T0212848

Practical Guide to Digital Manufacturing

Zhuming Bi

Practical Guide to Digital Manufacturing

First-Time-Right for Designs of Products,
Machines, Processes and System Integration

 Springer

Zhuming Bi ⓘ
Department of Civil and Mechanical Engineering
Purdue University Fort Wayne
Fort Wayne, IN, USA

With Contrib. by
Wen-Jun Chris Zhang ⓘ
Department of Mechanical Engineering
University of Saskatchewan
Saskatoon, SK, Canada

ISBN 978-3-030-70306-6   ISBN 978-3-030-70304-2   (eBook)
https://doi.org/10.1007/978-3-030-70304-2

This Springer imprint is published by the registered company Springer Nature Switzerland AG
The registered company address is: Gewerbestrasse 11, 6330 Cham, Switzerland

Preface

Many monographs and conference proceedings have been published on digital manufacturing and virtual manufacturing. Most of these resources focus on the introduction of new advancement on the theories, methodologies, and digitized tools made by the authors. Few books are available for readers to understand the thorough theoretical foundation and a broad scope of commercial computer-aided tools in digital manufacturing.

This book is unique and valuable to the readers in manufacturing in the sense that it will (1) introduce the thorough theoretical foundation, historical development, and enabling technologies of digital manufacturing; (2) serve as a practical guide for readers to use digital manufacturing tools in the designs of products, machines, processes, and integrated systems; (3) cover a broad range of computer-aided tools for geometric modeling, assembly modeling, motion simulation, finite element analysis, manufacturing process simulation, machining programming, product data management, and product lifecycle management; (4) use many real-world case studies to illustrate the applications of computer-aided tools to address various engineering design challenges in digital manufacturing, and (5) discuss the impact of cutting edge technologies, such as cyber-physical system, Internet of Thing (IoT), cloud computing, Blockchain Technologies, and Industry 4.0, on the advancement of digital manufacturing.

This book is written as a textbook for 3rd/4th-year undergraduates in Manufacturing Engineering, Mechanical Engineering, Industrial Engineering, and System Engineering and senior-level students in technical or engineering college, and as a reference book for Mechanical, Manufacturing, and Industrial Engineers who have the responsibilities of using computer-aided tools for the designs of products, machines, manufacturing processes, or system integrations and for the graduates who conduct researches relevant to manufacturing.

Zhuming Bi, Ph.D.
Professor of Mechanical Engineering

Wen-Jun Chris Zhang, Ph.D., P.Eng.
Professor of Mechanical and Biomedical Engineering

Contents

Chapter 1
Human Civilization, Products, and Manufacturing

Abstract The importance of manufacturing to the human civilization is discussed; main drivers to advance manufacturing technologies are discussed and classified. The technological evolution over the human history is examined to gain a good understanding of the needs, limitations, opportunities, and challenges of modern manufacturing technologies. Critical emerging technologies including *Internet of things* (IoT), *Big Data Analytics* (BDA), *Cloud Computing* (CC), *Blockchain Technology* (BCT), and *Rapid Technologies* (RP) are briefly introduced. The roles of computer-aided technologies in digital manufacturing are discussed. It is found that engineers should master digital skills to perform engineering practices in modern manufacturing. The organization of the book is presented as the conclusion of this chapter.

Keywords Human civilization · Human Development Index (HDI) · Gross National Income (GNI) · Primary industry · Secondary industry · Territory industry · Computer-aided technologies · Product Life Cycle (PLC) · Internet of Things (IoT) · Digital Manufacturing (DM) · Big Data Analytics (BDA) · Cloud Computing (CC) · Blockchain technology (BCT) · Rapid Prototyping (RP)

1.1 Human Civilization and Products

Human civilization reflects the advancement of human society in the forms of government, culture, industry, and common social norms (Sullivan 2020). Human civilization can be measured by *HDI*, which is a comprehensive measure of the averaged achievement of human development at the dimensions of *quality of life, education and knowledge*, and *standard of living*. Accordingly, HDI in these dimensions is quantified by three indices, i.e., *life expectation, average years of schooling*, and *GNI* per capita. The scores of these indicators are composed and normalized as an HDI (1) to reflect the advancement of human civilization at specific time and (2) to evaluate the advancement trend of human civilization within a specified period. For example, Table 1.1 shows the historical relations of life expectation, average years of schooling, and GNI and HDI from 1990 to 2018 by data (UNDP UNPD 2020a,

Table 1.1 The historical relation of GNI per capital and HDI by data (1990–2018) (UNDP 2020a, b)

Year	Life expectancy at birth	Expected years of schooling	Mean years of schooling	GNI per capita	HDI value
1990	75.2	15.4	12.3	37154	0.860
1995	76.1	16.0	12.7	39449	0.878
2000	76.8	15.1	12.7	45974	0.881
2005	77.7	15.7	12.8	49980	0.896
2010	78.7	16.2	15.4	50297	0.911
2015	78.9	16.2	13.3	54039	0.917
2016	78.9	16.3	13.4	54443	0.919
2017	78.9	16.3	13.4	55351	0.919
2018	78.9	16.3	13.4	56240	0.920

b), and it shows that the human civilization has been improved monotonously with the increases of three indices in our human society.

Three constitutive components of HDI are correlated to each other since all of them are determined by the availabilities of products and services. For example, a further look into the historical data in Table 1.1 shows the relation of HDI and GNI and that of GNI and the life expectance at birth in Figs. 1.1 and 1.2, respectively. While both of GNI and the life expectance at birth have contributed to the continuous improvement of HDI, the relations of GNI and life expectance at birth are very complex. In general, an increase of GNI leads to an increase of life expectance at birth. The greater a GNI is, the more income people can allocate to obtain products and services for more education and prolonged life spans, and the lower likelihood of diseases or premature deaths occur to people. Sufficient evidences have shown that in a nation, people at a lower level of incomes are less healthy than those at a

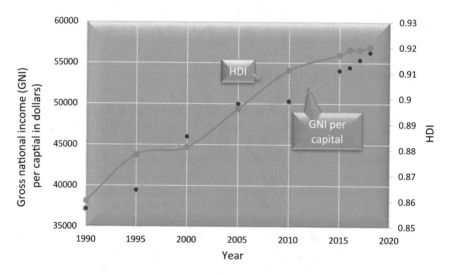

Fig. 1.1 The historical relation of GNI per capital and HDI from 1990 to 2018

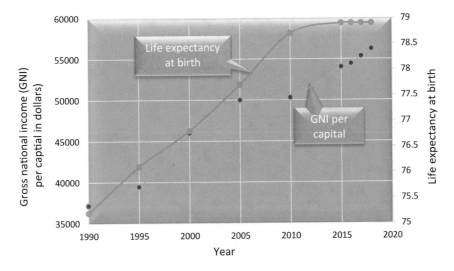

Fig. 1.2 Gross national income (GNI) per capital and life expectancy

higher level of incomes. The relationship of incomes and expected life experience is monotonously ascending; they are connected step-wise at every level of the economic ladder. For example, GNI can be interpreted into family incomes, which have been found to have the direct impact on the probabilities of occurrences of major diseases and illnesses that threaten humans' lives.

The survey by Woolf et al. (2015) gave the results in Figs. 1.3 and 1.4. It has showed that the reported rates of disease were higher for low-income Americans and that were accompanied by the higher rates of risk factors. For example, the obesity rates for rich people and poor people were 21.2% and 31.9%, respectively. It was found that such a difference was caused by the standards of living: people earned <$35,000 a year had three-time higher chance of smoking than those earned more than $100,000 a year. Only 36.1% of poor people had recommended levels of aerobic exercise in comparison with 60.1% of rich people. Similarly, Fig. 1.4 showed that incomes were associated with mental health: in comparison with people with an average family income for more than $100,000 a year, the people in a family with less $35,000 a year were four times or five times more likely *nervous* and *sad*. Similar trends occurred to somatic complaints such as pains and other ailments related to stress and depression).

The dependence of three indices for HDI is illustrated in Fig. 1.5. All aspects of GNI, education and knowledge, and standards of living are built on the availabilities of products. *Products* can be (1) *goods* for tangible things such as cars, phones, and computers and (2) *services* for intangible things such as education, banking, personal care, and entertainments (McCloy 1984). Products can be made in the ways of (1) *direct production* where people provide own goods and services or (2) *indirect production* where people cooperate with each other to produce the required goods and services. On the one hand, direct production is inefficient and only applicable to some

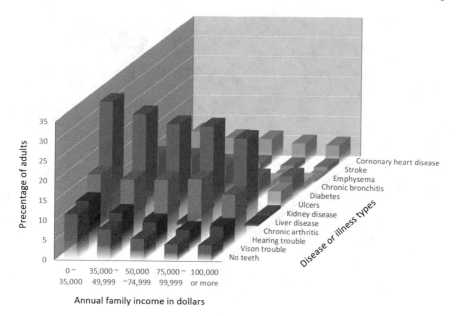

Fig. 1.3 The impact of family income on prevalence of physical diseases and illnesses (Woolf et al. 2015)

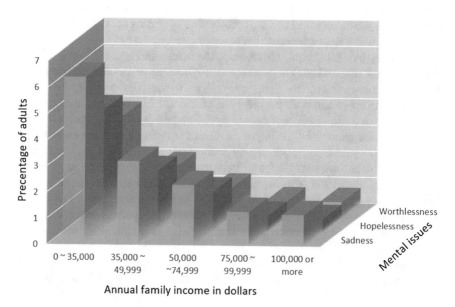

Fig. 1.4 The impact of family income on prevalence of mental illnesses (Woolf et al. 2015)

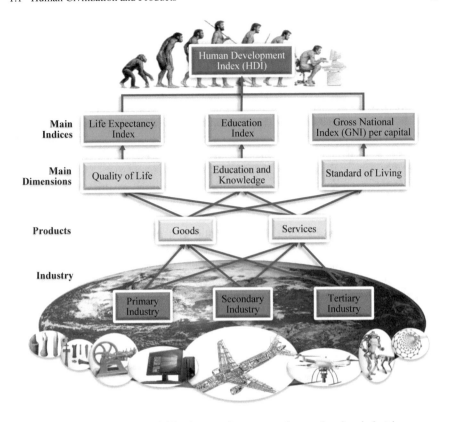

Fig. 1.5 Dependence of human civilization on primary, secondary, and tertiary industries

basic living needs. On the other hand, an indirection production led to the *division of labor* for the specializations in the modern society. Humans' wealth came from the primary, secondary, and tertiary industries with these specializations (McCloy 1984).

The standard of living is reflected by the value of goods and services that people can access and use; while GNI per capital determines what people can afford to buy. Alternatively, *gross domestic product* (GDP) is proposed as a measure of the market value of all of the products in a given time period. Furthermore, the products are made in three industries, i.e., *primary industry*, *secondary industry*, and *tertiary industry*. Note that *an industry* refers to a sector that applies the technologies to produce goods or services through a series of economic activities. In *primary industry*, natural resources are exploited in three different ways: (1) *extrusion* such as mining, quarry, and fishing to extract raw materials (such as oil, gas, metallic ores, diamonds, and sands) from the nature directly, (2) *cultivation* such as agriculture and forestry to obtain basic living supplies (fruits, vegetables, construction materials, timber, etc.), and (3) animal husbandry such as fish forming and cattle rearing for meat, milk, and furs. In *secondary industry*, the products from the primary industry are used as raw materials and converted into manufactured goods. Manufacturing goods are classified into (1) *capital goods* such as machinery and tooling which are used by other manufacturers and (2) *consumer goods* such as appliance and furniture which are used by end users. The secondary industry also include the businesses in *utilities* and *constructions* where manufactured goods are assembled as houses, factories, and plants. The *tertiary industry* provides services as products, and the services can be classified into (1) *commercial services* such as delivering goods to customers and providing financial supports which are crucial to the businesses in primary or secondary industries (2) *personal services* such as taxing, mail-delivering, and healthcare which are relevant to the quality and standards of living. Note that the majority of services in the tertiary industry rely on the wealth produced in the primary and secondary industries; therefore, it is worth to have a close look on the three industry sectors for their roles of wealth creation.

HDI is determined by the available wealth to earn incomes, gain knowledge, and improve standards of living, and it is interesting to see that the wealth of our human society is not just the accumulated commodities or resource reserves that we possess, the wealth includes the productive knowledge of humans to transform existing resources into desirable goods or services. Ruby (2003) argued that the wealth was an embodiment of physical and human capital, and it was used to generate the incomes to pay for desired goods and services X_i $(i = 1, 2, \ldots n)$,

$$X_i = f(L_i, K_i, M_i) \quad \text{for all} \quad i = 1, 2, \ldots n \tag{1.1}$$

where

X_i is the value of the ith of good or service;

L_i represents the quantity and ability of available labor;

K_i represents capital, machinery, transportation devices, and production infrastructure;

M_i represents available materials and other resources for production.

Since all of L_i, K_i, and M_i are manufacturing resources, the function $f(\cdot)$ in Eq. (1.1) represents a value-adding transformation in manufacturing. The amount of added value relates to the level of existing *manufacturing technologies* and *know-hows* used to convert inputs to outputs with better productivity. Therefore, other than a weighted average of the rate of population growth and the capital-accumulating rate, human civilization depends greatly on the growth rate of *manufacturing technologies* (Ruby 2003).

1.2 Drivers for Manufacturing Technologies

Manufacturing is to transfer raw materials into final products for users through a series of value-added and non-value-added processes. The transformation requires capital, people, machinery, tooling, processes, and systems to perform manufacturing businesses. Manufacturing contributes greatly to the value of national economies in the forms of assets, wealth, and strategic capabilities for defense and security, arts, literature, or other culture artifacts (NSTC 2008). Manufacturing has shown its importance to the national policies at the aspects of (1) providing employments, (2) attracting foreign direct investment (FDI), and (3) improving productivity and other economic measures (López-Gómez et al. 2017).

Manufacturing technology provides the tools to make products. Manufacturing technology covers *machine tools* (such as lathes and drill benches), *material removals* (such as turning and drilling) and *material forming* (such as pressing, stamping, and shearing), *additive processing* (such as 3D printing and laser sintering), *workholding* (such as clamps, blocks, and chucks)), *tooling* (such as drills, taps, and grinding wheels), *materials handling* (such as conveyors and automated guided vehicles), *automated systems* (such as flexible manufacturing systems and transfer machines), and *software tools* (such as computer-aided design, computer numerical controls, and computer-integrated manufacturing).

Figure 1.6 shows the main drivers to advance manufacturing technologies. These technological drivers are classified from the perspective of the cycle of continuous improvement (CI) of technological development. On the one hand, manufacturing technologies are *pushed* by the needs of the human society for further development and civilization. On the other hand, manufacturing technologies are *pulled* by the users' demands to make more and better products. In addition, for the transformation from raw materials to final products, manufacturing technologies are *evolved* by the integration of other relevant technologies in manufacturing businesses. Therefore, manufacturing technologies are gradually and iteratively advanced through CI cycles.

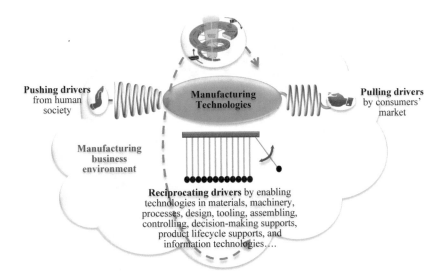

Fig. 1.6 Drivers of manufacturing technologies

1.2.1   Pushing Driver—Human Society

Manufacturing technologies are developed to meet the needs of the human society. Therefore, the evolution of human society is the primary driver to advance manufacturing technologies. Numerous researchers have discussed the evolving trends of the human society, and the identified main trends include (1) the globalization of the manufacturing business environment, which are featured by offshoring, outsourcing, and the growth of manufacturing capabilities, (2) the awareness of sustainability, which can be measured by carbon footprints, the efficiency of manufacturing operations, and quality of life and consumption, (3) the change of demographics such as the growing size of middle-class population and ageing workforce, (4) the urbanization of mobility, housing, infrastructure, and factories, (5) the threats to global stability from natural disasters and terrorism threats, (6) shortening of product lifecycles due to the advances of materials, processes, and technologies, (7) the change of customers' expectations on products such as personalization and fast rate of technology adoption. Accordingly, Fig. 1.7 shows six drivers to advance manufacturing technologies (López-Gómez et al. 2017). These drivers include the *ageing of workforce in developing countries, change of manufacturing skillsets, personalized products, increasing demands for products in cities, growing interest in technologies strategies,* and *reshoring manufacturing activities globally.*

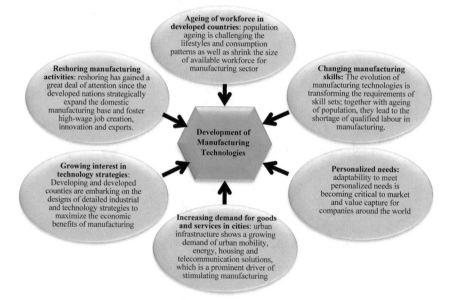

Fig. 1.7   Pushing driver of manufacturing technologies from human society

1.2.2 Pulling Driver—Consumer Markets

The added value of manufacturing operations is redeemed when products are delivered to meet customers' needs, and the customers' needs determine the functions, types, and volumes of products that manufacturers should make for. From this aspect, manufacturing technologies must be continuously advanced; since the demanded products become more and more complicated. Figure 1.8 shows that the customers' needs are continuously changing in terms of *spectrum, versatility, variety, life span, volume, lead time, integration, suppliers*, and *life supports* of products. The average volume of a specific product is gradually reduced to the extreme case of size-one, the average life span and lead time is continuously reduced. The measures of other aspects are increased.

The spectrum of products has been greatly expanded over time. At small scale, nano-machines were designed to operate in human in vivo environment for diagnosis, inspection, and medical treatment as shown in Fig. 1.9a (Amato 2012); at large scale, massive machines were expected to transport people to deep space as shown in Fig. 1.9b (Wikipedia 2021a). Manufacturing technologies must be advanced to explore new materials, and to design, fabricate, and assemble extreme small or large and complex products.

Taking an example of the iPhone products in Fig. 1.10, generation after generation, products become more and more *versatile* since new features will be added and the capacities will be expanded in each successive generation to make the products better. In each generation of products, customers have personalized choices of available features, and more product features imply more variations of customized products. For example, Fig. 1.10 shows the adoption of new features in iPhone series at the aspects of screen technology, 3D touch, networking support, cameras, video records, live photos, identification, charging, and water-resistance. Moreover, *product varieties* have been increased gradually due to an increase of product features.

An iPhone camera module has over 200 separate parts; the Boeing with 5400 supplier factories including secondary supplies; more than 750 million components

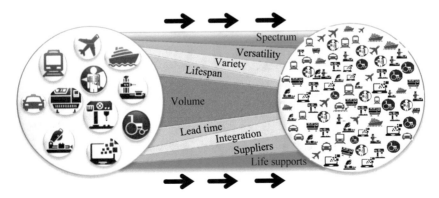

Fig. 1.8 Pulling driver of manufacturing technologies from consumer market

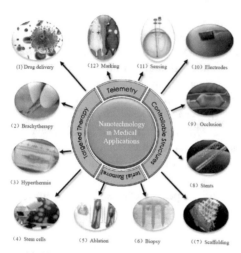

(a) Nano-machines in medical applications   (b) Humans mission to Mars

Fig. 1.9 Examples in expanded product spectrum

Fig. 1.10 Example of versatility and variety versus generation of products

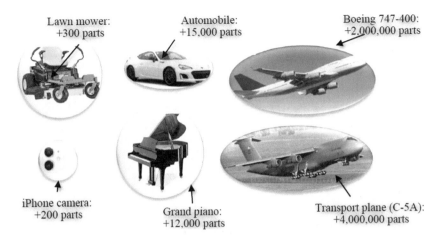

Lawn mower:
+300 parts

Automobile:
+15,000 parts

Boeing 747-400:
+2,000,000 parts

iPhone camera:
+200 parts

Grand piano:
+12,000 parts

Transport plane (C-5A):
+4,000,000 parts

Fig. 1.11 Examples of complex products

and assemblies were acquired by Boeing in 2012, around 500,000 people are hired by these suppliers. The program scales of the control systems for a Lockhead F-22 raptor, Boeing 787 Dreamliner, Airbus A380 and 2015 Ford F-150 were 2, 7, 100, and 150 million lines of code, respectively. Figure 1.11 shows some product examples, whole complexities can be defined mainly based on the number of parts and components in products (Digital Engineering 2016).

The complexity of products and processes determines the complexity of manufacturing technologies in making these products, and the five main factors to affect the complexity of manufacturing technologies are (1) functions of products, (2) number of complexities of manufacturing processes, (3) correlations of products and processes, (4) the interactions in enterprise information systems, and (5) the number of involved national and international standards, regulations, and specifications. As shown in Fig. 1.12, the product complexities by the five aforementioned factors have been continuously increased from *very simple* to *very complex* years over years. Taking an example of computers, based on today's standards, early computers were so simple in terms of its capabilities of computing and storage. A new-generation central processing unit (CPU) was developed at an average of *every 18 months*; it has *the double complexity* for *quadruple capabilities* at *roughly half price* in comparison to previous CPU. Other products have similar changing trends of complexities and lead times: products become old-fashioned, their attractions to customers are gradually vanished, and the market's demands decline in less and less time (Prasad 1997).

Correspondingly, *average life span* of products becomes shorter and shorter due to the early availability of new products with more functions and better performances. Taking the change of computer life spans in Fig. 1.13 as an example, according to the study at Arizona State University (ASU), the average life spans of personal

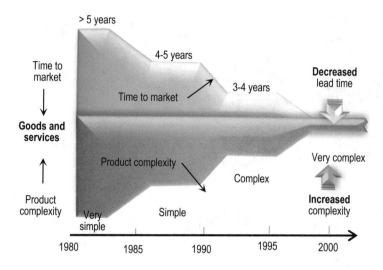

Fig. 1.12 Shortened lead time

Fig. 1.13 Example of shortened product lifespan—PCs at Arizona State University (ASU) (Babbitt et al. 2009)

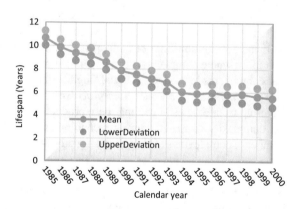

computers (PC) had been reduced greatly from 12 years in 1985 to <3 years in 2000 (Babbitt et al. 2009); the increase of computer ownership at a rapid rate was the main cause of such a change.

1.2.3 Reciprocating Driver—Relevant Technology Development

On the one hand, manufacturing technologies are the results of applying fundamental science and technology in solving problems in designing, making, and using

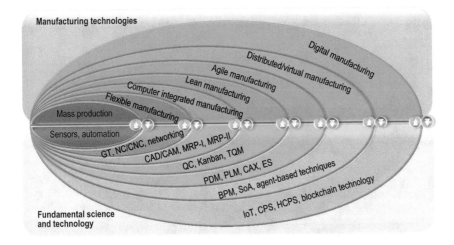

Fig. 1.14 Reciprocating drivers of manufacturing technologies by relevant technologies

products; on the other hand, new challenges of designing and making better products stimulate the further advancement of fundamental science and technology. From this perspective, technology development is a reciprocating driver to manufacturing technologies (Jiang et al. 2014). The impact on relevant technology development on manufacturing technologies has been clearly evidenced in the evolution of manufacturing technologies.

As shown in Fig. 1.14, every major shift of the manufacturing paradigm has been driven by the new development of relevant technologies (Cheng et al. 2000). For example, *sensors and automation* made the closed-loop automation in *mass production* possible. The emerging information technologies (IT) such as *Internet of Things* (IoT), *cyber-physical system* (CPS), *human cyber-physical system* (HCPS), and *blockchain technology* have laid the foundation to practice digital manufacturing paradigm (Bi et al. 2020). On the other hand, the demands for green, miniaturized, highly valued products have stimulated the studies in many emerging fields such as *nanoscience, environmental science, big data analytics, machine learning*, and *cybersecurity*.

1.3 Formulation of Engineering Design Problems

Figures 1.15 and 1.16 show that engineering designs are involved in all stages of product lifecycles (Wikipedia 2021b) and all aspects of corresponding manufacturing processes (Wang 2006), respectively.

A manufacturing system aims to make products, and numerous decision-making activities are involved in at different stages of product lifecycles. Therefore, any

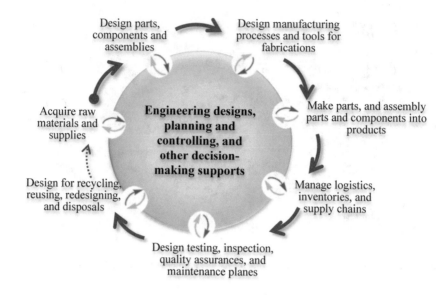

Fig. 1.15 Engineering designs in product lifecycle

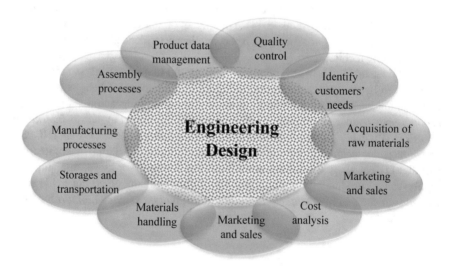

Fig. 1.16 Engineering designs in all aspects of manufacturing operations

decision-making activity in a manufacturing system can be formulated as an engi-neering problem, which consists of *inputs* (*I*), *outputs* (*O*), *purposes* (*P*), *variables* (*V*), and *constraints* (*C*). As shown in Fig. 1.17, to formulate an engineering problem, the scope of the problem should be defined to identify the set of aforementioned

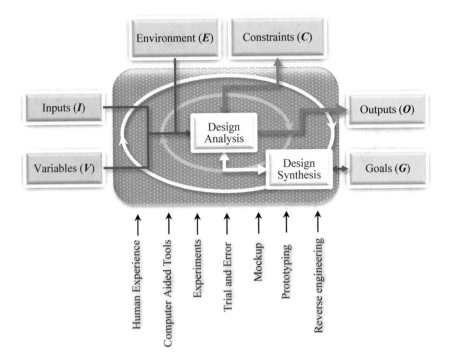

Fig. 1.17 Formulating an engineering problem

elements. Further, the relations of the system elements must be modeled and analyzed so that different solutions can be compared to each other to find an optimal solution to the formulated problem. With the trend of ever-increasing complexity, scale, and dynamics, various engineering approaches, such as computer-aided technique, prototyping, and reverse engineering are needed to assist humans in performing design analysis and synthesis in seeking engineering solutions.

Once an engineering problem is formulated, a general procedure shown in Fig. 1.18 can be followed to seek an optimal solution to the defined problem. The procedure consists of five design phases. At phase II, a design space is defined to include all of the possible solutions to be evaluated, and the relations of system elements with outputs (O) and constraints (C) are modeled. At phase III, possible solutions are continuously analyzed and evaluated against the expected design criteria; they are compared each other to identify an optimal solution through design synthesis. At phase IV, the detailed design is performed for the optimized solution, and the design is verified and validated to ensure no violence to the given constraints; if there is, then an iterative process is needed from phase II to IV until a valid solution is identified. At phase V, the physical solution is implemented as a practical solution to the stated engineering problem. It is clear that the information about an engineering solution is accumulated, while the design constraints must be satisfied.

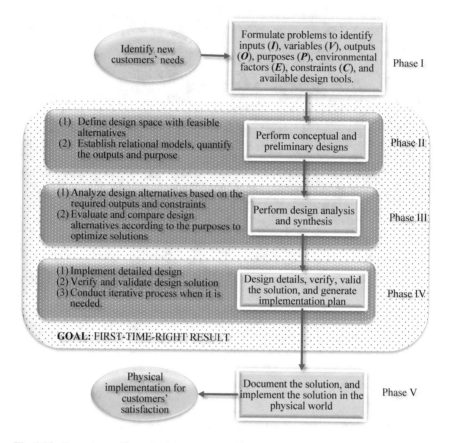

Fig. 1.18 General procedure of solving an engineering problem

1.4 Computers in Engineering Design

Regardless of the complexity level of a product or a manufacturing system, the engineering process follows similar process as discussed in Fig. 1.18. However, advanced techniques and tools are needed when the scale and complexity of an engineering problem is beyond of the capabilities of manual decision-makings. From this perspective, more and more human's roles are being replaced or assisted by computer-aided technologies such as computer-aided design (CAD), computer-aided engineering (CAE), and computer-aided manufacturing (CAM). Here, the characteristics of engineering design problems are discussed to understand the needs of using computer-aided techniques in engineering designs (Bi 2018). An engineering problem can be one of six typical types as shown in Table 1.2.

The typical engineering problems in Table 1.2 can be classified based on the levels of creativity of the corresponding engineering solutions. The level of creativity relies

Table 1.2 Classification of engineering designs

Type	Description	Example
Routine design	To find a design solution by following the standards and guidance that specifies the steps and calculations.	Follow ASTM fastener standards to select bolts, screws, or nuts for a fastener.
Redesign	To update an existing design when some functional requirements have been changed in its application.	Modify a robot program when the working points on a path have been changed in a new task.
Selection design	To select one of the pre-existing solutions for a given application.	Select an electronic device such as a computer at an office.
Parametric design	To optimize a set of discrete, continuous, or mixed design variables in a given conceptual structure.	Design a four-bar mechanism to operate a door subjected to the specified weight, operating range, space, and cost.
Integrated design	To build a product or system using existing components and modules.	Design a workcell for the identified machines and products in cellular manufacturing (CE).
Original design	To conceptualize a design from scratch to meet a set of newly identified functional requirements.	Propose a new solution to guarantee the security, trustiness, and responsiveness of data sharing in a large-scale distributed system.

to the characteristics of a 'solution space' and 'design variables' in a formulated engineering problem, and Table 1.3 shows a classification of engineering problems based on the levels of creativity. An engineering design can be a 'routine', 'innovative', or 'creative' design.

Both humans and computers are important in seeking the solutions of engineering problems, but they are competent to different levels and scopes of decision-making tasks. The differences of humans and computers have been widely discussed, and Table 1.4 gives a comparison of their strengths and weaknesses in engineering designs.

As shown in Fig. 1.19, in engineering practice, both humans and computers play their roles in accomplishing certain activities. Since computers and humans are good at different tasks as shown in Table 1.4, the strengths of computers and humans can be synergized by using computer-aided techniques in solving various engineering problems. Especially, computer-aided techniques can be utilized to automate design processes and improve effectiveness of engineering designs. Taking an

Table 1.3 Difference of routine, innovative, and creative designs

	Level of creativity	Routine design	Innovative design	Creative design
Solution Space	structure	Known	Known	Unknown
	Search procedure	Known	Unknown	Unknown
Design Variables	Types	Fixed	Fixed	Changed
	Ranges	Fixed	Changed	Changed

Table 1.4 Differences of humans and computers (Bi 2018)

	Human Designers	Computers
Strengths	• Identifying design needs, • brainstorming to think solutions 'out of box' • engineering intuition and big knowledge base, • selecting design variations, • the flexibility to deal with changes • qualitative reasoning, • psychologically, human decision is more trusted than artificial intelligence, • predict trends, patterns, or anomalies, and • learn from experience	• Fast speed, reliable, endurance, and consistent, • Capable of exploring a large number of options, • Carry out long, complex, and laborious calculations, • Store and efficiently search large databases, and • Provide information on design methodologies, heuristic data, and stored expertise.
Weaknesses	• easily tired and bored • cannot micro-manage • biased and inconsistent • prone to make errors • not good at quantified reasoning • incapable of utilizing the data presented in awkward manner	• difficult to synthesize new rules • limited knowledge base • no common sense

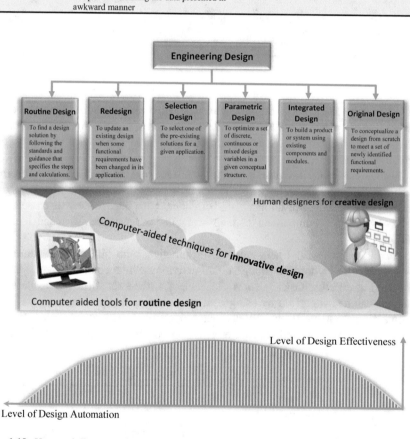

Fig. 1.19 Human designers and computers in engineering designs (Bi 2018)

example, Bi et al. (2006, 2010) automated the process of the configuration design for a modular robotic system, which was able to evaluate 10^4–10^5 robotic configurations automatically in finding the best configuration for the specified task.

1.5 Computers in Manufacturing

Due to the complexity of modern manufacturing systems, more and more manufacturing businesses rely on computers, and computer-aided techniques were widely applied to expand the scope of automation in manufacturing systems. The capability, the productivity, and the efficiency of a manufacturing system can be greatly improved by maximizing the degree of automation.

The dependence of the performance of a manufacturing system on the automation was explored by many researchers. For example, Williams (2000) analyzed the importance of manufacturing operations from the perspective of human-machine relations. As shown in Fig. 1.20, manufacturing activities are classified into two groups, i.e., the activities involved in the information flow, and the activities in the material flow. In the material flow, computers are used to mechanize and automate manufacturing operations; intelligent machines such as *computer numerical control* (CNC) machines and *automated guided vehicles* (AGVs) are widely used to replace human operations. In the information flow, computers are used to support decision-making activities; computer-aided techniques assist humans in making decisions at all phases of manufacturing systems' lifecycles from design, construction, operation, to the dissolve phase. While humans are still important in sustaining the agility and flexibility of a manufacturing system, it becomes more and more critical that humans and computers collaborate harmoniously to expand manufacturing businesses in a highly competitive environment.

A broad adoption of computer-aided techniques helps to maximize the scope of the automation. In particular, computer-aided techniques are taking over more and more human's efforts in decision-making supports. As a matter of fact, the advancement of manufacturing technologies was measured by the capabilities of computer-aided techniques in tackling with the growing *scale, complexity*, and *responsiveness* of manufacturing systems. Figure 1.21 shows the evolution of computer-aided techniques over time (Cheng and Bateman 2008; Bi and Cochran 2014). When the computer-aided tools such as *quality control* (QC), *total quality management* (TQM), *supply chain management* (SCM), *enterprise requirements planning* (ERP-I), *enterprise resources planning* (ERP-II), *product lifecycle management* (PLM), *software as a service* (SaaS), and *platform as a service* (PaaS), and *Infrastructure as a Service* (IaaS) became available, manufacturing processes were able to be automated by using CNCs, flexible manufacturing systems (FMSs), computer-integrated manufacturing (CIM), distributed manufacturing (DM), and predictive manufacturing (PM). On the other hand, computer-aided technologies were driven by the emerging needs of manufacturing systems to deal with the increasing volume, variety, velocity of the data in the globalized business environment.

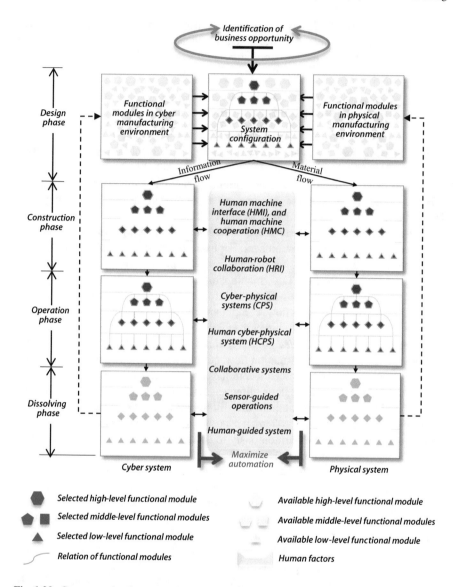

Fig. 1.20 Computers for the automation in modern manufacturing system

1.6 Emerging Digital Technologies

Digital manufacturing technologies refer to some computer-aided tools that are used in product designs, manufacturing processes, services, supply chain management, and system designs and operations. With the digital manufacturing, system elements

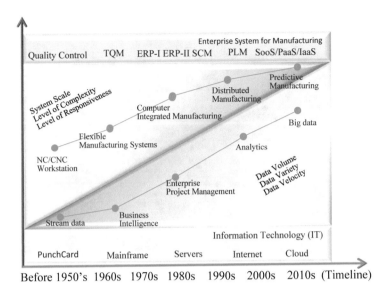

Fig. 1.21 Evolution of computer-aided technologies in manufacturing (Cheng and Bateman 2008; Bi and Cochran 2014)

and processes are networked and connected to fully utilize real-time data in optimizing a holistic manufacturing system. Digital technologies are widely adopted to enhance the competitiveness and in expanding the landscapes of manufacturing enterprises. In recent years, digital technologies have been progressing exponentially, especially by adopting new technologies at the following aspects.

1.6.1 Internet of Things *(IoT)*

Internet of things (IoT) is a network of smart things. A smart thing can be anything with sensing, data processing, or decision-making capabilities. IoT is the extension of the Internet with the connections to physical devices via transmission control protocol/Internet protocol (TCP/IP). As shown in Fig. 1.22, IoT connects the physical twin and digital twin seamlessly in digital manufacturing and IoT makes any interactions among the things possible. As shown in Fig. 1.23, an interaction possibly occurs to a *cyber and physical system (CPS)*, *machine-to-machine, human to machine, human to human, machine to human, machine to infrastructure,* and *machine to environment.*

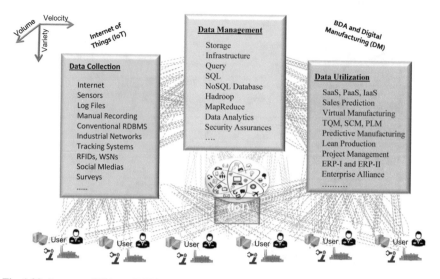

Fig. 1.22 Internet of Things (IoT) in enterprise systems (Bi and Cochran 2014)

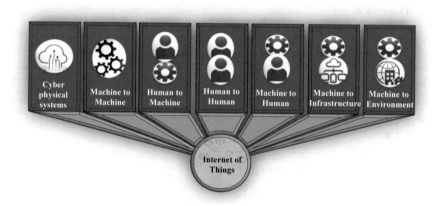

Fig. 1.23 IoT makes any interaction possible

1.6.2 **Big Data Analytics** *(BDA)*

Enterprises are forced to make more product variants in mass customization with an enhanced level of customers' satisfactions (Pasche 2008); their enterprise systems should be capable of dealing with the increasing volume, velocity, and variety (3 V) of data due to system uncertainties and complexity shown in Fig. 1.24.

Nowadays, enterprises are deluged in the Internet with the data scale from *terabytes* (2^{40} bytes), *petabytes* (2^{50} bytes) to *exabytes* (2^{60} bytes), and traditional big analysis methods became struggling in processing *heterogeneous*, *unstructured*, and

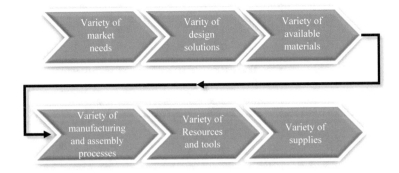

Fig. 1.24 Uncertainties and complexity in manufacturing

dynamic data. *Big data analytics* (BDA) aims to analyze a big data and explore information, knowledge, and wisdom at the unprecedented scale (Cloud Security Alliance 2013). As shown in Fig. 1.25, BDA is used in digital manufacturing to support real-time, short-term, and long-term decision-making activities over the product lifecycles. BDA is expected to achieve the goals of *computerization, connectivity, visibility, transparency, predictivity,* and *adaptability* based on the capabilities of BDA tools (KPMG 2018; Bi and Wang 2020).

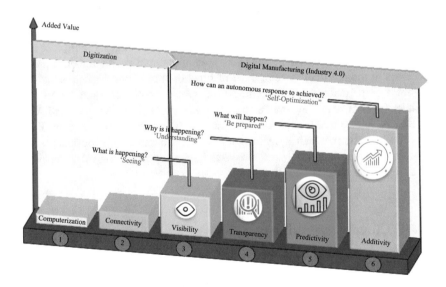

Fig. 1.25 The goals of BDA in digital manufacturing (KPMG 2018; Bi and Wang 2020)

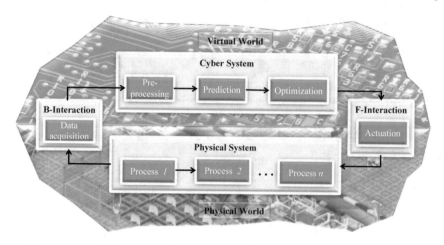

Fig. 1.26 Main components in CPS

1.6.3 Cyber-Physical Systems (CPS)

A cyber-physical system differs from stand-alone embedded systems, sensors or other conventional mechatronic systems in sense that the cyber electric and electronic system is fully coupled with the physical system to be controlled. The studies of CPS were prioritized to advance the manufacturing technologies for Industry 4.0. Existing works on CPS were at their infant stages and limited to the exploration of basic concepts, system architecture, technologies, and challenges. Figure 1.26 shows the main components of a cyber-physical system that include a number of physical processes in the physical system, a set of functional modules for data acquisition, pre-processing, prediction, and optimization in the cyber-system, and the functional modules for data acquisition and actuation as the interactions (Schmidt and Ahlund 2018). CPS adds the capabilities of computation, communication, and controls (3C) to a physical system, which allows a cyber-system to interact with physical processes directly. A CPS is featured by real-time sensing and dynamic controls to respond to the changes and uncertainties in applications.

1.6.4 Cloud Computing (CC)

An enterprise often faces some dilemmas in determining computing resources for its information system. On the one hand, the enterprise must optimize its system resources for core manufacturing businesses in the physical manufacturing system. On the other hand, products, manufacturing processes, and business environment become more and more complicated and dynamic, and high computing capabilities are expected to process big data for decision-making units to respond to changes

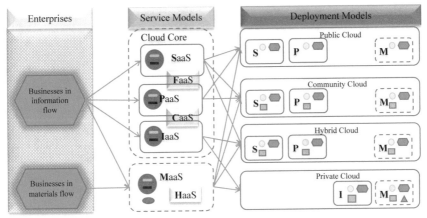

S - Service P -Platform I - Infrastructure M - Manufacturing F - Function C - Container H - Hardware

Fig. 1.27 Main components in CPS (Pedone and Mezgar 2018)

and uncertainties promptly. As shown in Fig. 1.27, *cloud computing* (CC) offers a technical solution for enterprises to utilize outsourcing computing resources over the Internet. The manufacturing businesses in both of the information and materials flows can be defined as services; and services will be provided in different service models such as *infrastructure as a service* (IaaS), *function as service* (FaaS), and *manufacturing as service* (MaaS). Services are available in a *public cloud*, *community cloud*, *hybrid cloud*, or *private cloud* over the Internet (Pedone and Mezgar 2018). For small and medium-sized enterprises (SMEs), CC allows to minimize the information infrastructure and reduce the cost of the enterprise system since CC is dynamic and flexible.

1.6.5 **Blockchain Technology** *(BCT)*

Manufacturing systems used to be closed, and system controls are centralized. With the emerging needs of (1) collecting and sharing data from smart things in distributed environment, (2) collaborating with business partners across system boundaries, and (3) utilizing external resources to enhance business capabilities, the controls of modern manufacturing systems are distributed and decentralized. Service-oriented architecture (SOA) allows enterprises to use distributed and external resources to expand manufacturing capabilities. However, some challenges of SOA have not been addressed satisfactorily (Viriyasitavar et al 2019a, b).

As shown in Fig. 1.28, blockchain technology (BCT) was proposed to assure the trustiness, security, and privacy of the services over the Internet. The performance of a decentralized system can be measured by some performance key indicators (PKIs)

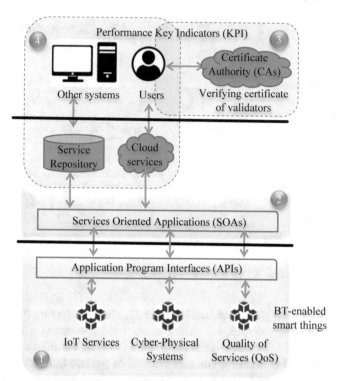

Fig. 1.28 Blockchain technologies in service-oriented architecture (Viriyasitavar et al. 2019a, b)

including *scalability*, *security*, *openness*, and *flexibility*. Three-layer system architecture shows an integration of blockchain technology (BCT) with IoT, SoA, PKIs, and other available services over the Internet. BCT-based applications at the first level ensure the validity of data from smart things such as sensors, CPSs, and quality of services (QoS). SoA at the second layer ensures that the manufacturing processes are executed and delivered in the form of services, and SoA deals with the consistency of data and interactions. PKIs at the third layer establishes the trustiness of validators. The blockchain validators should be clarified and endorsed by Certificate Authority (CA) for each system.

1.6.6   Rapid Prototyping *(RP)*

An engineering design process is generally iterative in which errors, defects, and omissions are gradually eliminated to find a right solution. A traditional engineering process involves in the validation where a physical product or system is built and tested against expected functionalities. This becomes cost-forbidden when

(a) Traditional material removal manufacturing

(b) Rapid prototyping (additive manufacturing) such as 3D printing

Fig. 1.29 Examples of material removal processes and rapid prototyping

the product or system is extremely complicated or the number of iterations is large. Rapid prototyping (RP) is a cost-effective technique for the fast fabrications of physical products, models, assemblies, or systems. RP is mostly implemented by an additive manufacturing process such as 3D printing. Rapid prototyping is referred as *solid freeform fabrication* (SFF), *additive manufacturing* (AM), or *direct manufacturing* (DM). In contrast to traditional manufacturing processes.

Figure 1.29 shows some example products from material removal processes and rapid prototyping. RP has exposed the following advantages: (1) there is no needs for tooling, forming, or fixturing, and RP can be operated more safely since no machining force is involved. (2) It uses layered fabrication, which is applicable to make products with any complex geometries. (3) RP is a near net shape process which has little waste. 4) RP can shorten the development time significantly, since no custom tools or programming is needed to make special products.

1.7 Digital Skills in Modern Manufacturing

With ever-increasing complexity of products and manufacturing processes, more and more manufacturing operations and decision-making support becomes beyond the reach of humans. Machines and computers are replacing humans not only for repetitive and boring tasks, but also for complicated decision-making supports at different levels and domains of manufacturing systems. On the other hands, humans and machines have to collaborate with each other closely to tackle with complex tasks (Bi et al. 2021), where operators, technicians and engineers are required to enhance their digital skills to work on machines and make decisions in various manufacturing businesses.

Muro (2017) discussed the trends of the requirements of digital skills in advanced manufacturing at the occupations from laborers and machinists to aerospace engineers. As shown in Fig. 1.30, the requirements for digital skills were increased from 2002 to 2016 with no exception. Taking the examples of production workers and tool and die makers, they used to heavily rely on their hand-on skills to fulfill their job responsibilities; however, more and more advanced machines demand higher and higher digital skills from human workers to operate, interact, and program machines. The average digital skill score of all major occupations was increased to 39 in 2016 from 24 in 2002.

In recent years, information technologies (IT) have been advanced rapidly; however, the adoption of new ITs has been lagged more or less. Figure 1.31 shows the relation of annual productivity changes and mean digital scores in different sectors (Muro 2017). The digital score of the occupations in advanced manufacturing was 39 in 2016, and the digital technology adoption in advanced manufacturing showed

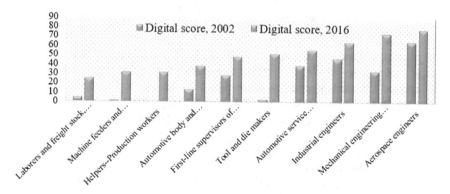

Fig. 1.30 Digital skill ratings for occupations in advanced manufacturing (Muro 2017)

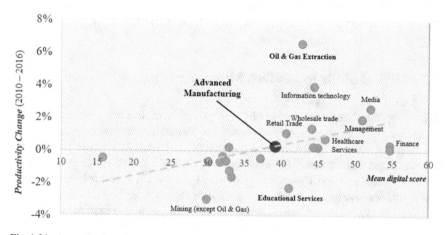

Fig. 1.31 Annualized productivity growth by sector versus digital score (2010–2016) (Muro 2017)

limited impact on the improvement of annual productivity change. The digital technologies have revolutionized some industry sectors especially for oil and gas extraction, since the new technologies such as big data analytics enhanced the exploitation capabilities to find new natural resources. It was surprising to note that the educational services were less benefited from digital technologies. Conventional designs of engineering curricula must be radically improved to fill the gap of the engineering education and emerging demands of digital skills in manufacturing sector (Bi and Wang 2020).

1.8 Organization of Book

Solving an engineering problem is generally an iterative process in the sense that the information of the final solution is gradually accumulated; it implies that design constraints in the formulated problem cannot be fully satisfied in the solving process until all the details of implementation are provided in the solution. This poses the challenge of using computer-aided technologies to practice the first-time right from the digital twin to physical twin due to (1) the uncertainties and complexity of products and systems and (2) the needs of verification and validation in the virtual environment.

This book emphasizes on engineers' knowledge and skills of using digital manufacturing tools to solve various engineering problems at different domains and levels of manufacturing systems. As shown in Fig. 1.32, both products and system resources have lifecycles that affect manufacturing systems. A manufacturing system involves a set of value-added and non-value-added manufacturing operations along the product lifecycle, and decision-making supports are needed to use various system resources in fulfilling these operations. In digital manufacturing, the decisions on system operations are made by functional units in the digital twin, and engineers should compete to utilize computer-aided technologies to make right decisions at the first time. Considering the time limit of a typical engineering course, this book selects the following four main subjects that are shown in Fig. 1.32.

Chapter 2 introduces *computer-aided design* (CAD). In CAD, engineers use computers to design, analyze, model, and evaluate products, processes, and systems. Computer-aided theory and methodologies are introduced to represent products, processes, and systems, and parametric modeling and knowledge-based engineering (KBE) are emphasized for knowledge reuse and exploration. Solid modeling techniques are introduced to create virtual models as the representations of geometric shapes, features, design intents, and topology and structures of products. Computer-aided kinematic and motion analyses are discussed for machine designs.

Chapter 3 introduces *computed-aided engineering* (CAE). In CAE, engineers use computers to model and analyze the responses of the product, process, or system that is subjected to *the specified loads and constraints* in applications. CAE is essential to eliminate the hidden mistakes and omissions in the virtual world before the digital twin is released to the physical world for implementation. The importance of CAE in manufacturing is discussed, different CAE tools are introduced, and the focus is put

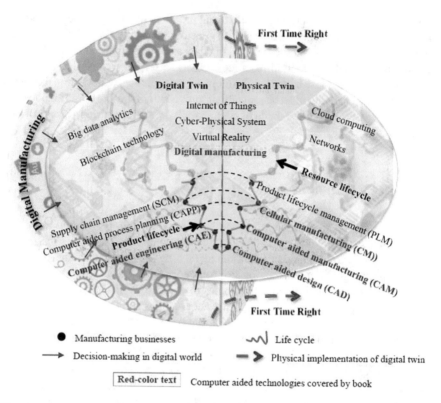

Fig. 1.32 Digital manufacturing makes the first time right from digital twin to physical twin

on *finite element analysis* (FEA). The theoretical fundamental of FEA is introduced, the general procedure of FEA modeling and simulation is presented. Some typical FEA applications in manufacturing are discussed.

Chapter 4 introduces *computer-aided manufacturing* (CAM). In CAM, engineers use a computer to design, model, analyze, optimize, and control manufacturing processes. CAM is complex since it involves *machines*, *tools*, *materials*, numerous *operating factors*, and multiple but conflicting performance criteria such as operating time, cost, accuracy, flexibility, and adaptability. CAM and relevant concepts are discussed, and computer-aided technologies are introduced for modeling and simulation of fixtures, mold and dies, and manufacturing processes. The programming tools are introduced for the controls of machine operations for specified products.

Chapter 5 introduces *computer-integrated manufacturing* (CIM). In CIM, engineers use computers to organize functional units and manufacturing operations at system level. A manufacturing system consists of various functional units for designing, manufacturing, assembling, transporting, managing, marketing, maintaining, and services at different levels and domains. The enabling technologies, such as group technology (GT), cellular manufacturing (CM) and discrete event dynamic simulation (DEDS) are discussed in detail. Computer-aided evaluation of manufacturing systems are introduced with the focus on cost and sustainability.

Chapter 6 introduces *digital manufacturing* (DM). In DM, engineers use computers to practice the first time right from *a digital twin* in the virtual world to a physical twin in the physical world. The functional requirements of DM are discussed, and new enterprise architecture is presented to deal with the complexity, uncertainties, and dynamics of system operations. Two digital technologies, reverse engineering (RE) and additive manufacturing (AM), are introduced as the examples of emerging digital technologies in digital manufacturing. There is no universal digital solution to any manufacturing system; therefore, applied studies are more or less required in developing a holistic digital solution to individual enterprises. Some research case studies are provided to explore the applications of digital technologies in various applications.

Design Problems

Problem 1.1. List some quantified measures to reflect the contributions of different industry sectors to human society, and discuss the importance of manufacturing to human civilization.

Problem 1.2. What are the main driving forces to develop new manufacturing technologies? Explain why.

Problem 1.3. What are the main pulling forces to develop new manufacturing technologies? Explain why.

Problem 1.4. What are the main reciprocating driving forces to advance manufacturing technologies? Explain why.

Problem 1.5. Formulate an engineering program for the following engineering projects, respectively.

(a) A local company produces various end-effectors for robots and other automated systems. A recent adventure is to develop new end-effectors for collaborative robots that work in an open environment with humans. The company meets the challenge to mount their end-effectors on collaborative robots due to different design standards at the interfaces. A mount of an end-effector on a collaborative robot must (1) have the flexibility of orientation shift between 0° and 90°, (2) meet the safety standards of coving all sharp edges and corners, and (3) minimize the impact on weight, robotic control, and cost.

(b) A truck assembly plant builds 1/2–1 t trucks. At the entrance of the assembly line, chassis frames are separated by spacers in stack. After a chassis frame is transported to the assembly line, human operators remove spacers from the frames and transport them to storage containers. The company has the difficulty in recruiting human operators at such workstations and the productivity of labors for such operations is low. The company expects to replace human operators by automation to locate four spacers on a chassis frame, to remove and transport spacers to the designated areas within the production cycle of one truck per minute.

(c)   The precision cooling business unit of a client company has a production line with brazing furnace for heat treatment processes of heat sinkers. One human worker uploads parts at one end and unload parts at the other end of the heat treatment line. The company has some used robots that have the full capabilities of loading and unloading heat sinkers. Develop an automated solution of using used robots to replace human operators.

Problem 1.6.   Have an example of emerging digital technologies of your interest, debrief the concept and relevant methodologies and tools, and discuss its role in addressing the challenges in modern manufacturing.

Problem 1.7.   Have a discussion of the importance of digital skills in engineering practice, and have a wish list of the digital skills you would like to master

References

Amato P (2012) Swarm-intelligence strategy for diagnosis of endogenous diseases by nanobots. https://pdfs.semanticscholar.org/58cf/550b97fe50ca8718c31c98bcdecbb9e06aef.pdf

Babbitt CW, Kahhat R, Williams E, Babbitt G (2009) Evolution of product lifespan and implications for environmental assessment and management: a case study of personal computers in high education. Environ Sci Technol 43(13):5106–5112

Bi ZM (2018) Finite element analysis applications: a systematic and practical approach, 1st edn. Academic Press, ISBN 10 018099526

Bi ZM, Cochran D (2014) Big data analytics with applications. J Manag Anal 1(4):249–265. https://doi.org/10.1080/23270012.2014.992985

Bi ZM, Wang XQ (2020) Computer aided design and manufacturing, 1st edn. Wiley-ASME Press Serious. ISBN 10 1119534216

Bi ZM, Lin Y, Zhang WJ (2010) The general architecture of adaptive robotic systems for manufacturing applications. Robot Comput-Integr Manuf 26(5):461–470

Bi ZM, Gruver WA, Zhang WJ, Lang SYT (2006) Automated modeling of modular robotic configurations. Robot Auton Syst 54 (12):1015–1025

Bi ZM, Miao ZH, Zhang B, Zhang WJ (2020) The state of the art of testing standards for integrated robotic systems. Robot Comput Integr Manuf 63(June 2020):101893

Bi ZM, Luo M, Miao Z, Zhang B, Zhang WJ, Wang L (2021) Safety assurance mechanisms of collaborative robotic systems in manufacturing. Robot Comput Integr Manuf 67 (February 2021):102022

Cheng K, Bateman RJ (2008) e-Manufacturing: characteristics, applications and potentials. Prog Nat Sci 18(2008):1323–1328

Cheng K, Pan PY, Harrison (2000) The Internet as a tool with application to agile manufacturing: a web-based engineering approach and its implementation issues, Int J Prod Res 38(12):2743–2759

Cloud Security Alliance (2013) Big data analytics for security intelligence. https://downloads.clo udsecurityalliance.org/initiatives/bdwg/Big_Data_Analytics_for_Security_Intelligence.pdf

Digital Engineering (2016) By the numbers: product complexity. http://old.digitaleng.news/de/by-the-numbers-product-complexity/

Jiang H, Zhao S, Yin K, Yuan Y, Bi ZM (2014) An analogical induction approach to technology standardization and technology development. Syst Res Behav Sci 31(3):366–382

KPMG (2018) Supply chain big data series part 1. https://advisory.kpmg.us/content/dam/advisory/en/insights/pdfs/2018/supply-chain-big-data-part-1-shaping-tomorrow.pdf

López-Gómez C, Leal-Ayala D, Palladino M, O'Sullivan E (2017) Emerging trends in global advanced manufacturing: challenges, opportunities and policy responses. http://capacitydevelop ment.unido.org/wp-content/uploads/2017/06/emerging_trends_global_manufacturing.pdf

McCloy D (1984) Industry: structures and operation. Technology 1984:286–304

Muro M (2017) Get with the program digitalizing America's advanced manufacturing sector. www. nacfam.org/wp-content/uploads/2017/03/Digitalizingusmfg_090717_v3.pptx

National Science and Technology Council (NSTC) (2008) Manufacturing the future. https://nifa. usda.gov/sites/default/files/resource/nanotech__manufacturing_rd.pdf

Pedone G, Mezgar I (2018) Model similarity evidence and interoperability affinity in cloud ready industry 4.0 technologies. Comput Indus 100:278–286

Prasad B (1997) Analysis of pricing strategies for new product introduction. Pricing Strat Pract 5(4):132–141

Pasche M (2008) Product complexity reduction—not only a strategy issue. In: 11th QMOD conference. quality management and organizational development attaining sustainability from organizational excellence to sustainable excellence, 20–22 August 2008, Helsingborg, Sweden. http:// www.ep.liu.se/ecp/033/080/ecp0803380.pdf

Ruby DA (2003) The creation of wealth and economic growth. http://www.digitaleconomist.org/wth_4020.html

Schmidt M, Ahlund C (2018) Smart buildings as cyber-physical systems: data driven predictive control strategies for energy efficiency. Renew Sustain Energy Rev 90(2018):742–756

Sullivan N (2020) What is a civilization?—definition & common elements. https://study.com/aca demy/lesson/what-is-a-civilization-definition-common-elements.html

UNPD (2020a) United nations development programme—human development reports. http://hdr. undp.org/en/content/human-development-index-hdi

UNDP (2020b) Inequalities in human development in the 21st century briefing note for countries on the 2019 human development report, USA. http://hdr.undp.org/sites/all/themes/hdr_theme/country-notes/USA.pdf

Viriyasitavar W, Xu LD, Bi ZM, Sapsomboon A (2019a) New Blockchain-Based Architecture for Service Interoperations in Internet of Things. IEEE Internet Things J 6(7):739–748

Viriyasitavar W, Xu LD, Bi ZM, Pungpapong V (2019b) Blockchain and Internet of things for modern business process in digital economy—the state of the art. IEEE Trans Comput Soc Syst 6(6):1420–1733

Wikipedia (2021a) Human mission to Mars. https://en.wikipedia.org/wiki/Human_mission_to_ Mars

Wikipedia (2021b) Product lifecycle. https://en.wikipedia.org/wiki/Product_lifecycle

Williams T (2000) The Purdue enterprise reference architecture and methodology (PERA) institute for interdisciplinary engineering studies. Purdue University. http://citeseerx.ist.psu.edu/viewdoc/download?doi=10.1.1.194.6112&rep=rep1&type=pdf

Wang G (2006) Introduction to CAD/CAE. http://www2.ensc.sfu.ca/~gwa5/index_files/25.353/ind exf_files/1Introduction-06-1.pdf

Woolf SH, Aron L, Dubay L (2015) How are income and wealth linked to health and longevity? Available online, https://community-wealth.org/sites/clone.community-wealth.org/files/downlo ads/paper-woolf-et-al.pdf (accessed on April 18, 2021)

Chapter 2
Computer-Aided Design

Abstract In computer-aided design (CAD), engineers use computers to design, analyze, model, simulate, and evaluate a product, process, or system. In this chapter, the application of CAD in the creations of virtual models of products or systems is focused on. CAD was originally developed to represent geometric information of objects by using computer-drawing software. With the continuous evolution in several decades, the capabilities of CAD tools have been expanded significantly beyond computer graphics. In this chapter, various CAD techniques are introduced to model products, processes, and systems at different levels and aspects, *parametric modeling,* and *knowledge-based engineering* (KBE) are specially introduced, and basic solid modeling techniques are discussed to model *geometrics, features, design intents*, and *assembling relations*; finally, kinematic and motion analyses are explored to machine designs.

Keywords Computer-aided designs (CAD) · Computer-aided geometric modeling (CAGM) · Parametric modeling · Motion analysis

2.1 Introduction

Product or system development involves an iterative design process, from (1) the identifications of *customers' requirements* (CRs), (2) interpreting CRs into design specifications, defining design space for different design concepts, (3) evaluating and comparing design concepts, (4) implementing the selected design concepts, and finally to (5) verifying and validating design solutions (Bi 2018). Modern products or systems are mostly complex sufficient which are far beyond designers' capabilities to fulfill all design tasks manually. Computer-aided technologies, such as Computer-Aided Design (CAD), Computer-Aided Manufacturing (CAM), Computer-Aided Process Planning (CAPP), and Computer-Integrated Manufacturing (CIM), can relief designers from manual tasks for repetitive, routine, error-prone calculations, analyses, and drawings. Advanced computer-aided tools, products, or systems with better performances can be designed in shortened lead times and at a reduced cost (Lyu et al. 2017).

Computer-aided design (CAD) is the technology of using computers to design, model, analyze, and evaluate a product or system and document the design process. CAD is a powerful and effective tool for engineers to conduct research, innovation, new product design, and development. CAD software has been widely used in the manufacturing sector for the purposes of (1) accelerating and optimizing product or system designs, (2) facilitating collaboration and cooperation via graphic communication, (3) developing and maintaining the databases of products, processes, and systems, (4) reducing the cost and lead time of product development, and (5) improving the productivity of designers (Wikipedia 2020a). A CAD tool offers the flexibility for engineers to lay out and create virtual models for products and systems and ensure them to fulfill the anticipated functions even before their physical replica is made. CAD has become an essential technology that is used almost in every application such as *aerospace, automotive, shipbuilding, military, healthcare, industrial and architectural design,* and *textile* and the *fashion world* (Jhanji 2018).

CAD is used to automate *engineering design processes* by which human intelligence, knowledge, innovation, and creativity are utilized in developing products or systems to meet *engineering specifications* and *customers' requirements* (CRs). Generally speaking, *an engineer design process* corresponds to a series of loosely structured, open-ended activities to (1) define a problem with design needs, criteria, and constraints, (2) generate, model, and evaluate design options, and (3) determine, implement, and validate the design solution (Chang 2014). *Computer-aided geometric modeling* (CAGM) is a branch of CAD that mainly concerns the representation of geometries and shapes of objects. Figure 2.1 shows that CAGM is the foundation of CAD since the design of a product or system generally starts from

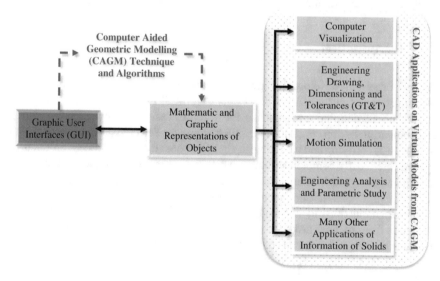

Fig. 2.1 Role of CAGM in computer-aided systems (CAD) (Bi and Wang 2020)

Table 2.1 Seven components of a CAGM system

Component	Description
Input (I):	Gives the information of *design needs* such as *specifications*, *topologies*, *features*, *dimensions*, and *design intents*.
Variables (V):	Specifies the information of *the design space* such as (1) the representations of design intents, parameters, features, and assembly mates, (2) the modelling strategies, techniques, and topologies and sequences of graphic entities.
Outputs (O):	Are the outcomes be generated in *the design process* such as *solid models*, *engineering drawings*, *rendered scenes*, *animations*, and *the physical data for sequential engineering analyses*.
Goal (G):	Is the set of soft objectives to be optimized in the design process such as *customer satisfactions*, *productivity*, *time*, *cost*, *flexibility* to adopt the changes, and *manufacturability*.
Computer Environment (E):	The physical entities used to execute the design process such as computing hardware, software, storages, and the devices for inputs, outputs and interfaces.
Constraints (C):	Is the set of hard objectives that *a design solution* must satisfy such as *the functionalities*, *compliances*, *standardization*, *accessibility*, *ergonomic needs*, and *free of illegality*, *redundancy*, *conflicts or interferences*.
Resources (R):	Include other relevant resources to support and facilitate the design process such as *designers' expertise*, *design templates*, *libraries and utilities*, *toolboxes*, *add-ins tools*, *parametric modelling tools*, *macro supports*, *reverse engineering* (RE), and *machine learning*.

creating a virtual geometric model of objects. CAGM aims to represent the geometries, shapes, dimensional and spatial relations of objects. Once an object is virtually modeled, any associated information can be obtained and utilized to support relevant decision-making activities such as design optimization. For examples, (1) a solid model is often used to visualize and evaluate the conceptual design before the physical product is actually made; (2) engineering drawings are created from a virtual product model to facilitate its manufacturing processes, and engineering drawings are generated directly from solid models with all relevant information such as *geometric dimensioning and tolerances* (GD&T); (3) other engineering analyses, such as *computer-aided engineering* (CAE) and *computer-aided manufacturing* (CAM), can be conducted at any design stage when the virtual model is ready.

In CAGM, *applied mathematics* and *computational geometry* are integrated to represent the shapes of objects, associated behaviors, and properties. From a system perspective, a CAGM consists of the following seven components, which are described in Table 2.1 and Fig. 2.2.

2.2 Computer-Aided Design (CAD) System

A CAD system deals with data that is intangible, and it is helpful to understand a CAD system by comparing it with an equivalent system that deals with tangible materials. From this perspective, a CAD system is equivalent to a manufacturing system in terms of having a series of transferring processes from inputs to outputs (see Fig. 2.3). In

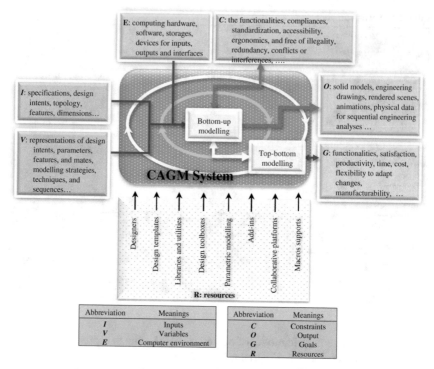

Fig. 2.2 Description of computer-aided geometric modeling (CAGM) system

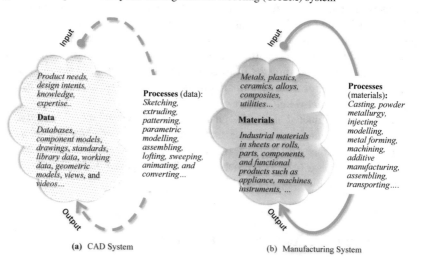

Fig. 2.3 Transformation similarity of a CAD system and manufacturing system

a CAD system, certain types of data inputs are *processed* and *transferred* into other types of data outputs through a series of computing processes (Bi and Cochran 2015). Types of data can be *product needs, design intents, expertise, databases, component models, drawings, standards, library data, working data, geometric models,* and *views, animation,* and *videos.* Types of computing processes can be performed by *sketching, extruding, parametric modeling, assembling, lofting, sweeping, animating, and converting,* and more *add-ins functions* for different applications.

System architecture describes constitutive components, their relations, and the rules for the changes of these system factors over time (Niu et al. 2013). Figure 2.3 shows a comparison of the architecture of a CAD system and a manufacturing system. Similar to a manufacturing system architecture that consists of *machine tools, controllers, material handling facilities,* and other manufacturing resources that are required to transfer raw materials to finished products, a CAD system architecture consists of *computing hardware, software,* and *input and output devices* that are required to transfer, share, process, and produce output data.

As shown in Fig. 2.4, *the hardware* of a CAD system is one or a number of networked computing resources. *A computer resource* typically consists of (1) *a central processing unit* (CPU) to process data, (2) *memory and storage* to keep data, and (3) an *operating system* to manage the resource. The hardware determines the capacities of the CAD system in terms of *processing speeds,* and *volume and types of data.* The computing capacity used to be measured by the number of transistors on an integrated circuit and it was doubled every 18 months according to *Moore's law.* Since the advancement of *classic digital computing* (CDC) in the forms of binary digital electronics seemed to approximate its limits, new computational models such as *analog computing* (AC), *neuro-inspired computing* (NC), and *quantum computing* (QC) were expected to continuously improve the computing capabilities of hardware systems exponentially (Intel 2002; Shalf and Leland 2015). Moreover, the computing capacity of a CAD system can be increased by *parallel*

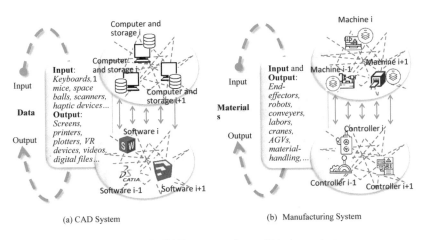

(a) CAD System   (b) Manufacturing System

Fig. 2.4 Architecture similarity of a CAD system and manufacturing system

computing by a number of CPUs or computers in the distributed environment or *cloud computing* by the cloud services over the Internet (Zissis et al. 2017; Wang et al. 2017). An *input device* is to input raw data and the designer's intents in a CAD system; common types of input devices *keyboards, mice, space balls, haptic devices, barcode readers,* and *scanners.* An *output device* is to output design outcomes for their usages; common types of output devices are *screens, printers, plotters, Virtual Reality* (VR) devices, *videos,* and *digital files.* Finally, the software determines the functionalities of a CAD system. CAD software can be either *open-source, non-commercial free* or *commercial.* Popular open-source or free CAD software packages are *Autodesk 123D, freeCAD, SketchUp, Onshape, TinkerCAD, 3D System Cubify, OpensCAD, RepoCAD,* and *Blender* (Junk and Kuen 2016). Examples of popular commercial CAD systems are *Solidworks, AutoCAD, Creo, Inventor, Solid Edge, Catia 3D Experience,* and *NX for Product Design* (Hooper 2020). Without losing the generality, Solidworks is selected in this chapter to illustrate the functionalities of CAD system.

2.3   Properties of Solid Models

The geometric representation of three-dimensional (3D) solids plays important role in making discrete products. A system for the geometric representation of objects consists of four components (1) symbol structures of solid objects, (2) processes in which geometric information such as volume and dimensions are defined, (3) input mechanisms for creating and editing representations in an iterative process, and (4) the devices to output the graphic results. As shown in Fig. 2.5, the sub-system for entering, processing, modifying the graphic representations of objects is called computer-aided graphic modeling system (Requicha 1980).

In CAGM, solids are postulated as the abstract geometric entities that are bounded, closed, regular, and semianalytic in the subset of 3D *Euclidean space* (E^3). A very

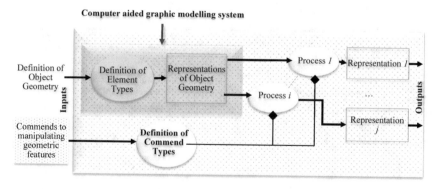

Fig. 2.5 Four components of a computer graphic modeling system

small subset of E^3 can be used to represent physical solids since an abstract solid must possess the properties of *rigidity, homogeneity, finiteness, closure, finite describability,* and *determined boundaries* that are described in Table 2.2 (Requicha 1980). The properties in Table 2.2 can be used to justify if a computer model is a legal solid model or an operation over the solids is legal or illegal. The example in Fig. 2.6 is not a legal solid model; it does not possess the properties of homogeneity and determined boundaries due to the existence of dangling line and face. The example in Fig. 2.7 shows that the different algorithms for the same Boolean operation may produce a legal or illegal solid model.

Table 2.2 The properties of solid models

Properties	Description
Rigidity:	A solid model has a given shape or configuration that is independent of the orientation and location of solid.
Homogeneity:	Any position in E^3 can be defined as one of '*interior*', '*exterior*', and '*boundary*' unambiguously; the solid cannot have dangling or isolated elements.
Finiteness:	A solid must have a finite volume of space.
Closure:	Rigid motions and the Boolean operations of solids must sustain all properties of solids.
Finite describability:	A solid has a finite set of elements such as '*faces*' to ensure that the solid is representable as a computer model.
Determined boundaries:	A solid has a set of clearly defined boundary faces to determine the *interior* volume.

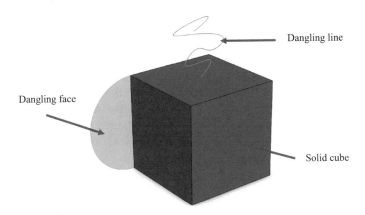

Dangling line

Dangling face

Solid cube

Fig. 2.6 Example of illegal solid model

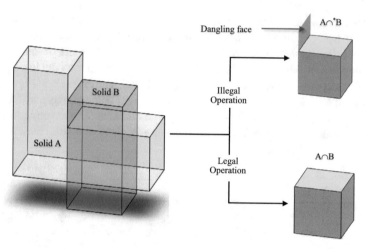

Fig. 2.7 Example of illegal solid operation

2.3.1 Legality of Solid Models

A physical body (B) can be represented by (1) a set of graphic elements such as *vertices* (V), *edges* (E), loops (L), genus (G), and *faces* (F) and (2) their dependent relations. The meanings of a loop and genus are illustrated in Fig. 2.8.

As a pyramid example shown in Fig. 2.9, the finite volume is bounded by five faces (F_1, F_2, F_3, F_4, and F_5), each face is formed by a number of edges (e.g., F_1 is formed by the edges of E_1, E_2, E_3, and E_4), and each edge is formed by two vertices (e.g., E_1 is formed by the vertices of V_1 and V_2). The graphic elements of vertices, edges, faces, and bodies are at different levels but with some dependences. To have the complete in a solid model, a data structure is required to represent the topological relations of different graphic elements. Figures 2.10 and 2.12 show a hierarchical and network data structure for a pyramid model, respectively.

Note that the hierarchical data structure in Fig. 2.10 includes some redundant information since a graphic element at a low level has its relations to multiple elements at a high level. For example, a vertex can be an end point of two edges, and an edge can be a boundary line of two faces. A network data structure in Fig. 2.11 can eliminate such redundancies since data pointers are used to represent the topological relations

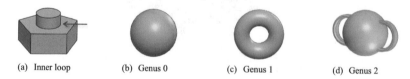

(a) Inner loop   (b) Genus 0   (c) Genus 1   (d) Genus 2

Fig. 2.8 The meanings of loops and genus

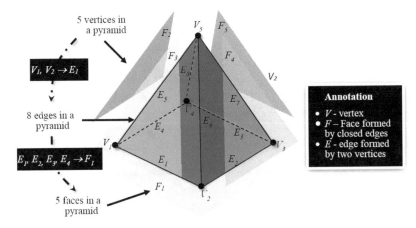

Fig. 2.9   Vertices, edges, and faces of a pyramid

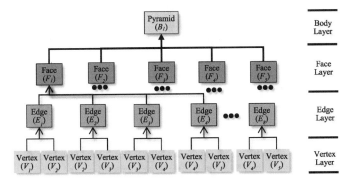

Fig. 2.10   Hierarchical structure of pyramid object

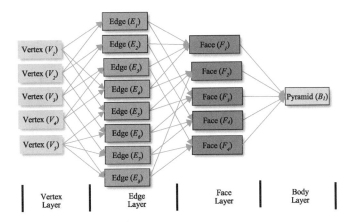

Fig. 2.11   Network structure of pyramid object

Fig. 2.12 Example of solid
object in Example 2.1

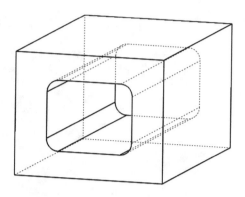

of graphic elements. Moreover, the number of data pointers for a certain graphic
element can be determined by the number of relations of this element with others.

The Euler-Poincare equation can be applied to justify if a solid model is legal
(Sanchez-Cruz et al. 2013). Euler and Poincare proved that a polyhedral is homo-
morphic to a sphere; thus, for a model with the given numbers of faces (F), edges
(W), vertices (V), bodies (B), loops (L), and genus (G), it is topologically valid if
one of the first two of the following three conditions is satisfied:

$$
\left.
\begin{array}{ll}
F - E - V - L = 2(B - G) & \rightarrow \; General\,object \\
F - E + V = 2 & \rightarrow \; Simple\,solid \\
F - E + V - L = B - G & \rightarrow \; Open\,object
\end{array}
\right\}
\qquad (2.1)
$$

Based on which condition has been satisfied, the corresponding object type is
general object, *simple solid*, and *open object*, respectively.

Example 2.1 Determine the numbers of different graphic elements, and use the
Euler-Poincare equation to classify the object type.

Solution For the object in Fig. 2.12, $V = 24$, $E = 36$, $F = 1\,4$, $G = 1$, $L = 2$, and B
$= 1$. Accordingly, the first conditions in the Euler-Poincare equation is satisfied, i.e.,

$$
24 - 36 + 14 - 2 = F - E - V - L \Leftrightarrow 2(B - G) = 2(1 - 1)
$$

Therefore, the represented model is a *general object*.

2.3.2 Geometric Constraints

Depending on the physical process in generating the shape of solid, the corresponding
solid model is valid when it meets the geometric constraints such as the following.

Legality. A solid model must be legal that has all of the properties in Table 2.2.
It should meet one of the first two conditions in the Euler-Poincare equation. Two

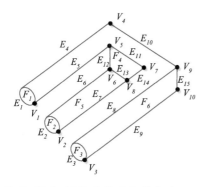

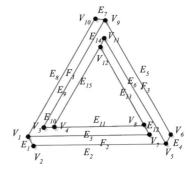

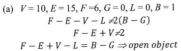

(a) $V = 10, E = 15, F = 6, G = 0, L = 0, B = 1$
$$F - E - V - L \neq 2(B - G)$$
$$F - E + V \neq 2$$
$$F - E + V - L = B - G \Rightarrow open\ object$$

(b) $V = 12, E = 14, F = 3, G = 0, L = 0, B = 1$
$$F - E - V - L \neq 2(B - G)$$
$$F - E + V \neq 2$$
$$F - E + V - L = B - G \Rightarrow open\ object$$

Fig. 2.13 Illegal model examples (*open objects*)

well-cited examples in Fig. 2.13 are illegal solid models since the third condition in the Euler-Poincare equation is satisfied.

Free of Interference. In representing a physical solid, no overlapped material is allowed at any position within the solid volume. Therefore, no interference is allowed in the corresponding solid model. Figure 2.14 shows the representation of a tapered spring. When the wire diameter is smaller than the pitch of the spring, the solid model is valid (see Fig. 2.14a). When the wire diameter is large than the pitch of the spring, the interference occurs between two neighboring rings, the solid model becomes invalid; unless that the overlapped materials are merged at the interfered volume (see Fig. 2.14b).

Accessibility. When the shaping process of a solid object is concerned, the finite volume for a solid must be accessible in the manufacturing process. While different

(a). no interference when the size of wire is smaller than pitch

(b). an interference when the size of wire is larger than pitch

Fig. 2.14 Interference example

manufacturing processes shape the geometric shape of solid in different ways, the accessibility to the finite volume can be different. In other words, an accessible solid volume for one manufacturing process does not mean that it will be accessible to another manufacturing process. Accessibility and accessibility have been specially considered in creating an assembly model of products. Figure 2.15 shows an example of a combined solid model with a total of four objects, and three objects are within the enclosed cavity of the fourth object. The virtual model is valid if the combined solid is made by additive manufacturing; otherwise, the solid volumes of three objects are inaccessible.

Manufacturability. A geometric constraint might be relevant to the required tooling in a shaping process of solid object. Taking an example of a product from the casting process in Fig. 2.16, the geometric shape of the product is determined by

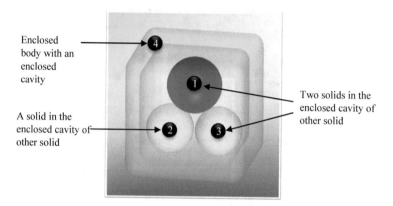

Fig. 2.15 Example of accessible/inaccessible solid volume

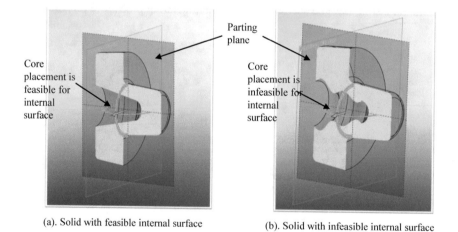

(a). Solid with feasible internal surface   (b). Solid with infeasible internal surface

Fig. 2.16 Consideration of manufacturability of solids

the cavity formed in the mold assembly. An internal surface of casting is defined by a core, and one has to ensure the core is feasibly removed from the mold assembly without causing damage to internal surface. From this perspective, one of the solid models of casting in Fig. 2.16a is valid since it is feasible to place a removable core to generate an internal surface.

2.4 Graphic Modeling Techniques

As shown in Fig. 2.17, the representation of a physical object consists of the graphical elements at different levels from *points*, *lines*, *surfaces*, *volumes*, *features* to *design intents*. Accordingly, the modeling techniques can be classified based on their capabilities in creating and manipulating these graphical elements into *wireframe modeling*, *surface modeling*, *solid modeling*, *space decomposition*, *feature-based modeling*, and *knowledge-based modeling*. The higher level of graphic elements is, the more advanced the computer-aided graphic modeling (CAGM) tool is required.

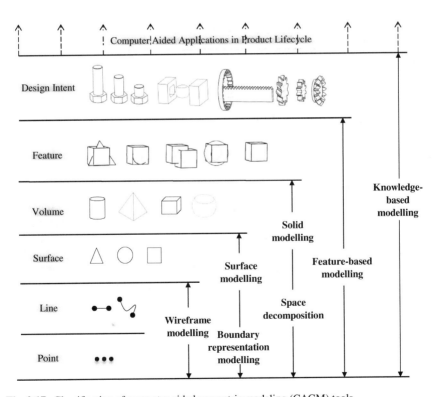

Fig. 2.17 Classification of computer-aided geometric modeling (CAGM) tools

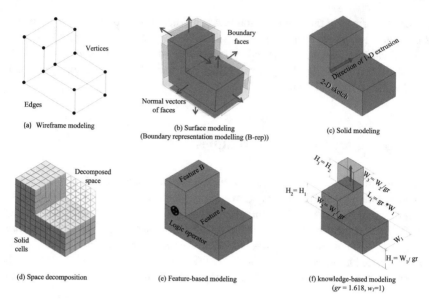

Fig. 2.18 Modeling techniques versus types of graphic elements

The example in Fig. 2.18 shows the differences of using these modeling techniques to model a physical object formed by two or more rectangular blocks in a golden ratio, i.e., *Width*: *Length* = 1: 1.618.

2.4.1 Wireframe Modeling

In *wireframe modeling*, graphic elements are *points* for vertices and *lines* for edges, and an object is represented by its boundary edges. Edges can be *circles, arcs, straight* or *curvy lines*. Since no upper level graphic information such as *faces* or *volumes* is available, the visibility of a graphic element is not detectable, and the information of surface areas or masses is not available. Since creating a wireframe model requires to input the coordinates of vertices, wireframe modeling is ineffective to model complex shapes due to a large number of data points. Moreover, a wireframe model may cause *the ambiguity* of the represented geometry. However, since *points* and *lines* are two primary types of graphic elements, wireframe modeling can be used as the supporting technique of other modeling methods.

2.4.2 Surface Modeling and Boundary Representation Modeling (B-Rep)

In *surface modeling*, a physical object is modeled as a set of finite surface patches in free forms. A surface model has sufficient information to determine the visibility of graphic elements. However, a surface model does not include the information of volumes or masses; since no thickness of volume information is available.

Surface modeling can be advanced to be *boundary representation modeling* (B-Rep). B-Rep further defines a finite and closed cover of a solid over a surface model. To define the volume of a solid, each constitutive surface patch is treated as *a half space* to separate *inside* space from *outside* space in a Cartesian space:

$$H = \{P : P \in E^2 \ and \ f(P) < 0\} \tag{2.2}$$

where H is the set of all points in the inside half space; P is an arbitrary point in the Cartesian space E^3; $f(P) = 0$ is the function of surface patch; $f(P) < 0$ is the constraint for any point at the negative direction of normal vector of the surface patch.

As shown in Fig. 2.17(b), the normal vector of a surface patch defines the inside half space of the patch; each surface patch divides the Cartesian space into two regions of infinite extension. Furthermore, if a point in the Cartesian space is at the inside half space of all boundary faces, it is in the volume of the solid. Accordingly, the volume of the solid S is the intersection of the half spaces H_i where $(i = 1, 2, ...N)$ of all surface patches as,

$$S = \bigcap \left(\sum_{i=0}^{N} H_i \right) \tag{2.3}$$

where S is the solid pace, N is the number of boundary surface patches, and H_i is the half space of the i-th surface patch $(i = 1, 2, ...N)$.

2.4.3 Solid Modeling and Space Decomposition

In *solid modeling* or *space decomposition modeling*, the volumetric information is defined by specifying all geometric dimensions in a 3D space.

Figure 2.19 shows three ways to specify 3D volumetric dimensions in solid modeling. *The first way* (see Fig. 2.19a) gives the dimensions of a solid in a 3D space by a 2D sketch and 1D depth along the direction of extrusion. *The second way* (see Figs. 2.19b and 2.18c) gives the dimensions of a solid in a 3D space by one 1D path, 1D thickness and 1D width over the cross-section. *The third way* (see Fig. 2.19c) gives the dimensions of solid in a 3D space by one 1D path and 2D profile.

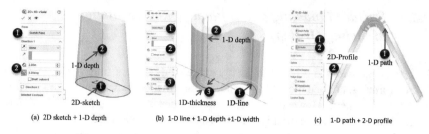

(a) 2D sketch + 1-D depth   (b) 1-D line + 1-D depth +1-D width   (c) 1-D path + 2-D profile

Fig. 2.19 Three ways of specifying 3D dimensions in solid modeling

In space decomposition modeling, a solid is represented by a set of isomorphic 3D cells. The cell size determines the precision of the geometric representation, and it should usually be several-order smaller than the dimensions of the solid. As shown in Fig. 2.18d, a continuous space where the solid locates is discretized into a 3D array of isomorphic cells, the state of each cell is checked to see if it locates in the finite volume, and the solid model is the collection of all cells in the volume of solid. Space decomposition modeling is widely used to represent a continuous domain by a set of discretized elements and vertices in numerical simulations such as *finite element analysis* (FEA).

2.4.4 Feature-Based Modeling

In *feature-based modeling*, basic graphic elements such as *drawing references* and *solid primitives* are treated as *features,* and a solid object is composed of a set of constitutive solid features by the composition operations so-called *Constructive Solid Geometry* (CSG). Note that a feature is unnecessary a solid element; a feature such as a reference coordinate system is defined to *create, position,* and *dimension* solid features efficiently.

Figure 2.18e shows that the solid is formed by *uniting* feature-A and feature-B at the locations these two features are placed. Feature-based modeling is very flexible in the sense that any pre-built graphic element can be defined as a feature for the purpose of reusing (Fig. 2.20). shows 8 most commonly solid primitives which include *cube, rectangular block, prism, sphere, cylinder, tapered cylinder, cone,* and *torus.* In creating a complex geometry, solid primitives can be tailored by some solid modeling tools such as *extruding, revolving, sweeping,* and *lofting.* These tools will be discussed in the coming sections Table 2.3.

As shown in Fig. 2.21, three basic Boolean operations of CSG for the composition are *union* (∪), *intersection* (∩), and *difference* (\). Note that (1) the operation of difference depends on the order of features, and (2) the operations take place when the positions and orientations of the features are given.

Table 2.3 Common features and illustrations

Type	Operation	Illustration
Sketched	Extrude (2D-sketch + 1D depth)	
	Revolve (2D-sketch + 1D axis)	
	Sweep (2D enclosed profile + 1D path in 2D or 3D)	
	Boundary boss/base (Two 2D profiles + path normal controls)	
	Loft (a set of 2D profiles + the ordered connections of vertices on the profiles)	

(continued)

Table 2.3 (continued)

Type	Operation	Illustration
Build-in	Hole wizard (the set of hole positions + hole type + dimensions)	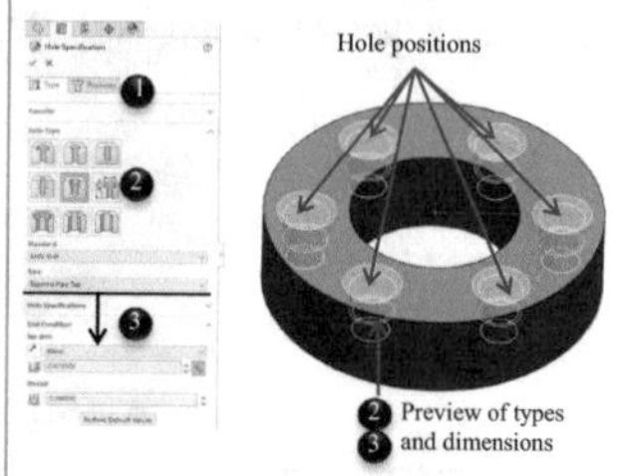
	Thread (Thread edge + end-conditions + thread dimensions)	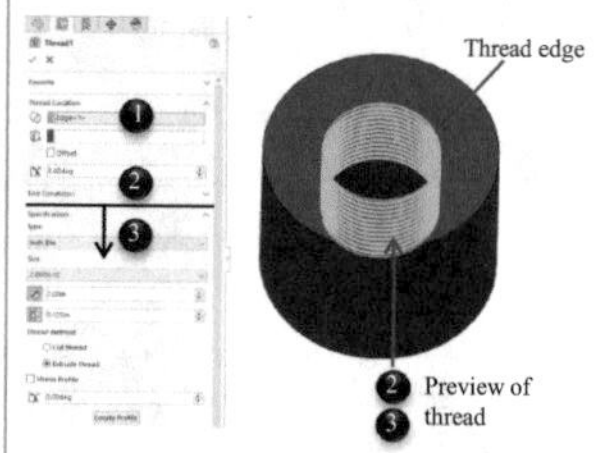
	Fillet/Chamfer (a set of selected edges or faces + size of fillet or chamfer)	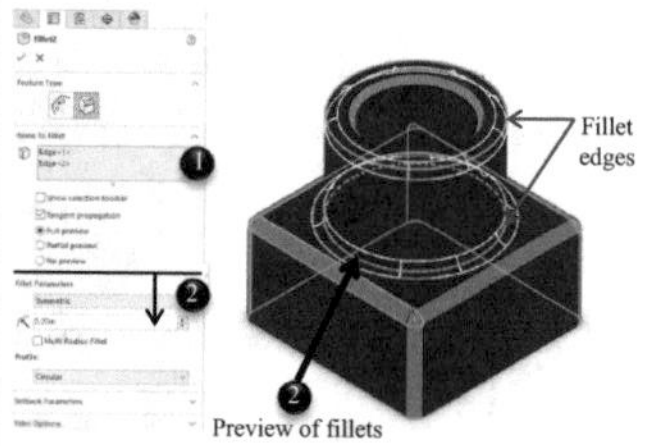
	Pattern (linear or circular) (the direction of pattern + the instance (features, solids, or other entities) + other parameters such as spacing, and number of instances)	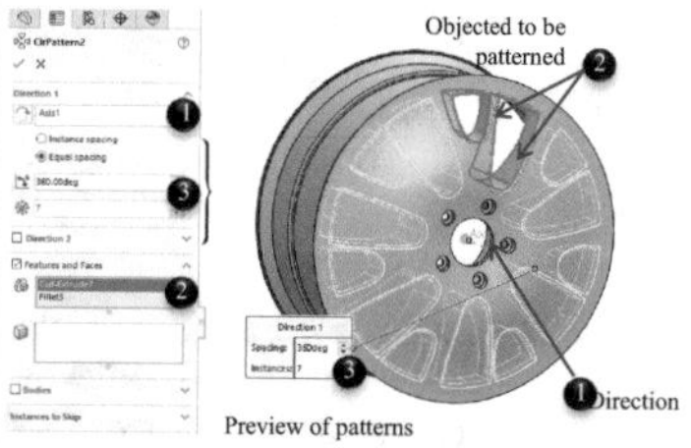
	Rib (an open 2-D sketch + rib thickness + the direction of material)	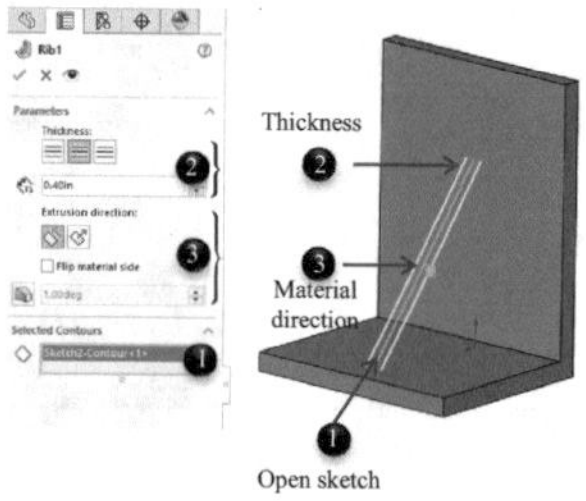

(continued)

Table 2.3 (continued)

Type	Operation	Illustration
	Draft (a set of surfaces to be drafted + the draft angle + the direction of drafting)	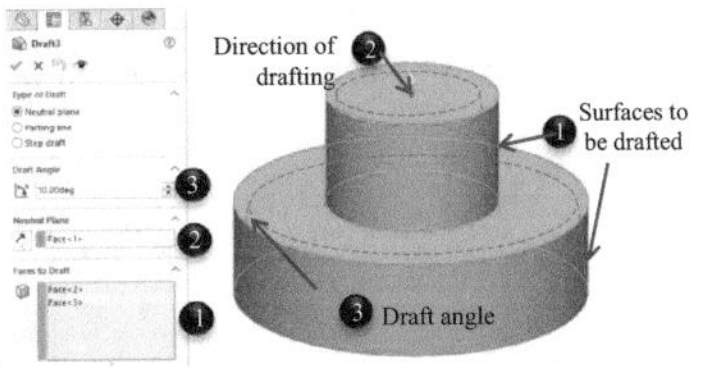
	Shell (the face to create the shell + the wall thickness)	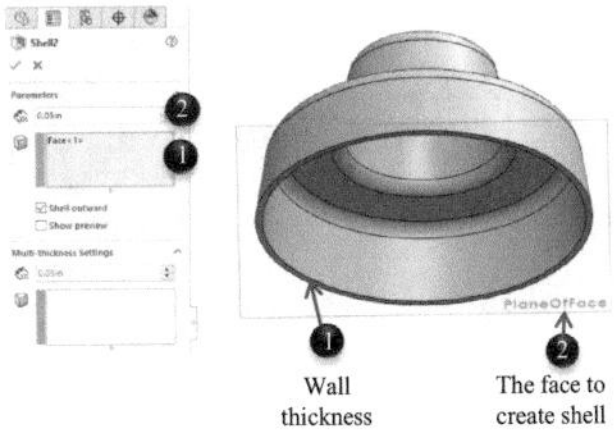
	Wrap (Wrap type + warp method + wrap sketch + wrap face + warp thickness)	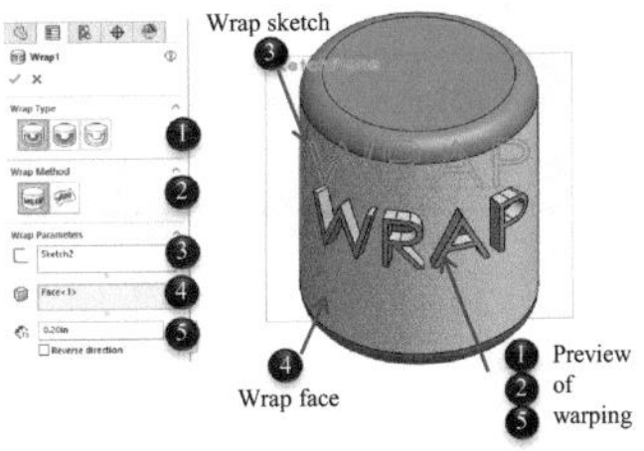
Reference	Point (the relation of the selected entity with the point + the selected entity)	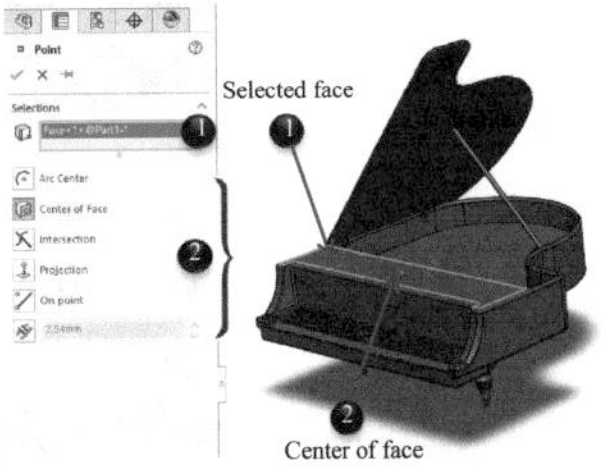
	Axis (the relation of the selected entity with the axis + the selected entity)	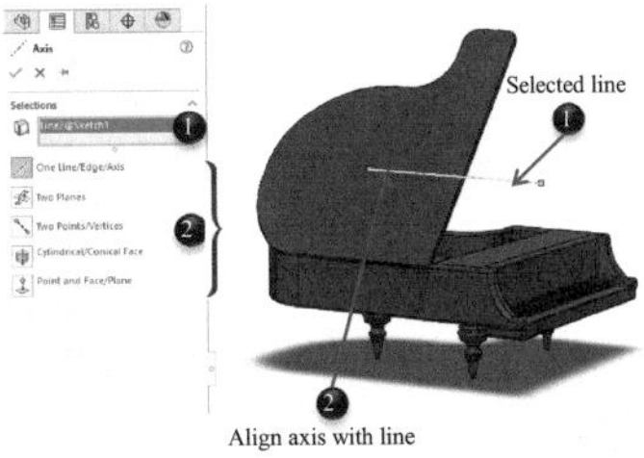

(continued)

Table 2.3 (continued)

Type	Operation	Illustration
	Plane (the relation of the selected entity with the axis + the selected entity)	 Selected face ① ② Align plane with the selected face
	Coordinate system (the origin + two of three axes)	 Z-axis ④ Y-axis ③ Origin ① X-axis ①

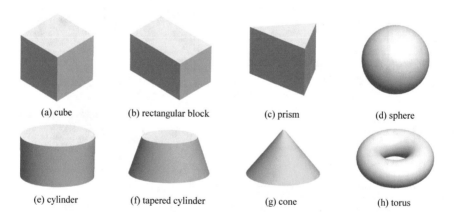

(a) cube   (b) rectangular block   (c) prism   (d) sphere

(e) cylinder   (f) tapered cylinder   (g) cone   (h) torus

Fig. 2.20 Eight types of commonly used solid primitives

To place a 3D feature at the right position and orientation for a CSG operation in a given coordinate system, one or more coordination transformations can be applied. A *coordinate transformation* (CT) is a mathematical process of transforming a set of graphic features into a new position and orientation coordinate system by *translating*, *copying*, *rotating*, *mirroring*, and *scaling* these features. Figure 2.22 shows the example of the transformed object using these four operations, respectively.

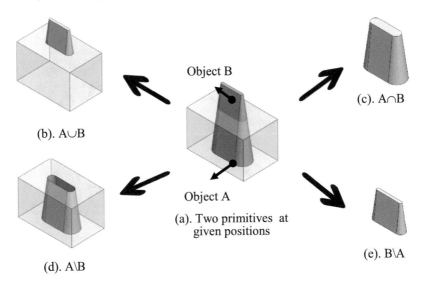

(b). A∪B

Object B

(c). A∩B

(d). A\B

Object A

(a). Two primitives at given positions

(e). B\A

Fig. 2.21 Three basic Boolean operations of 3D features

Fig. 2.22 Four operations of coordinate transformation (CT)

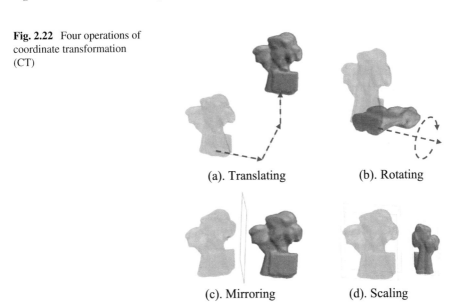

(a). Translating

(b). Rotating

(c). Mirroring

(d). Scaling

Assume that (1) a solid is constructed from a set of solid primitives $S_{i,pi}$ at the respective positions p_i ($i = 1, 2, 3,\ldots, N$), where N is the number of solid primitives; (2) a Boolean operation ('∪', '∩' or '\') over two primitives is denoted as $\otimes$, then the solid is represented as a series of composition operations in CSG as

$$S_c = ((S_{1,p1} \otimes S_{2,p2}) \otimes S_{3,p3} \ldots) \tag{2.3}$$

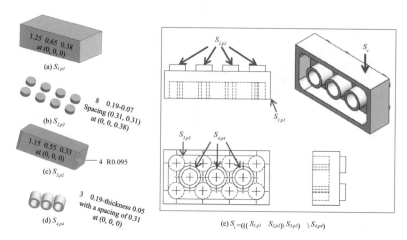

Fig. 2.23 Using CSG to build a model for a Lego piece

where S_c stands for the composed solid, $S_{i,pi}$ ($i = 1, 2, 3, ...N$) is solid primitive i at its respective position p_i, and $\otimes$ is one of the Boolean operations ('∪', '∩' and '\'). Figure 2.23 shows the series of logical operations to model a Lego piece S_c from $S_{i,pi}$ ($i = 1, 2, 3, 4$); note that a solid primitive, such as $S_{2,p2}$ and $S_{4,p4}$, can be a composed solid itself.

In feature-based modeling, a solid is modeled by a set of solid primitives and the corresponding logical operations over these primitives. Accordingly, the data structure of the solid is a graph for the order and dependences of the Boolean operations of solid primitives. Therefore, a CSG model consists of

(1)   A binary tree for the Boolean operations,
(2)   A set of the links to solid primitives as the outer leaf nodes of tree, and
(3)   A set of intermediate components as the interior nodes of tree as the results of the Boolean operations over solid primitives or components.

Figure 2.24 shows a data structure example. The model has two data types for (1) solid primitives and (2) composition operations, respectively. *For the types of solid primitives*, the model includes three features, i.e., two blocks and one cylinder. *For the types of logical operations*, the model includes one *union* (∪) operation of two blocks and one *difference* (\) operation of the united component and the cylinder.

Since feature-based modeling creates a model for an object from solid primitives. Since solid primitives include volumetric information as well as graphic elements at low-levels such as vertices, edges, and boundary faces, a CSG model has the complete solids. Therefore, featuring-based modeling requires less steps than wireframe modeling and surface modeling to create a virtual model. However, feature-based modeling specifically emphasizes the representation of graphic contents of physical objects, it can be further improved as knowledge-based modeling to incorporate design intents in modeling practices.

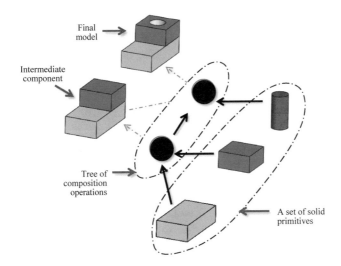

Fig. 2.24 Example pf CSG data structure

2.4.5 Knowledge-Based Modeling

Products are designed and manufactured with purposes; so are the geometries and shapes of products. Even though there are numerous ways to create a virtual model for a given object, it is very helpful for a modeler to take into account *design knowledge* in modeling graphic objects. Incorporating design knowledge in a solid helps in (1) highlighting main attributes of products, (2) simplifying the modeling processes, (3) facilitating product evolution and the reuse of product designs, (4) reducing product design times and improving the productivity of modelers, and (5) sharing and exchanging design concepts in communication and design collaboration.

The mechanism of incorporating design knowledge in a solid model is the use of design intents. *Design intent* is to describe how a design feature is represented and how it is guided when it needs changes. Design intents are embedded in a graphic model and corresponded to certain *dimensions* and *relations*; a change made on a design intent updates these graphic contents (i.e., dimensions and relations) automatically. Since design knowledge is far beyond graphic contents themselves, design intents are not just sizes and shapes of features, the scope of design intents covers *tolerances, requirements or constraints of manufacturing and assemblies, relationships of features* and *dimensions,* and other design factors in product lifecycles (Bi and Wang 2020).

From the perspective of solid modeling, design intents are represented by various types of design parameters in solid models. Using design parameters ensures that the model can always meet the requirements specified by design intents. Taking an example of the sketch in Fig. 2.18c, the design intent of the golden ratio of length and width of the rectangle was represented as a design equation in a parametric model;

no matter what the length or width is, they are in a golden ratio. From the perspective of product use, if the design intents of a product cannot be adequately represented, the product model might be misleading and sometimes even useless.

2.5   Design Parameters, Features, and Intents

Either feature-based modeling or knowledge-based modeling is a type of *parametric modeling method*. In parametric modeling, various *parameters* are defined to represent *positions, dimensions, geometries*, and *relations* of *features*. Updating a parameter changes all of the associated features of a solid model automatically. From this perspective, a parameterized model is embedded with design knowledge and intelligence so that design features can be defined and modified with the minimized manual efforts.

2.5.1   Design Features

Design features are high-level representations of building blocks of a solid by which graphic elements, engineering knowledge, and design intents are structured. To represent a solid model, the modeler should firstly understand the object to be modeled in terms of its constitutive building blocks and the corresponding composition operations. The modeling procedure follows after these building blocks are identified, these building blocks are models as design features, respectively, and then structured as a solid model based on the specified Boolean operations. Figure 2.25 illustrates the directions of understanding and modeling procedures of an exemplified solid.

As shown in Fig. 2.25, graphic features can be classified into *sketched features* and *build-in features*; in addition, the modeling references or any reusable entities can be treated as features. Table 2.3 shows some common features and examples under each catalogue.

As shown in Table 2.3, (1) the examples of sketched features are *extrude, revolve, sweep, boundary boss/base*, and *loft*; in a sketched feature, the main attributes are defined in two-dimensional (2D) or 3D sketch(s). (2) The examples of the build-in features are *hole wizard, thread, fillet, chamfer, pattern, draft, rib*, and *shell*. In a build-in feature, the main attributes are determined by computer algorithms based on users' input(s). From a manufacturing perspective, most of the features are associated with certain manufacturing processes, for example, a fillet on a shaft shoulder relates to a turning operation, a counter-bore hole relates to a drilling operation, and a draft angle relates to a variety of the manufacturing processes; the part geometry is determined by that of molds, dies or tools. (3) Examples of reference features are *points, axes, planes*, and *coordinate systems*.

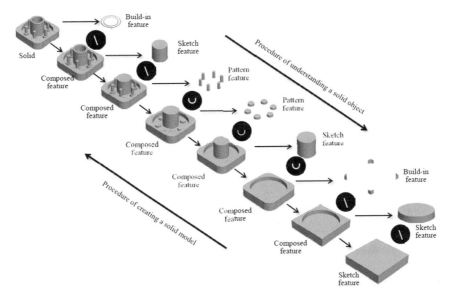

Fig. 2.25 Understanding and modeling procedures in feature-based modeling

2.5.2 Design Parameters

Design parameters are at the lowest level of a solid model, and design parameters include everything that are required to represent design features and their relations. Table 2.4 shows some common types of design parameters and examples in feature-based modeling.

Table 2.4 Common types of design parameters and their applications

Parameter	Data Type	Example
Dimensional variables	*Numerical*	(1) the height and diameter of cylindrical feature, (2) the length, the height, and the width of a rectangle block, and (3) the radius of a fillet feature.
Constraints	*Boolean*	(1) parallel, perpendicular, angular relations of two lines, (2) equal radii of two arcs, (3) convert entities in sketching, and (4) the symmetric relations of two objects about a plane.
Operations	*Mixed data and logic*	(1) the direction and instances of a pattern feature, (2) drafting over existing face, and (3) the direction and the thickness of a rib feature.
Equations	*Mathematic expression*	(1) the link value of multiple dimensions, and (2) the dependent relation of two or more dimensional variables.

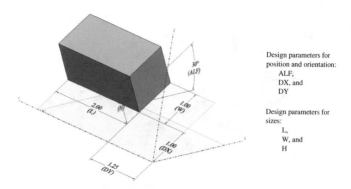

Fig. 2.26 Example of defining design parameters for sizes and positions of object

2.5.2.1 Dimensional Variables

Generally, a graphic feature has a number of dimensions to describe (1) the geometric size and (2) the relative position and orientation with respect to a reference coordinate system. By defining a *design parameter* to a dimension, the design parameter applies a *hard constraint* to the dimension; since the dimensional value will be fully controlled by the design parameter. Figure 2.26 shows an example of the fully constrained object where the design parameters have been defined for its position, and orientation, and sizes in three dimensions.

A constraint is *hard* implies that the change of the corresponding dimension has to be made by updating the value of the design parameter. Note that the *default setting* of a dimension in a modeling process is a soft constraint; in other words, the dimensional value can be changed by an undergoing algorithm in computer-aided modeling tool. Figure 2.27 shows the difference of the hard and soft constraints; a fillet radius in the case of (a) hard constraint is fully controlled by its design parameter.

It is a good practice to define a design parameter to a critical dimension as a hard constraint in modeling, especially at the stage of the sketching process of 3D features. In Solidworks, the dimensions in a sketch are generally created by using the *smart-dimension tool* in sketching, and additional dimension(s) of a 3D feature are automatically generated by the software based on the modeler's inputs when the feature is created.

2.5.2.2 Constraints

A *constraint* refers to a dependent relationship of two or more geometric entities. Constraints are mostly used to represent the design intents for the relations of geometric entities in sketching processes. Figure 2.28 shows a list of common relations of graphic entities such as *axes, line segments,* and *planes.*

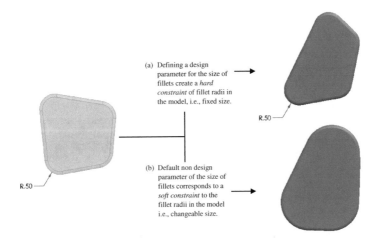

(a) Defining a design
    parameter for the size of
    fillets create a *hard
    constraint* of fillet radii in
    the model, i.e., fixed size.

R.50

(b) Default non design
    parameter of the size of
    fillets corresponds to a
    soft constraint to the
    fillet radii in the model
    i.e., changeable size.

R.50

Fig. 2.27 Difference of hard and soft constraints on a dimension

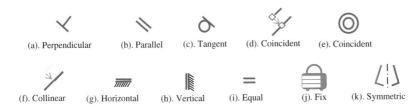

(a). Perpendicular   (b). Parallel   (c). Tangent   (d). Coincident   (e). Coincident

(f). Collinear   (g). Horizontal   (h). Vertical   (i). Equal   (j). Fix   (k). Symmetric

Fig. 2.28 Common constraints for geometric relations of graphic entities in sketching

In the sketching process, a modern CAD tool can automatically detect applicable constraints of a graphic entity with others. Once a constraint is detected and becomes active, the system gives the feedback, and the modeler has the option to apply this constraint to the selected entities. A constraint will be automatically accepted when no action is taken when the constraint is active; otherwise, the modeler needs further actions to remove or change a constraint.

A well-structured sketch usually involves in many constraints to represent different design intents. Constraints in a sketch can be created, modified, or removed by the *Display/Delete Relations* tool in Solidworks. Figure 2.29a is the icon for the Display/Delete Relations for activation. Figure 2.29b shows a scenario when the selected entities are *a point* and *a line*. The system identifies three applicable constraints: (1) the point is a *midpoint* of the line segment, (2) the point is *coincident* with the line segment, or (3) both point and line entities are *fix* in the sketch. Figure 2.29c shows the case when the selected entities are two line segments. The applicable constraints identified by the system include *horizontal, vertical, collinear, perpendicular, parallel* relations of two lines, an equal length of two lines, and both line segments are fixed in the sketch. Figure 2.29d illustrates the case when two points are selected, the applicable constraints identified by the system include the

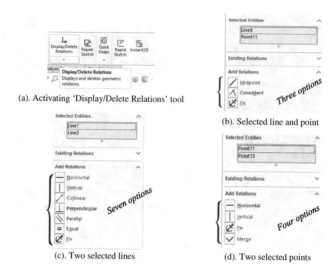

Fig. 2.29 Examples of applicable constraints of selected entities

relations of *horizontal, vertical, merge* of two points, or both of the points are *fix* in the sketch. Therefore, the system determines the types of applicable constraints based on the selected number and types of graphic entities.

2.5.2.3   Operations

An operation is to create a graphic feature from a sketch and add one or two dimensions to the features. Certainly, design parameters should be defined to represent these dimensions and other associated options related to the operation. Figure 2.30 shows the setup of a linear pattern operation. The design parameters involved in the operation include *the direction, spacing,* and *the number of instances* of the first and the second directions of patterning, and the feature(s) or bodies to be patterned.

Figure 2.31a shows the setup of a surface thickening operation. The design parameters involved in the operation include *the selected surface, the direction,* and *the dimension* of thickening. Figure 2.31b shows the setup of a revolving operation. The design parameters involved in the operation include *the axis, the angle,* and *the direction of revolving,* and *the selected 2D sketch.*

2.5.2.4   Design Equations

A *design equation* is used to express the dependence of one design parameter on other design parameters. Using design equations is effective when a set of design parameters are associated with each other by certain rules or standards; therefore, design equations are widely used to model machine elements and other standardized

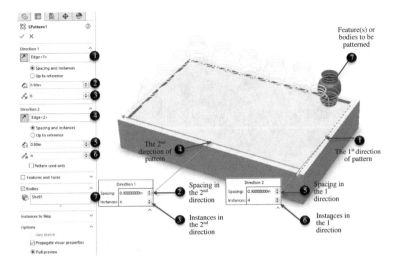

Fig. 2.30 Design parameters in the *operation* of linear pattern

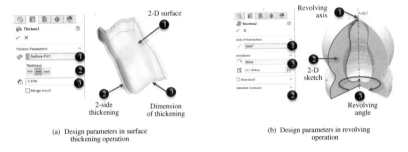

(a) Design parameters in surface thickening operation

(b) Design parameters in revolving operation

Fig. 2.31 Design parameters in the *operations* of *surface thickening* and *revolving*

products. Note that standardization is critical to reduce product cost, and standardization makes independent parameters dependent. Therefore, design equations can be used to represent the design intents where geometric dimensions are adhered to a specified set of rules. Figure 2.32 shows the models of a gear or a nut where a number of design equations have been defined to represent the dependences of these parameters in two models, respectively.

A design equation is a mathematic expression that follows the syntax of the programming languages of a CAD system. In such an expression, a dependent variable on the left side of the assigning symbol '=', and its value is determined by the expression on the right side of '='. Variables in the expression of a design equation can be either of a *global* or independent variable. The design variables are mostly numeric for dimensioning; however, design variables can also be defined in other

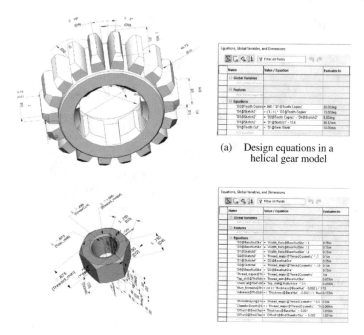

(a)   Design equations in a
helical gear model

(b)   Design equations in a nut model

Fig. 2.32 Machine element models with design equations

formats such as the states of features, material, and document properties. In addition, a CAD system supports most common operations over design variables. For example, Solidworks supports a list of the legal operations as shown in Table 2.5.

Figure 2.33a shows how to access the *equations* tool in Solidworks, which is listed in the *insert* menu. Once the equations tool is activated, the window of *managing equations* is opened as shown in Fig. 2.33b. It consists of three fields for *global variables*, *feature variables*, and *dimensional variables*, respectively. A variable on the left side is determined by the expression on the right side of the equation, and its evaluated value is seen in the last column.

Table 2.5 Legal mathematic operations in *design equation tool* of Solidworks

Operator or Function	Meanings	Operator or Function	Meanings
+	Addition	arccos(x)	find the angle of the cosine ratio x
−	Subtraction	atn(x)	find the angle of the tangent ratio x
*	Multiplication	arcsec(x)	find the angle of the secant ratio x
/	Division	arccosec(x)	find the angle of the cosecant ratio x
^	Exponentiation	arccotan(x)	find the angle of the cotangent ratio x
sin(x)	fine the sine ratio of angle x	abs(x)	find the absolute value of x
cos(x)	fine the cosine ratio of angle x	exp(x)	find e raised to the power of x
tan(x)	fine the tangent ratio of angle x	log(x)	find the natural log of x to the base of e
sec(x)	fine the secant ratio of angle x	sqr(x)	find the square root of x
cosec(x)	fine the cosecant ratio of angle x	int(x)	find the integer of x
cotan(x)	find the cotangent ratio of angle x	sng(x)	find the sign of x as '-1' or '1'.
arcsin(x)	find the angle of the sine ratio x	pi	refers to the ratio of the circumference

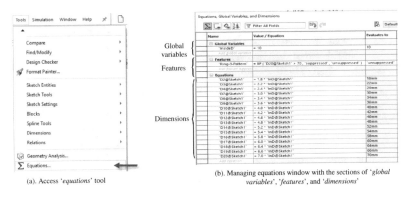

(a). Access 'equations' tool

(b). Managing equations window with the sections of 'global variables', 'features', and 'dimensions'

Fig. 2.33 Using the *design equations tool* in Solidworks

2.5.2.5 Design Tables

A *design table* is used to represent a set of configurations in one part or assembly model. Using a design table is able to represent a part or product family by a single model. One part or assembly is referred to as a *configuration* in the family. Typically, a model of a part family consists of (1) a set of *design parameters* for major dimensions and the states of optional features (2) a *design table* that includes a set of configurations with the assigned dimensions or states of these design parameters.

A design parameter in a design table differs from that in a design equation; in sense that the former is discrete or logical variable, while the latter is continuous. In addition, a number of design parameters can be dependent variables; while a design equation can only deal with one dependent variable.

A design parameter can be any graphic entity that has an associated parameter identity (*ID*) in a solid model. Common types of design parameters are *constraints*, *dimensions*, and *states* of *features*, *parts*, and *mates*. Therefore, the design table tool is capable of (1) including a variety of sketches, features, parts, or assemblies in a single model, (2) creating new components from existing components, and (3) incorporating the modeling intelligence such as parametrical constraints and design equations in a design table. Figure 2.34 shows some part and product families, which can be modeled by a design table with a set of the design parameters for *features*, *parts*, *components*, and *assembling mates*. To sustain the consistency of data sources, only one design table is allowed in one part or assembly model.

In a design table, the variants of part or product are represented by the configurations, and the configurations can be modeled manually or directly generated by computers. As shown in Fig. 2.35a, a design parameter can be assigned with different values manually to create different configurations. Alternatively, a design table can be defined to assign different values to design parameters in order to create new configurations automatically (see Fig. 2.35b).

When a design table is defined for a model with some existing configurations, these configurations will be automatically included in the table. Available configurations

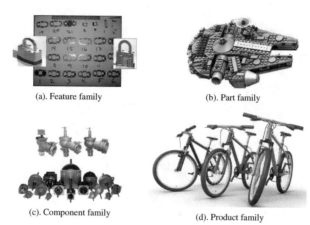

(a). Feature family   (b). Part family

(c). Component family   (d). Product family

Fig. 2.34 Examples of using a design table for part families

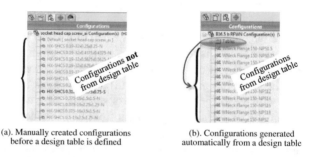

(a). Manually created configurations   (b). Configurations generated
before a design table is defined   automatically from a design table

Fig. 2.35 Example of configurations in a design table

can be viewed and edited under the *configurations* tool of the model tree as shown in Fig. 2.35. Finally, if a model has a design table already, no new design table is allowed, and the changes on the configurations can only be made by editing the existing design table.

The design table tool in Solidworks is implemented by the external Microsoft Excel program. Therefore, Microsoft Excel must be available to apply the design table tool. A design table is directly associated to its model in sense that all of the controlled attributes in the table must have internal links to the corresponding design parameters such as dimensions or feature states in the model.

Figure 2.36 illustrates the procedure to create a design table in a part or assembly model. *Firstly*, all controllable dimensions and features must be defined in the model. *Secondly*, the modeler has the options of (1) manually define a few configurations or (2) proceed to create a design table directly. *Thirdly*, the design table tool under the *insert* menu is activated, and the modeler then selects and adds the dimensions and features as controllable attributes into the design table. Table 2.6 shows the types of design parameters which can be controlled by the design table. *Fourthly*, a number of

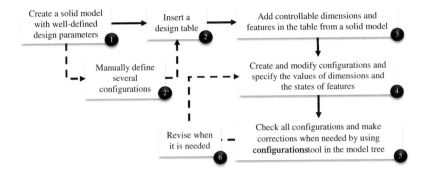

Fig. 2.36 The procedure of defining a design table in a model

Table 2.6 Types of legal design parameters in a design table (Solidworks 2018)

	In either of a part model or an assembly model	Only in Part Model	Only in Assembly Model
Design variables	▪ Tolerance type ▪ Configuration Specific Properties ▪ Model color ▪ Linear and Radial Pattern ▪ Spacing and Instances Derived Configurations ▪ Lighting state ▪ Equation state ▪ Sketch relationship state ▪ Mass Properties ▪ Center of Gravity	▪ Feature state ▪ Configuration of base or split part ▪ Dimension values	▪ Component state ▪ Mate state ▪ Referenced Configuration ▪ Display State ▪ Assembly feature state (cuts) ▪ Dimension and Mate values ▪ Bill of materials (BOM) part number ▪ Expand in BOM

configurations are defined by assigning different values to design parameters. *Fifthly*, the newly created configurations are previewed and checked under the configurations tool of the model tree. If something is wrong with certain configuration, it has to be fixed by editing the design table.

Figure 2.37 shows the procedure of defining a design table in Solidworks.

As shown in Fig. 2.37a, the design table tool can be accessed by clicking '*Insert*' in the menu bar → '*Tables*' in the expended list → '*Design Table*' in the expended list. A popup window in Fig. 2.37b shows that there are three options in creating a design table: (1) the *blank* option is for an empty template, (2) the *auto-create* option is to provide guides in adding dimensional variables into the table template, and (3) the *from file* option provides the flexibility to import a pre-defined design table. Figure 2.37c shows the result of selecting the *auto-create* option, a pop-up window will list the majority of legal design parameters such as dimensional variables and other model attributes. The modeler can choose the design parameters with variants and insert them in the design table. The modeler can also insert other design parameters that are not listed at the phase of editing. As shown in Fig. 2.37d, once all the controllable attributes are added, the modeler can create a new configuration one by one by adding a new row with specified values for controllable attributes in the design table. After all, configurations are inserted, they can be previewed and checked in the *configurations* of the model tree as shown in Fig. 2.37e.

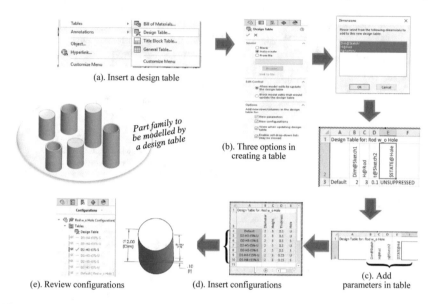

(a). Insert a design table

Part family to be modelled by a design table

(b). Three options in creating a table

(c). Add parameters in table

(d). Insert configurations

(e). Review configurations

Fig. 2.37 Interface of defining a design table in Solidworks

Example 2.2 Create a model for an optical encoder family with six levels of resolutions as shown in Fig. 2.38.

Solution To plan the modeling process, two identified design intents are to (1) use design equations to represent the dependences of dimensional variables and (2) use a design table to represent all variants (e.g., configurations) of the encoder model. Accordingly, Fig. 2.39 shows the three major steps of the modeling process. *Firstly*, the basic model of an optical encoder is created. The base feature is a solid circle with a central hole, and it is then added six pairs of features (see Fig. 2.39a). Each pair involves a cut feature and corresponding circular pattern, and its radical dimension and the number of the pattern instances will be renamed and treated as design parameters later. *Secondly*, design equations are defined to represent the dependences of the dimensions of the cut features in these pairs as shown in Fig. 2.39b. *Thirdly*, a design table is created that includes six configurations as the variants of the encoder model.

Figure 2.39c shows the layout of a design table: (1) the first row is the table title that indicates the associated ID of the model. (2) *The first field of the second row* (e.g., the highlighted *black* field) must not be edited since it tells the contents in the second row are design parameters, which include the internal links to the corresponding dimensions, features, or other properties in the model. In this example, the second row includes (*a*) the dimensional parameters for inside and outside diameters and (*b*) the states of cut and pattern features. (3) The fields in the first column (e.g., the highlighted *blue* fields) lists the configurations. In this example, it lists six configurations for different levels of resolutions. (4) The rest fields of the table (e.g., the highlighted

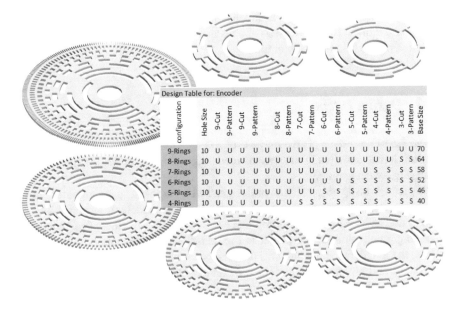

Design Table for: Encoder

configuration	Hole Size	9-Cut	9-Pattern	9-Cut	9-Pattern	8-Cut	8-Pattern	7-Cut	7-Pattern	6-Cut	6-Pattern	5-Cut	5-Pattern	4-Cut	4-Pattern	3-Cut	3-Pattern	Base Size
9-Rings	10	U	U	U	U	U	U	U	U	U	U	U	U	U	U	U	U	70
8-Rings	10	U	U	U	U	U	U	U	U	U	U	U	U	U	U	S	S	64
7-Rings	10	U	U	U	U	U	U	U	U	U	U	U	S	S	S	S	S	58
6-Rings	10	U	U	U	U	U	U	U	U	U	U	S	S	S	S	S	S	52
5-Rings	10	U	U	U	U	U	U	U	U	S	S	S	S	S	S	S	S	46
4-Rings	10	U	U	U	U	U	U	S	S	S	S	S	S	S	S	S	S	40

Fig. 2.38 Design table for optical encoder family

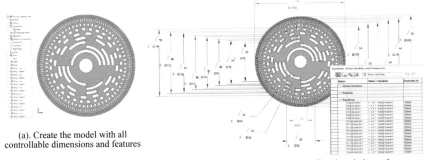

(a). Create the model with all controllable dimensions and features

(b). Use design equations to define the relations of dimensions and features

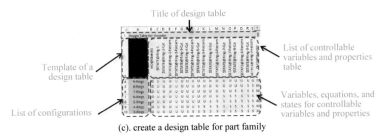

Title of design table

List of controllable variables and properties table

Template of a design table

Variables, equations, and states for controllable variables and properties

List of configurations

(c). create a design table for part family

Fig. 2.39 Example of creating a part model with a design table

yellow fields) assign a value to a dimensional parameter or a state (e.g., '*U*' for *unsuppressed* or '*S*' for *suppressed*) of a feature. In addition, the design table can be formatted by the Microsoft Excel tool.

2.5.3 Design Intents

Design intent represents (1) the purpose of a design feature, (2) the constraints that a feature has to satisfy, (3) the way of modeling a feature, or (4) the guidance of modifying such a feature.

Any feature on an object should be designed with purposes, so is a graphic feature of a solid object. In feature-based modeling, it is the modeler's responsibility to identify design intents and the appropriate ways of representing these design intents.

Taking an example of the model in Fig. 2.18f, one design intent is the golden ratio of the length (L) and width (W) of rectangle, which is widely used to represent the beauty and ergonomic appearance of real-world objects (see some examples in Fig. 2.40). Therefore, a design intent of $L/D = 1.618$ has been identified and represented as a design equation for the relation of L and D in Fig. 2.40f. Such a design intent should be considered when the object examples in Fig. 2.40 are modeled.

Design intents are the modeler's understanding of the object to be modeled. From the perspective of modeling, there are numerous methods to create the same graphic feature, and design intents are (1) the methods selected to create certain features and

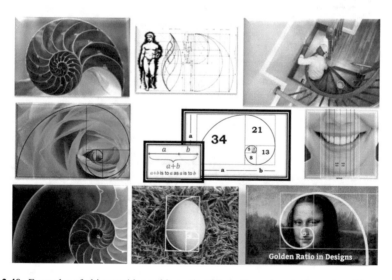

Fig. 2.40 Examples of objects with a golden ratio of their dimensions in the real world

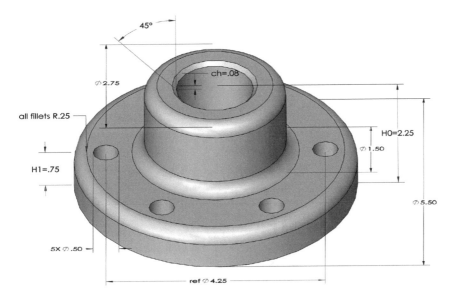

Fig. 2.41 Example of considering design intents in modeling process

(2) the rationales why such selections are made. Therefore, design intents should be considered at every step of modeling procedure.

Example 2.3 Create a model for a flange in Fig. 2.41 and discuss some exemplifying design intents in the modeling process.

Solution Other than a solid primitive, a solid consists of multiple features that will be created, respectively. Therefore, the modeling process depends on the decomposition of the solid into the features. There are many ways to decompose a solid into features, and it will be very helpful to incorporate a design intent in identifying features and planning the steps of creating these features.

Figure 2.42a shows the design intent *to minimize the number of features*; accordingly, the modeling steps are minimized. It is appropriate when the information of the solid is detailed and finalized, and the future change occurring to the solid will be rare. The main disadvantages of such a design intent are (1) the complexity of features, (2) the coupling of dimensional variables of multiple graphic attributes, and (3) the inefficiency in making the changes on an existing model. Figure 2.43a shows the implementation of the modeling process using this design intent. The solid model consists of two features in the model tree; e.g., one extrude bass (*F1*) and one extrude cut (*F2*). The right side of Fig. 2.43a shows the logical operations of these two features.

Figure 2.42b shows the design intent to *maximize the flexibility of the modeling process*; accordingly, the features are decomposed to the lowest layer. The number of features, or the number of the steps in the modeling process, will be maximized; in return, the advantages of such a design intent are (1) the intuitive way to plan

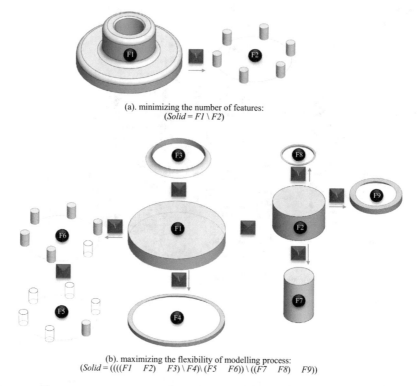

(a). minimizing the number of features:
($Solid = F1 \setminus F2$)

(b). maximizing the flexibility of modelling process:
($Solid = ((((F1 \quad F2) \quad F3) \setminus F4) \setminus (F5 \quad F6)) \setminus ((F7 \quad F8) \quad F9))$)

Fig. 2.42 Basic design intents in planning modeling process

the modeling process, which is extremely productive in the conceptual design of products, (2) the simplicity of creating features, (3) the highest level of reusability of created features, (4) the independence of design features which allows to make a change on an individual feature with the minimized influence on others, and (5) the flexibility and expandability of making changes on an existing model. Decomposing a solid into the features at the lowest level provides the best resolution of modeling; therefore, we can call it the strategy for *precise modeling*, and we strongly recommend the readers to consider the flexibility and reusability in planning the modeling process. Figure 2.43b shows the implementation of the modeling process using this design intent. The solid model consists of nine features in the model tree; e.g., two extrude basses (*F1* and *F2*), two extrude cuts (*F5, F7*), three fillets (*F3, F4, F9*), one chamfer (*F8*), and one circular pattern (*F6*). The right side of Fig. 2.43b shows the logical operations of these nine features to represent the solid with all these features.

Whenever there are some options in the modeling process, design intents should be considered to select an optimal option. Taking an example of creating the first solid feature from a 2D sketch in Fig. 2.43a or Fig. 2.43b, the modeler has to select the sketch plane, it will be helpful to consider that the first sketch-plane determines the orientation of the solid with respect to the default world coordinate system.

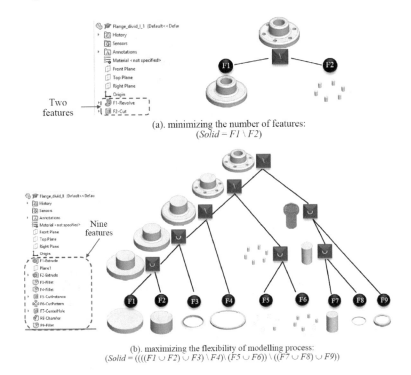

(a). minimizing the number of features:
$(Solid = F1 \setminus F2)$

(b). maximizing the flexibility of modelling process:
$(Solid = ((((F1 \cup F2) \cup F3) \setminus F4) \setminus (F5 \cup F6)) \setminus ((F7 \cup F8) \cup F9))$

Fig. 2.43 Implementations of modeling process

Accordingly, it determines the contents of the standardized *front view*, *right view*, and *top view*. Note that the front view is usually the main view to include the main dimensions of the solid. Therefore, the design intent at the first step of modeling should ensure that the first sketch plane includes the most significant dimensions of the solid.

In a graphic model of solid, the positions of orientations of graphic features are determined with respect to a specified coordinate system. A default coordinate system in Solidworks is shown in Fig. 2.44. It consists of an origin (*O*), and three planes for

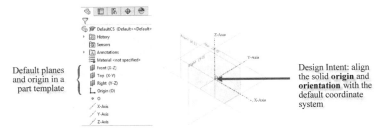

Fig. 2.44 Default coordinate system to locate and orientate a graphic feature

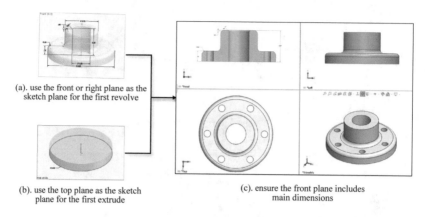

(a). use the front or right plane as the sketch plane for the first revolve

(b). use the top plane as the sketch plane for the first extrude

(c). ensure the front plane includes main dimensions

Fig. 2.45 Design intent in selecting the first sketch plane

front (X-Z), top (X-Y), and right (Y-Z) views. Three axes (X, Y, Z) of the coordinate system are dependent, which can be derived from the aforementioned information. In general, it is beneficial to align the original and orientation of the solid with the default coordinate system; so that the position and orientation of the solid can be known and easily determined at the beginning of modeling process. To implement this design intent, Fig. 2.45a shows that a right or front plane should be selected as the sketch plane when the first feature is revolved. Figure 2.45b shows that a top plane should be selected as the sketch plane when the first feature is extruded. In addition, the sketch should be centralized at the origin.

2.6   Modeling Procedure

In feature-based or knowledge-based modeling, a solid is decomposed as a set of features, which can be modeled, respectively. The attributes of a feature to be defined are *geometry and shape*, *dimensions*, *constraints*, and *operator*. The solid model is then created by applying logical operations (e.g., union ∪, intersection ∩, and difference \) on the graphic features. As shown in Fig. 2.46, the constitutive features of the solid are identified, created, and then added or removed from the solid volume using logical operations. For the geometry and shape, a graphic feature can be either of (1) a built-in feature *fillet*, *chamfer*, *rib*, or *hole or thread wizard* or (2) a sketched feature such as *extrude*, *revolve*, *loft* or *sweep*. The modeling tool for a built-in feature provides a *wizard* for the modeler to input geometric data; while a sketched feature requires the modeler to define 1D or 2D dimensions in sketches. Note that dimensioning a graphic feature requires to select or create certain references such as *points*, *axes*, *planes*, and *coordinate systems*. The corresponding tools for feature-based modeling are grouped in Fig. 2.46 and illustrated in Fig. 2.47 correspondingly.

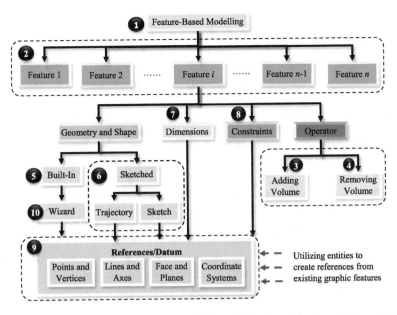

Fig. 2.46 The classification of feature-based modeling tools based on their roles on feature attributes

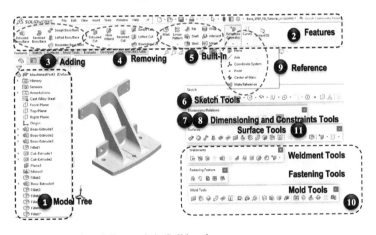

Fig. 2.47 Feature-based modeling tools in Solidworks

Figure 2.47 shows some major feature-based modeling tools in Solidworks, and these tools are associated with the need of creating and modifying different features and attributes in Fig. 2.46. Modelers should be aware of these tools and know-how to access when they are needed. As shown in Fig. 2.47, (1) *the model tree* is used to represent the features and their relations in a solid; the information of any entity can be accessed using the interface in the model tree; (2) *the feature manager* in the second row shows available tools to create different features; (3) logical operators in Fig. 2.46 are functioned by selecting feature types and specifying appropriate references; (4) different graphic features (e.g., built-in or sketched) in Fig. 2.46 correspond to different modeling tools in the feature manager; (5) the sketches, the dimensions, and the constraints in Fig. 2.46 are defined mainly by the tools in *the sketch manager* and other tools for parametric modeling such as *linked values, design equations*, and *design tables*; (6) the references in Fig. 2.46 are defined and managed using *the reference geometry manager*; in editing a model, any existing graphic elements (vertices, edges, and planes) can be used as the references to modify or create new features; (7) to create high-level built-in features, the Solidworks includes many adds-in tools such as *weldment, fastening, mold,* and *surface tools* as the tool wizards.

Feature-based or knowledge-based modeling is very effective to deal with the complexity and changes in product designs. The principal strategy is '*divide and conquer*'; a solid is modeled as a set of graphic features, and each feature is modeled or modified at a time to minimize its impact on other features. Figure 2.48 shows a routine procedure to define a graphic feature in the modeling process. It begins with the determination of *design intents*; note that a design intent represents the selection of modeling strategies for the given feature. For example, a cylinder can be created

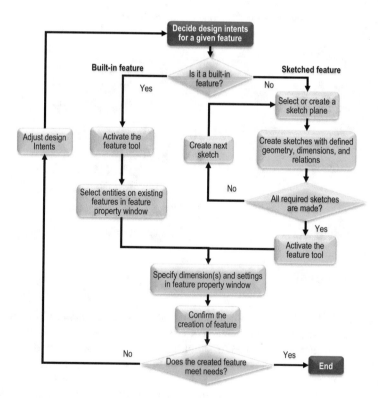

Fig. 2.48 Divide and conquer: *one feature at a time* in knowledge-based modeling

by *extruding*, *revolving*, even *lofting* or *sweeping*, the modeler should choose the simplest tool in modeling. Note that a design intent is the modeler's choice when there are a number of modeling options. As discussed early, a graphic feature can either be of *built-in feature* or *sketched feature*. A built-in feature such as a fillet and surface thickness can be created upon an existing feature, and the inputs can be specified directly in the *feature property* window. A *sketched feature* such as an e*xtrusion* requires to create one or more sketches to define the geometry and dimensions. For complex solid, specifying an appropriate reference for sketching and dimensioning is not always a trivial task; in many cases, new references should be specially defined to define the position and orientation of the feature subjected to certain constraints.

2.7 Assembly Modeling

With few exceptions of simple products, discrete products are assembled from parts and components due to many different reasons: (1) parts or components are from different materials, through different processes, or made at different locations or facilities; (2) sizes of parts are too large to be processed as a whole on machine tools; (3) finish products are large-scale that are inappropriate for transportations; (4) products are machines that have relative motions of their components; (5) product variants are made by selecting different modules and assembling them in different ways (see Fig. 2.49 for examples). Therefore, engineering products are mostly assembled units.

On average, 50% of the total manufacturing cost is tied to the assembly processes of products; manufacturing enterprises must know how to reduce the cost of assembly

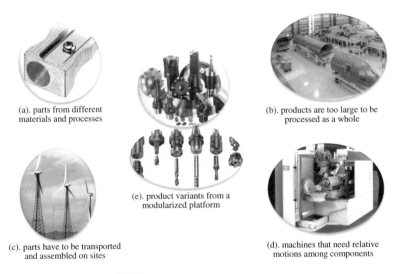

(a). parts from different materials and processes

(b). products are too large to be processed as a whole

(e). product variants from a modularized platform

(c). parts have to be transported and assembled on sites

(d). machines that need relative motions among components

Fig. 2.49 Products are assembled for many reasons

processes to gain their competitiveness. In early time, the designs of assembly processes were less emphasized than product designs. This might lead to the situation that some problems relevant to assembly processes were discovered too late and too expensive to be fixed. If an omission or error relating to assembly processes can be identified and fixed at the product design stage, the enterprise can save significant costs.

Design for assembly (DFA) is a design methodology by which the constraints and implementation of assembly processes are taken into consideration at the design stages of products. The primary goal of DFA is to reduce the cost, the time, and the complexity of assembly processes, and two basic approaches of DFA are (1) to simplify product structures by reducing the total number of parts in a product and (2) optimize the features on the product for easy grasping, moving, orienting, inserting, and fitting (Wikipedia 2020b). As shown in Fig. 2.50, Boothroyd (1994) gave the examples of using DFA for the motor-drive assembly and reticle assembly; the numbers of parts and assembly operations in these products were significantly reduced. Therefore, the costs of these products were greatly reduced due to the simplification of assembly processes via DFA.

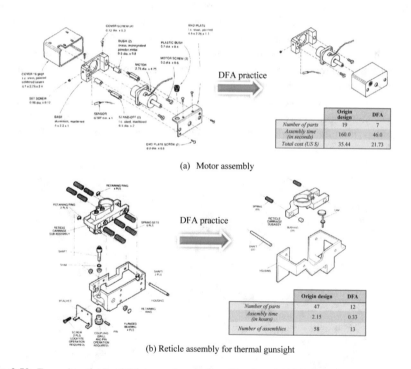

(a) Motor assembly

	Origin design	DFA
Number of parts	19	7
Assembly time (in seconds)	160.0	46.0
Total cost (US $)	35.44	21.73

(b) Reticle assembly for thermal gunsight

	Origin design	DFA
Number of parts	47	12
Assembly time (in hours)	2.15	0.33
Number of assemblies	58	13

Fig. 2.50 Examples of using DFA in product designs (Boothroyd 1994)

2.7.1 Terminologies in Assembly Modeling

Other than geometric and feature modeling of products, the functional under-standing of assemblies is crucial to clarify product designs (Gui and Mantyla 1994). *Assembly modeling* aims to represent and visualize the assembly relations of parts and components in a product model; the parts and components are presented as solid models.

An assembly model consists of a set of *components* for the desired functions and the *connections* that apply constraints on components. Generally, an assembly model uses a hierarchical structure for the connections of components. To model the assembly of a product, one has to understand the following terminologies.

(1) **Degrees of Freedom** (DOF)

Degrees of freedom are a number of independent variables to describe the position and orientation of a physical body in space. As shown in Fig. 2.51a, a free body in a 3D space has six DOF since three independent variables (T_x, T_y, and T_z) are used to describe the translations along three axes (X, Y, and Z), and other three independent variables (R_x, R_y, and R_z) are used to describe the rotations around three axes (X, Y, and Z). As shown in Fig. 2.51b, a constraint will reduce DOF of a body since it confines one or a few DOF of the body. For example, if the body has to be on a plane (e.g., the XY-plane), the constraint will apply to the free motions of R_x, R_y and the remained DOF of the body on the plane will be three DOF (e.g., T_x, T_y, and R_z).

(2) **Assembly Mates and Constraints**

Assembly mates create geometric relations of components. When an assembly mate is added to two components, their relative motions along one or a few directions are confined, and the rest of the unconfined motions are referred to as allowable motions. Therefore, the mates in an assembly model determine the DOFs of the components. A product without any allowable relative motion is called a *structure*; a product with some allowable relative motion is called a *machine*, and the number of allowable DOF is referred to as the DOF of the machine.

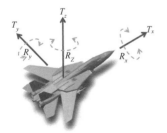

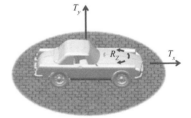

(a) 6 DOF in three-dimensional space   (b) 3 DOF in two-dimensional space

Fig. 2.51 DOF of a free body in 3D or 2D space

Table 2.7 Possible mating relations of two components with points, edges, and faces

Component II \ Component I	Point	Edge	Face
Point	A	D	E
Edge	D	B	F
Face	E	F	C

An assembly mate involves two or more graphic elements from two components. Note that three basic types of graphic elements are *points*, *edges*, and *faces*; Table 2.7 shows six possible mating types (*A*, *B*, *C*, *D*, *E*, and *F*) of the graphic elements from two components (components I and II).

Assume that component I is fixed, Table 2.8 shows the numbers of confined and allowable DOF of component II when the aforementioned mating types are applied, respectively.

In a computer model, a mating relation is unnecessary to have a physical contact. Therefore, a broad scope of mating relations has been introduced to confine the relative motions of two or more components. The types of mates are classified as *standard mates*, *advanced mates*, and *mechanical mates*. Tables 2.9, 2.10 and 2.11 show the types of standard mates, advanced mates, and mechanical mates, respectively.

(3)   **Root Components**

In an assembly model, *a root component* determines the position and orientation of the assembly in a reference coordinate system. For simplicity, a root component is 'fixed' in the reference coordinate system by defaults. A root component usually has a relatively large size, and importantly, has more mates with others in comparison to other objects in an assembly. Therefore, a root component should be selected based on the significance of a component in defining assembling relations with others. Figure 2.52 shows some examples of root components in respective products. In these examples, root components are highlighted by 'blue' color.

2.7.2   Modeling Methods

Products can be modeled in two methods, i.e., (1) *bottom-up methods* and (2) *top-down methods*.

Bottom-up methods are more traditional. Firstly, parts are designed and modeled; secondly, parts are inserted into an assembly model, and mates are defined among the parts to determine the spatial relations of parts in the assembly model. Individual parts should be edited at the part level, and the changes occurring to parts to be updated in the assembly model automatically.

Table 2.8 Confined and allowable DOF of components subjected to different mating relations

Mating type	Degrees of Freedom (DOF)		Illustration
	Confined	Allowable	
A (point-point)	T_x, T_y, T_z	R_x, R_y, R_z	
B (edge-edge)	T_x, T_z, R_x, R_z	T_y, R_y	
C (face-face)	T_y, T_z, R_x	R_y, R_z, T_x	

(continued)

Table 2.8 (continued)

Mating type	Degrees of Freedom (DOF)		Illustration
	Confined	Allowable	
D (point-edge)	*The point must be one edge, which confines two of* (T_x, T_y, T_z)	T_x, R_x, R_y, R_z	
E (point-face)	*The point must be one face, which confines one of* (T_x, T_y, T_z)	T_x, T_y, R_x, R_y, R_z	
F (edge-face)			

Table 2.9 The types of standard mates

Type	Symbol	Applicable graphic elements
Coincident		Point-point, point-line, point-plane, line-plane, plane-plane
Parallel		Line-line, line-plane, plane-plane
Perpendicular		Line-line, line-plane, plane-plane
Tangent		Arc-line, arc-arc, arc-cylinder, cylinder-cylinder, plane-cylinder
Concentric		Arc-arc, arc-cylinder, cylinder-cylinder
Lock		Any graphic element

Table 2.10 The types of advanced mates

Type	Symbol	Applicable graphic elements
Symmetric		Two points/lines about a line, two points/lines/planes/any objects about a plane
Width		One or two planes in the middle of two other planes
Path mate		Point-path, point-trajectory
Line/line coupler		Two parallel planes

Table 2.11 The types of mechanical mates

Type	Symbol	Applicable graphic elements
Cam		Cam: a set of faces cam follower: a face or vertex
Hinge		Arc-arc, arc-cylinder, cylinder-cylinder
Gear		Axis-axis, axis-cylinder, cylinder-cylinder
Rack pinion		Rack: edge, sketch line, centerline, axis, cylinder pinion: cylinder, arc, axis, revolved surface
Screw		Axis-axis, axis-cylinder, cylinder-cylinder
Universal joint		Axis-axis, axis-cylinder, cylinder-cylinder

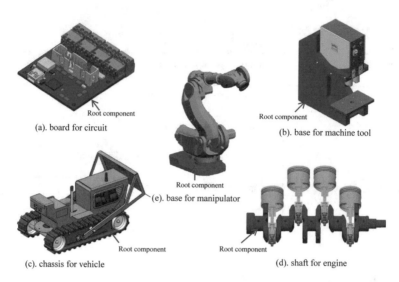

(a). board for circuit

(b). base for machine tool

(e). base for manipulator

(c). chassis for vehicle

(d). shaft for engine

Fig. 2.52 Examples of selecting a root component

A bottom-up design method is suitable for the scenario that a product is made of previously constructed, off-the-shelf parts, and standard components like hardware, pulleys, motors, etc. These parts are standardized; they should not be changed in engineering designs; instead, engineers select different components when the requirements are changed. In using a bottom-up method, *firstly*, the parts at the lowest level are modeled or obtained. *Secondly*, an assembly model is created and the root component is selected, inserted, and placed in the reference coordinate system. *Thirdly*, other part models are inserted in sequence; when a part is inserted, its spatial relations and constraints with existing parts are defined by mates. The assembly model is completed when all parts are inserted and the mates among the parts are fully defined. Using a bottom-up method allows engineers to focus on the details of the assembly relations since parts and components are modeled individually.

Example 2.4 Create a 6-DOF modular robot configuration from a modular robotic system. As shown in Fig. 2.53, the modular robotic system consists of standardized modules including rotary joints, linear joints, links, wrists, and grippers (Bi et al. 2008).

Solution As shown in Fig. 2.54, three critical design issues are involved in the application of a reconfigurable system. These design issues are *architecture design*, *configuration design*, and *control design*.

Architecture design determines the types of system modules and the connections of system modules. Modules in a reconfigurable system are encapsulated; in other words, the internal implementation of a module does not affect the interfaces of the module with others. The connections of a module are the options by which the module can be interacted with others. The design of system architecture aims to produce

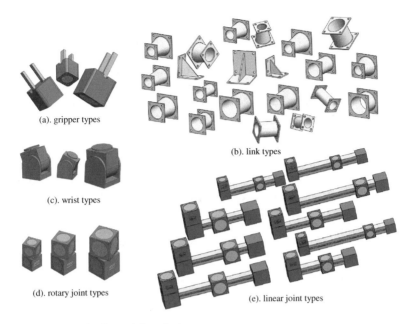

(a). gripper types

(b). link types

(c). wrist types

(d). rotary joint types

(e). linear joint types

Fig. 2.53 An example of a modular robotic system

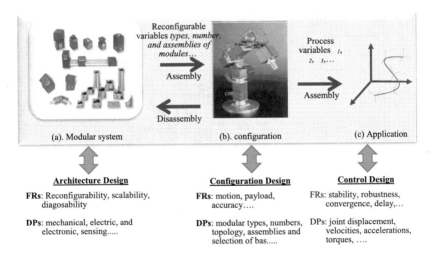

Fig. 2.54 Task-oriented design of modular robotic systems (Bi et al. 2008)

as many system configurations as possible with a given set of modules. Note that different configurations can be used to fulfill different tasks; the more configurations a system can generate, the better capabilities of the reconfigurable system can deal with changes and uncertainties in a dynamic environment. Architecture design is involved at the phase of reconfigurable system design. In *configuration design*, it is assumed that the reconfigurable system is given; for example, the types and the numbers of robotic modules in Fig. 2.52 are given. Configuration design is to select a set of modules and configure modules into a robot to fulfill the functional requirements of a given task optimally. Configuration design is involved at the phase of system application. Active modules in a reconfigurable system have their local controls; however, there are system-level goals when modules are assembled into a robot. Therefore, control design is to coordinate system modules, so that these modules can collaborate with each other to fulfill given tasks satisfactorily. The control design is involved at the phase of system operation (Bi et al. 2008).

Since the problem does not specify the functional requirements of the 6-DOF robot in terms of the workload, trajectory, velocity, and acceleration of toolpath, arbitrary three 1-DO joints are selected to build a 6-DOF model. As shown Fig. 2.53, the demonstrated robot has 6 DOF; it consists of 1 R-90 rotary joint, 1 L-70 linear joint, 2 R-70 rotary joints, 1 W-70 wrist, 1 link module for the connection of R-90 and L-70 modules, and 1 angled link module for the connection of L-70 and R-70. Note that a W-70 has 2 DOF. In the robot assembly, each active module has one or 2 DOF which are driven by respective motors (Fig. 2.55).

In a top-down assembly method, parts and components are created during the course of assembly modeling. The details of part models are not available when the assembly relations are defined, and constitutive parts or components are created one by one based on the conceptualized structure and assembly relations of products. Note that even though parts are created during the course of assembly modeling, they can be saved, either internally or externally, as individual models. Top-down modeling allows engineers to utilize geometric relations and constraints in the

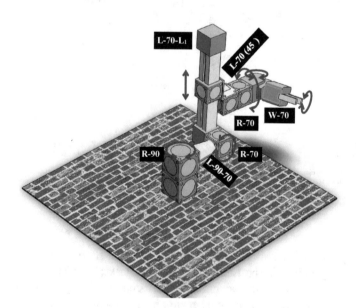

Fig. 2.55 A 6-DOF configuration of the modular robot system

high-level structure for newly created parts. In such a way, geometric modeling and assembly modeling can be proceeded simultaneously. Engineers can view the assembly relations of the part when the part is modeled.

A top-down method reduces the rework when the assembly relations are changed in a product design since the derived parts are associated with the assembly constraints when they are modeled. Therefore, top-down modeling is very useful at the conceptual design stage; it is widely used in tooling design since the geometries and spatial arrangement of tooling depend on the parts to be manufactured. In practice, a top-down method can be used to create an assembly model partially, i.e., a few critical parts in assembly or some key features of parts. Engineers can use the top-down modeling method to layout an assembly including key parts customized to assembly relations (Dassault Systems 2020).

Example 2.5 A customer wants to have a dining table with the features shown in Fig. 2.56: (1) the table takes a footprint of a circular area of Ø 2000 mm; (2) the table is supported by a hexagonal frame that serves for six peoples; (3) the table has an overall height of 1200 mm and stands on four legs; (4) each side of the table has a drawer with the height of 200 mm; (5) steel materials are used for all support parts and wood materials are used for the parts with a large surface and the parts for six drawers. Use top-down modeling to design a dining table to meet the aforementioned requirements.

Solution The top-down method is used to create six parts for a dining table. The main dimensions of the parts are determined based on the drawing of the conceptual design

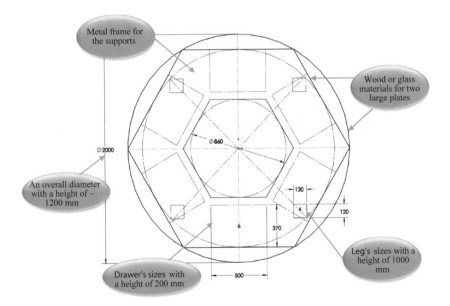

Fig. 2.56 Requirements of new dining table in top-down design example

in Fig. 2.56. As the result, Figs. 2.57 and 2.58 show the exploded and collapsed views of the assembly model, respectively. The bills of materials (BOM) table in Fig. 2.57 shows the dining table is assembled from 17 parts.

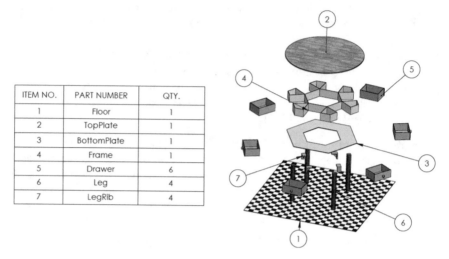

ITEM NO.	PART NUMBER	QTY.
1	Floor	1
2	TopPlate	1
3	BottomPlate	1
4	Frame	1
5	Drawer	6
6	Leg	4
7	LegRib	4

Fig. 2.57 Top-down modeling for the design of a dining table

Fig. 2.58 Collapsed assembly model of dining table in Example 2.6

2.7.3 New Assembly-Level Features

In assembly modeling, mates are defined to represent the spatial relations of parts in upper level components. To meet assembly constraints, new geometric or reference features can be created for parts or components at the assembly level; such features are called assembly-level features. *An assembly-level feature* differs from a feature in a part model in sense that the references, sketches, and dimensions of an assembly-level feature are associated with the assembly model. Assembly modeling supports the creation of assembly-level features. Figure 2.59 shows the modeling tools of SolidWorks in *the assembly commend group*, and the following tools relate to the assembly-level features: (1) **Insert Component**. A part or component is added into the assembly model; the part can be newly created in the assembly model using the top-down method; in other words, the new part at least takes a reference plane from the assembly model as its sketch plane for the first solid feature as the correspondence. (2) **Component Pattern**. A linear or circular pattern is defined by selecting one or a group of parts and specifying the number and spacing of the pattern. (3) **Smart Fasteners**. A standard faster is defined for a pair of selected parts at their contact surfaces. (4) **Assembly Features**. The tools are used to (i) create geometric features relevant to assemblies such as *hole series, hole wizard, simple hole, extruded cut, revolved cut, swept cut, fillet*, and *chamfer*, and (bi) define an assembly relation such as *weld bead* and *belt/chain*.

Example 2.6 Create a new part in Fig. 2.60a to connect the handle to the body of the light.

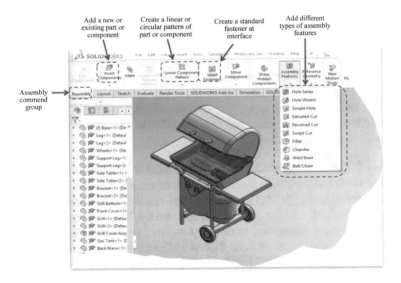

Fig. 2.59 Interface for adding assembly-level features

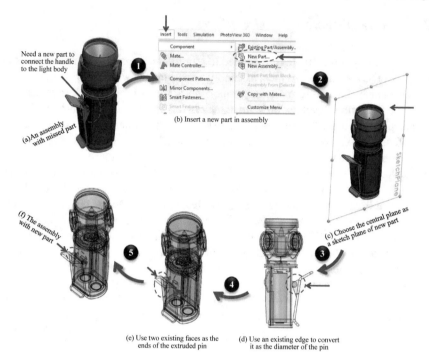

Fig. 2.60 The procedure of adding a new part in the assembly process

Solution Figure 2.60b–f illustrates the steps to insert a new part using the information of other parts in the assembly model (Fig. 2.60a). *Firstly*, the *new part* option under *the insert tool* is activated to begin the process of creating a new part in the assembly model. *Secondly*, the symmetric plane of the light body is selected as the associated sketch plane of the first feature of new part. *Thirdly*, the hole profile of the handle is projected on the sketch plane and converted as a circle in the sketch. *Fourthly*, an extruded feature is created by specifying the newly created sketch and two sides on the handle as the bounds on two extrude directions, respectively. *Finally*, a new part is fully defined, and the inserting process is terminated and returned to the assembly model with a newly created pin.

Example 2.7 A characterized human chest model is modified to investigate the impact of sternotomy wires on the transmission of ultra-wideband (UWB) in a wireless body area network (WBAN). Figure 2.61a shows the original model that consists of the layered models of cloth, skin, fat, muscle, sternal, bone, and heart in chest; note that different layers correspond to different dielectric properties in the simulation. Figure 2.61b and c show the models of wires and heart valve that will be mounted on sternal and heart, respectively. Modify the simulation model by inserting wires in the chest model so that it can be used for the numerical simulation of signal transmission (Särestöniemi et al. 2019).

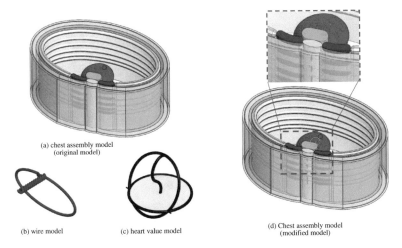

(a) chest assembly model
(original model)

(b) wire model

(c) heart value model

(d) Chest assembly model
(modified model)

Fig. 2.61 Example of eliminating interferences in assembling (Särestöniemi et al. 2019)

Solution In an assembly model, the interference of solid objects must be eliminated. When a new part is inserted, possible interferences should be detected. When an interference is identified, it can be eliminated by adding a cavity feature at the assembly level. Figure 2.61d shows the assembly model is modified in such a way that the cavity features are created in the part models of heart, sternal, fat, bone, and muscle; these cavities are used to accommodates the heart valve and wires without interference.

2.7.4 Exploded Views and Bill of Materials

An assembly model represents constitutive parts and their spatial relations in a product or system. When the number of constitutive parts increases, it becomes difficult to view the assembly relations in one configuration. Defining exploded views provides a useful way for engineers to look into the assembly relations in multiple configurations. In each configuration, parts can be moved and placed at exploded positions, and the movement traces can be modeled as exploded lines to represent assembling paths. Moreover, the exploding steps can be animated and recorded so engineers can review the assembling processes of a product vividly. As shown in Fig. 2.62, the tools in assembly modeling also provide the function of creating *the bills of materials* (BOM) automatically, The BOM list all types of parts as well as the numbers of each part type. Exploded views help engineers to visualize the assembly plan for verification and validation purpose effectively.

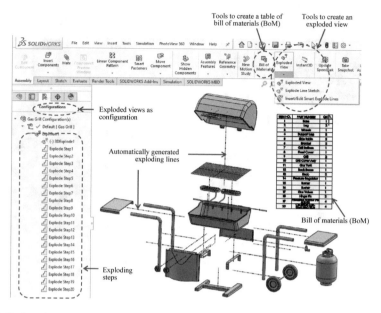

Fig. 2.62 Interface to create an exploded view

Figure 2.63 shows an example of an engine assembly model. Figure 2.63a is a collapsed configuration with all mates among the parts, and Fig. 2.64b is an exploded view to show all parts as well as the parts for assembling processes. The exploded view helped to make an automated assembly plan (Yu et al. 2014).

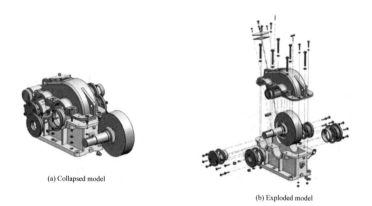

Fig. 2.63 Example of exploded view—cylinder gear reducer (Yu et al. 2014)

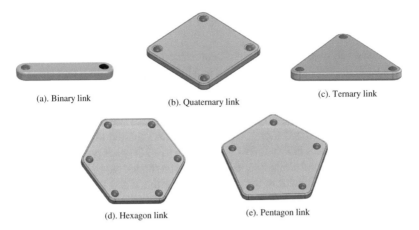

(a). Binary link

(b). Quaternary link

(c). Ternary link

(d). Hexagon link

(e). Pentagon link

Fig. 2.64 Classification of links

2.8 Kinematic and Dynamic Modeling

Modern machines or products are mostly mechatronic systems; however, they are built upon mechanical systems. In designing a mechanical system, mechanics investigates the kinematic and dynamic behavior of a mechanical system, which is subjected to mechanical loads such as displacement constraints and driving forces.

A mechanical system consists of a set of *links* that are connected by *joints*. A mechanical system has a special link called *an end-effector* to perform tasks. Assembly of links and joints ensures that the end-effector can move with specified degrees of freedom (DOF). To model a mechanical system, links and joints are firstly represented adequately.

2.8.1 Link Types

In a mechanical system, any rigid body in the structure is modeled as a conceptual link, and the mechanical structure is the assembly of links and joints. In the machine design theory, links are classified based on the number of connections to other objects. As shown in Fig. 2.64, a conceptual link is called a (a) binary link, (b) ternary link, (c) quaternary link, (d) pentagon link, and (e) hexagon link if the link has the number of connections as 2, 3, 4, 5, and 6, respectively. Note that a connection of a link to other objects implies that additional motion constraints are added to constrain the motion of the link relative to others.

2.8.2 *Joint Types and Degrees of Freedom (DOF)*

The motion of an object or a system is represented by *degrees of freedom* (DOF) of motion. DOF are the number of minimal and independent variables that are required to describe the position and orientation of an object or system in space at any instant of time.

As shown in Fig. 2.65a, a free body in the 3D space possesses six DOF; since the body can be translated along X-, Y-, Z-axes and rotated along X-, Y-, Z-axes, respectively. As shown in Fig. 2.65b, a free body in a 2D space possesses three *DOF* since it can be translated along X- and Y-axes and be rotated along Z-axis; the rotating axis is always perpendicular to the plane *O-XY* for the X and Y translation.

When two links are joined, some constraints apply to the jointed links. The type of joint determines the degrees of motion to be confined. Figure 2.66 shows six joint

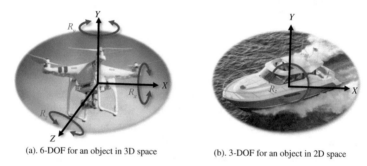

(a). 6-DOF for an object in 3D space   (b). 3-DOF for an object in 2D space

Fig. 2.65 Free object and motion degrees of freedom (DOF) in 3D and 2D spaces

(a). Prismatic joint   (b). Revolute joint   (c). Screw joint

(d). Cylindrical joint   (e). Universal joint   (f). spherical joint

Fig. 2.66 Classification of joints and the degrees of freedom of motion

types where the direction(s) of unconstrained motion(s) are illustrated where the rest of the directions are constrained motions.

A *prismatic joint* in Fig. 2.66a has one translational motion; a translation or rotation along any one of five other directions is fully constrained. A *revolute joint* in Fig. 2.66b has one rotational motion. A *screw joint* in Fig. 2.66c allows the translation and rotation along the same axis simultaneously, but these two motions are coupled and the joint only has one DOF. Differing from a screw joint, a *cylindrical joint* in Fig. 2.66d has one transition and one rotation along the same axis, while these two motions are independent; the joint has two DOF. A *universal joint* in Fig. 2.66e and a *spherical joint* in Fig. 2.66f have two and three rotations without any translations, respectively.

2.8.3 Kinematic Chains

A mechanical system without the consideration of its energy source and ground component is called a kinematic chain. A *kinematic chain* specifically refers to the topology of the assembly of rigid bodies (or links) by joints.

Figure 2.67a shows *an open-loop kinematic chain* where links carry one upon another in a series. Due to a few constrained motions occurring to links, an open-loop kinematic chain usually has a large motion range but with the limited rigidity to carry external loads. In addition, any link or joint brings new sources of error; the errors from these links and joints are stacked up linearly in an open-loop kinematic chain. Theoretically, an open-loop chain has a relatively low accuracy.

Figure 2.67b shows *a closed-loop kinematic chain* that includes two or several special links with more connections to others. These links are connected to a group of others in parallel. The load on a multi-connected link is shared by a set of connected links. Therefore, a closed-loop kinematic chain has the better capability to carry the external load. In addition, since the same link is connected with others in parallel, the errors at joints are averaged instead of stacked. Therefore, a closed-loop kinematic

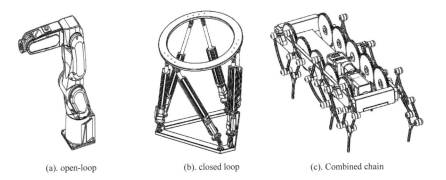

(a). open-loop   (b). closed loop   (c). Combined chain

Fig. 2.67 Three types of kinematic chains

chain is expected to have a better motion accuracy; however, it has a relatively small range of motion due to the constraints by multiple links.

Figure 2.67c shows *a hybrid kinematic chain* that includes both open-loop or closed-loop chains. A hybrid chain makes the tradeoff between the loading capability and the range of motion. The closed-loop sub-chains are used where the loads are large to the machine, and the open-loop sub-chains are used where the machine needs a large accessible space for given tasks.

2.8.4 Mobility of Mechanical Systems

When the links, the joints, and the assembly topology are given, the mobility of a mechanical system can be determined. *The mobility* of a mechanical system's mobility (M) is quantified as the number of DOF of system.

DOF of a mechanical system is defined with respect to a selected reference frame called *a ground reference frame* as,

$$M = \lambda(l - j - 1) + \sum_{i=1}^{j} f_i \tag{6.1}$$

where

M be degrees of freedom (DOF) of system,

l is the total number of links, including fixed link,

n is the total number of joints,

f_I is the degree of freedom of relative motion between element pairs of i-th joint, and

λ is an integer $\lambda = 3$ for a plane mechanism and $\lambda = 6$ for a spatial mechanism.

If a mechanism is planar, and all of the joints are low-pair joints (prismatic joint or revolute joint), Eq. (6.1) can be simplified as the Gruebler's equation as

$$M = 3(l - 1) - 2j \tag{6.2}$$

where

M be degrees of freedom (DOF) of a planar mechanism,

l is the total number of links, including the fixed link, and

j is the total number of low-pair joints.

In a machine, independent motions occur to active joints, and active joints are driven by motors. The motions of active joints are transferred to the end-effector link where the task is performed. Figure 2.68 shows some simple machines with the driving motion (in red) and the driven motion at the end (in black). All of the example machines except Fig. 2.68g have 1-DOF input and output. It implies that

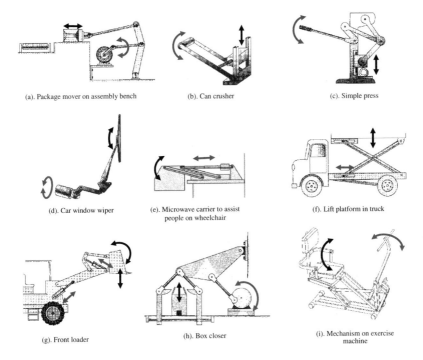

(a). Package mover on assembly bench   (b). Can crusher   (c). Simple press

(d). Car window wiper   (e). Microwave carrier to assist people on wheelchair   (f). Lift platform in truck

(g). Front loader   (h). Box closer   (i). Mechanism on exercise machine

Fig. 2.68 Examples of simple machines

many feasible solutions are available to satisfy the same motion requirement in the application.

At the conceptual design stage, the designer should be able to analyze degrees of freedom of motion when the assembly model of a machine is given.

Example 2.8 Evaluate the DOF of the mechanism shown in Fig. 2.68a.

Solution As shown in Fig. 2.68a, the mechanism is a planar mechanism. Let the DOF of a link in a plane be $= 3$. The mechanism includes six links, and seven joints, i.e., $l = 6$ and $j = 7$. All joints are either of 1-DOF rotational or translational. Accordingly, $f_i = 1$ for $(i = 1, 2, \ldots 7)$. Using Eq. (6.1) finds that,

$$M = \lambda(l - j - 1) + \sum_{i=1}^{j} f_i = 3(6 - 7 - 1) + 7(1) = 1 \qquad (6.3)$$

Machine design is a complicated process. With the computer implementation of the aforementioned kinematic and dynamic modeling methods, computer-aided design tools are able to assist engineers to design and optimize a machine for the expected functional requirements. Figure 2.69 shows that engineers could interact and fully utilize the capabilities of computer-aided design tools in virtual machine

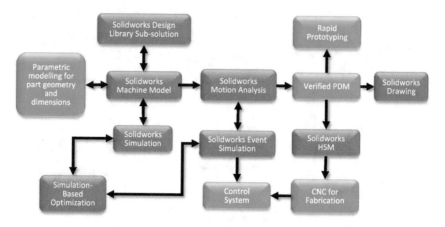

Fig. 2.69 Main functional modules for virtual machine design in SolidWorks

design (Markkonen 1999). *Firstly*, a design process begins with the preparation of a CAD model of a machine including geometries, dimensions, material properties of parts and components, mating relations in the assembly model, and boundary conditions in the application environment. *Secondly*, a simulation model is defined to establish a mathematical model for kinematic and dynamic behaviors of system. *Thirdly*, model parameters are specified; main modal parameters include the properties of motors, the profiles of expected motions, the duration of simulation, and design variables to be investigated. *Fourthly*, the simulation is performed to find the solution to the formulated mathematic model numerically. *Fifthly*, the simulation result is analyzed and verified, and an iterative process is repeated to precedent steps until the design and analysis goal is achieved. It should be noted that even CAD tools are available to take over many critical tasks, engineers' involvement in the computer-aided design process is essential to a successful machine design. For a machine design at different stages, design scopes and goals are different, and different design tools are needed at different stages. SolidWorks provides a comprehensive toolset to support virtual machine design.

2.8.5 *Motion Simulation*

SolidWorks *Motion Analysis* is to create a simulation model to study the position, velocity, acceleration, and torque of a mechanism, which is subjected to external loads. Virtual motion analysis brings significant benefits to machine design. (1) The number of physical prototypes can be minimized since the simulation will be able to identify most of the potential design errors and omissions. (2) Virtual analysis takes much less time than the experiments. It supports parametric studies to optimize the design before it is prototyped; this allows exploring more design options at a very early stage. (3) Simulation-based optimization can be utilized to improve machine design. (4) The simulation is integrated with all the other engineering analyses; it becomes possible to gain the quantified insights for additional engineering analysis and investigate the design feasibility over the entire product lifecycle. In the following, the main steps of a motion simulation are discussed.

2.8.5.1 Model Preparation

The first step for motion simulation is to prepare a CAD model of machine. The machine model includes all constitutive parts, and the connections of the parts are presented in a simplified form. In addition, the materials are specified for all solid bodies, so that the strengths of parts are determined in terms of allowable stresses. The forces acting on parts in the process of machine operation are then determined based on the required motion of machine.

When all parts are modeled, the next step is to model the assemblies of parts. If two parts have a static spatial relation in the machine, these two parts should be grouped since no relative motion is allowed between them. If two parts involve in a relative motion with each other, a correct type of joint must be selected. The assembly relations of parts are modeled as *mates* in assembly modeling. As shown in Fig. 2.70, the mates in the SolidWorks are catalogued into (a) *standard mates*, (b) *advanced mates*, and (c) *mechanical mates*. If two parts are bonded, the mates of these two parts are likely a combination of a few of standard mates in (a). Advanced mates are applied in a scenario where more than two entities are involved (e.g., symmetric and width) or a coupling of two motions occurs (e.g., path mate and linear coupler). Mechanical mates are special mates to represent the motions of typical machine elements.

Figure 2.71 shows an assembly model example of a Yumi robot for motion simulation. The robot has two mechanical arms, and each arm has a 7-DOF motion. Each DOF is enabled by an active rotary joint. In the assembly model, each DOF is modeled as a combination of one '*coincident*' mate of two planes and one '*concentric*' mate of two cylindrical surfaces from two joined parts.

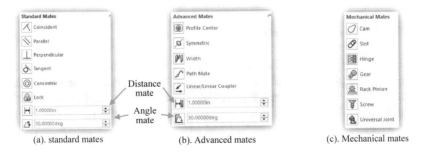

Fig. 2.70 Mates in assembly modeling of SolidWorks

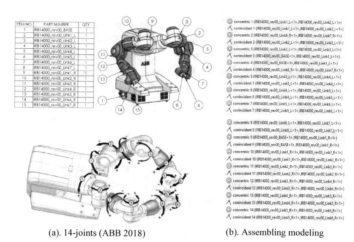

(a). 14-joints (ABB 2018)   (b). Assembling modeling

Fig. 2.71 Assembly modeling of ABB yumi robot

2.8.5.2 Creation of Simulation Model

SolidWorks has three functional modules for a motion study: *Animation, Basic Motion*, and *Motion Analysis*. The animation tool simulates the kinematic behaviors of models without the consideration of dynamics. Users can use the animation to visualize possible motions using an assembly model. Both of the basic motion and motion analysis tools are used to simulate kinematic and dynamic behaviors with the consideration of dynamic properties and driving forces but with different levels of calculation accuracies. Galliera (2010) gave the comparison of three simulation tools as shown in Table 2.12.

As shown in Fig. 2.72a, *the Motion Analysis tool* is included in 'Add-Ins' in 'Options' of SolidWorks. It is not loaded as default; therefore, a user has to activate the tool before it can be accessed. A new motion study can be created by right-clicking the motion study tab as shown in Fig. 2.72b. After the motion analysis has been activated, the list of the options under 'animation' includes Animation, Basic Motion, and Motion Analysis.

2.8.5.3 Define Motion Variables

As shown in Fig. 2.73, a motion study includes a set of motion variables for *motor, spring, damper, force, contact*, and *gravity*. These motion variables should be defined for the machine to be simulated. For example, a set of motors must be defined for all the active joints in a machine. In addition, quite a few properties have to be specified when a motor is defined in the motion analysis.

Figure 2.74 shows the interface in defining a motor in the motion study. *Firstly*, the motion of a motor can be *translational* or *rotational*. *Secondly*, the motion is

Table 2.12 Comparison of SolidWorks Animation, Basic Motion, and Motion Analysis

Types	Solvers	Description
Animation	3D Dimensional Constraint Manager (3DDCM) by D-Cube	The 3DDCM solver is capable of positioning parts in an assembly model or in a mechanism. The animation can be used to build, modify and animate the assembly model mainly for the visualization of the changes occurring to geometries, appearances, dimensions, and constraints such as mates. An animation can be defined as a smooth interpolation of multiple static views, or referred as *keys*, in a given animation time.
Basic Motion	Ageia PhysX	The Ageia PhysX is a physics solver primarily for the animations in games. The basic motion tool simulates how objects behave, move and react for life-like motion and interaction. Adopting the Ageia PhysX in the basic motion makes the simulation look realistic but the motion is not precise. The basic motion tool is capable of approximating the functions of motors, springs, collisions, and gravity effect. It is physics-based, which allows updating the simulation quickly with less computation. It suits the best for the presentation-worthy animation.
Motion Analysis	ADAMS solver	The ADAMS solver is a sophisticated tool to analyze the kinematic and dynamic behaviors of mechanical systems. The motion analysis tool aims to analyze the forces, torques, contact forces, and power consumption accurately. The simulation result over time can be exported for other engineering analysis after the simulation is completed. The motion analysis tool is used to simulate and analyze a machine with the consideration of driving forces, springs, dampers, and frictions. The kinematic solver takes into account of motion constraints, material properties, mass, and component contacts.

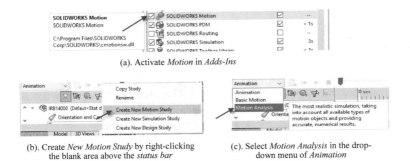

(a). Activate *Motion* in *Adds-Ins*

(b). Create *New Motion Study* by right-clicking the blank area above the *status bar*

(c). Select *Motion Analysis* in the drop-down menu of *Animation*

Fig. 2.72 Create a Motion Study for a machine model

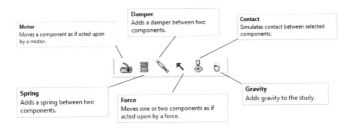

Fig. 2.73 Types of motion variables in a Motion Study

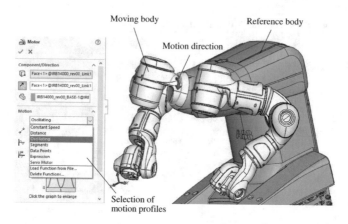

Fig. 2.74 Defining a motor in a SolidWorks motion study

associated with a moving body, and the motion is relative to a reference body along a specified direction; therefore, the moving body, the reference body, and the motion direction must be specified. *Thirdly*, the profile of a motion can be one of *Constant Speed*, *Distance*, *Oscillating*, *Segments*, *Data Points*, *Expression*, or *Servo Motor* from the drop-down list of *Motion*. *Fourthly*, the direction of the motion must be specified.

2.8.5.4 Setting Simulation Parameters

Other than motion variables, the motion study allows to customize the properties of a simulation model shown in Fig. 2.75a. The user can (1) specify the number of frames per seconds in calculation; (2) decide whether or not the simulation can be visualized in the course of calculation; (3) refine the accuracy of 3D contact or of the representation of solid geometry; (4) specify the cycle settings; (5) specify the solving algorithm and the tolerance as the criterion of termination as shown in Fig. 2.75b.

2.8.5.5 Motion Simulation

The motion analysis does not initialize the calculation automatically when some changes are made in the simulation model; therefore, the user has to accept the changes for a new simulation by clicking the 'calculate' icon to run the simulation shown in Fig. 2.76. After the calculation is completed, the user can use the animation tool to review the motion of the machine over time. In addition, the simulation result can be saved and exported to external sources in an .AVI or other formats.

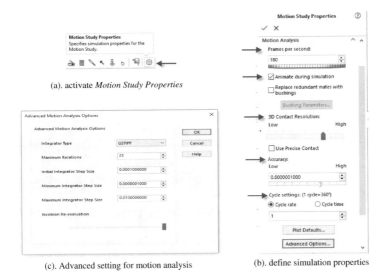

(a). activate *Motion Study Properties*

(c). Advanced setting for motion analysis

(b). define simulation properties

Fig. 2.75   Interface to set simulation properties

Fig. 2.76   Run a motion study simulation

2.8.5.6   Analyze Simulation Data

A Motion Analysis model involves a large number of motion variables and simulation parameters. It should be a very rare case that a user defines all of the simulation parameters appropriately at his or her first iteration. The user should know what kinematic and dynamic properties are expected from the simulation to be capable of making engineering judgments to see if the simulation result is reasonable. Figure 2.77 shows the selection of kinematic and dynamic variables relevant to a motion study. Any of these variables can be selected to investigate how it is changed over time in simulation. Figure 2.78 shows an example plot of the change of torque over time for a specified motor in the machine.

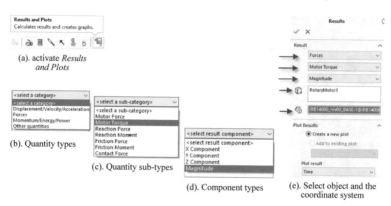

Fig. 2.77 Define a plot for change of kinematic or dynamic variable over time

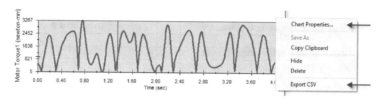

Fig. 2.78 The change of driving torque over simulation time

2.8.5.7 Mechanical Event Simulation

For a machine with motion, the forces exerted on solid bodies are varied over time; accordingly, stress distributions over the bodies are changed over time. For a safe design, it is helpful to determine when and at what amplitude of the maximum stress occurs to the bodies in a machine. The time-dependent loads from a motion analysis simulation can be utilized for the structural analysis of a solid body in a machine.

By incorporating a structural analysis in a motion study, the distributions of stress, a factor of safety, or deformation over solid bodies can be analyzed directly without manually setting up boundary conditions and loads; since the loads are imported automatically from the motion study results. The user can investigate the effect of dynamic motion load on stress or deformation distribution over one part or component. Figure 2.79 shows an example where the stress and deformation of a part in a robot have been analyzed at the specified timeframe. In this simulation model, the loads and boundary conditions are automatically defined in the motion study model.

A motion study can be performed on a specified time and time range. In addition, the data flow from a motion study to structural simulation is one-directional. In other words, the results of the stress analysis do not affect the motion study model. Note that detailed stress analysis can be performed in SolidWorks Simulation which will be discussed in detail in Chap. 3.

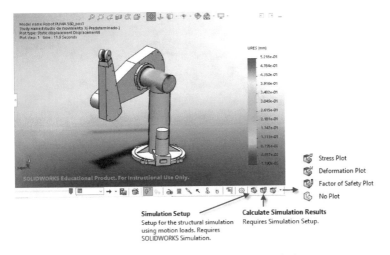

Fig. 2.79 Incorporating mechanical event simulation in motion analysis

Example 2.9 The modular robotic system in Fig. 2.53 is used to create an assembly model of 3-DOF parallel kinematic machine (PKM). The diameters of base and end-effectors are set as 600 mm and 200 mm, respectively. Each branch of the 3-DOF parallel kinematic machine consists of one 70-mm active rotary joint, two 70-70-110 type-A links, and two 70-mm passive rotary joints. Taking the reference home position of the assembly model, the motions of three active joints are given in Table 2.13.

Assume that the materials are set as 1060 alloy for all components in the assembly model, and the direction of gravity force is along negative Z_b. (1) Create the assembly model for motion study. (2) Define and analyze the motion of the end-effector based on the given joint motion. (3) Visualize the trace of the reference (O_e) of the end-effector platform. (4) Export the results of displacements and driving torques of active joints. (5) Export the results of the displacements of O_e (x_e, y_e, θ_e) with respective time.

Solution Hybrid modeling is adopted to (1) build three kinematic branches from the modules in Fig. 2.53 by the bottom-up method and (2) create the ground and top plates based on the specified diameters by the top-down method. Figure 2.80 shows the assembly model of 3-DOF parallel kinematic machine. It consists of a ground platform, three 3-DOF branches, and an end-effector platform.

To create a motion study model, all active joints are defined as 'motors'. In the given 3-DOF parallel kinematic mechanism, each branch is built from two passive rotary joints and one active rotary joint. The first rotary joint in each branch is selected to define the motor for this branch as shown in Fig. 2.81. The motion properties of the three motors are defined based on Table 2.13 accordingly. The completed model includes three motors; in addition, the direction of gravitational acceleration is specified to take into consideration the effect of the weights in motion.

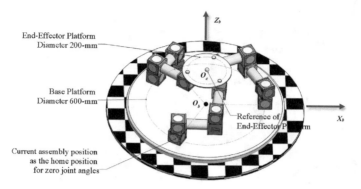

Fig. 2.80 Assembly model of 3-DOF parallel kinematic machine

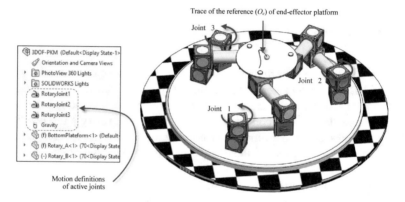

Fig. 2.81 The trace of the reference O_e subjected to the given motions of three active joints

Table 2.13 Specified joint motions

Joint No.	Motion profile	Motion range (°)	Frequency (hertz)
1		40	1
2	*Oscillation*	35	0.5
3		20	1

The simulation is calculated, and the simulation results can be visualized, analyzed, and exported. Figures 2.82, 2.83, and 2.84 show the changes of joint displacements, driving torques, and end-effector motions over time, respectively.

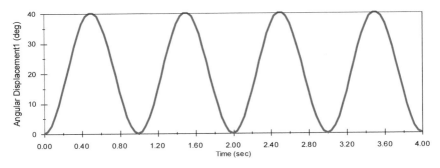

(a) joint angle (θ_1) over time (second)

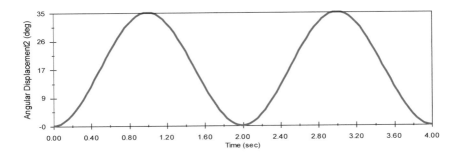

(b) joint angle (θ_2) over time (second)

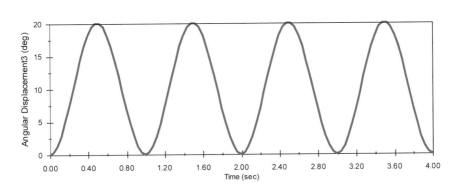

(c) joint angle (θ_3) over time (second)

Fig. 2.82 Joint displacements over time

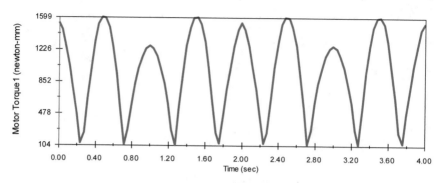

(a) joint torque (τ_1) over time (second)

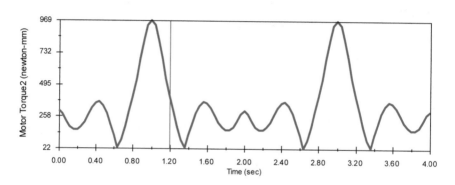

(b) joint torque (τ_2) over time (second)

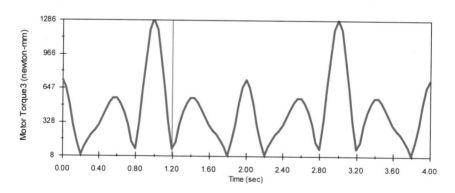

(c) joint torque (τ_3) over time (second)

Fig. 2.83 Joint torques over time

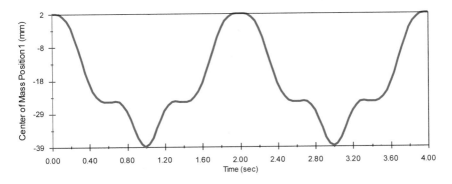

(a) end-effector platform x_e (mm) over time (second)

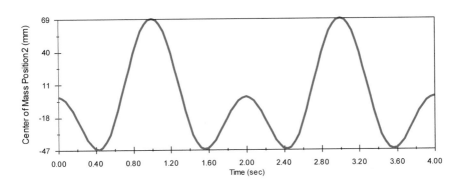

(b) end-effector platform y_e (mm) over time (second)

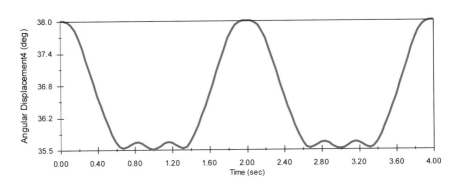

(c) end-effector platform θ_e (mm) over time (second)

Fig. 2.84 The displacements of end-effector over time

2.9   Summary

Engineers should master the skills of using CAD tools to design, analyze, model, simulate, and evaluate a product, manufacturing process, or system. In this chapter, various CAD techniques are introduced to represent products, processes, and systems at different levels and aspects. To improve the productivity of engineering designs, engineers should understand the theory, methods, and tools of parametric modeling and knowledge-based engineering (KBE), and use computer-aided tools efficiently to model the geometrics, features, design intents, and assembling relations. For machine design, engineers can benefit greatly by using the motion simulation tools for kinematic and dynamic analysis of machines in an integrated computer-aided environment.

Design Problems

Problem 2.1 Identify the number of vertices, edges, faces, loops, and genus of the following objects, and use the Euler-Poincare law to justify if they are legal solid models.

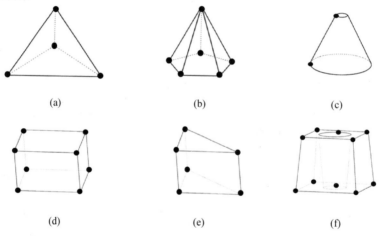

(a)   (b)   (c)

(d)   (e)   (f)

Problem 2.2 Identify the number of vertices, edges, faces, loops, and genus of the following objects, and use the Euler-Poincare law to determine object types.

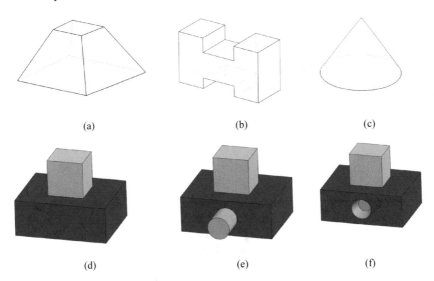

(a)   (b)   (c)

(d)   (e)   (f)

Problem 2.3 Using the Euler-Poincare Law to justify the validity of open objects in the first column of Table 2.14.

Table 2.14 Examples of objects for problem 2–3

Example	F	E	V	L	B	G	F-E+V-L=B-G

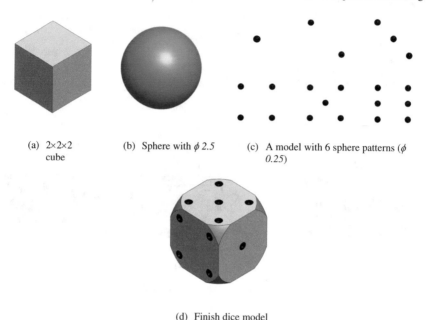

(a) 2×2×2
     cube

(b) Sphere with ϕ 2.5

(c) A model with 6 sphere patterns (ϕ
     0.25)

(d) Finish dice model

Fig. 2.85 A dice model using CSG composition

Problem 2.4 Create a dice model as shown in Fig. 2.85d from the given three features using the CSG composition.

Problem 2.5 Review the drawing in Fig. 2.86, identify the set of solid primitives, and determine the composition operations to create the represented solid model.

Problem 2.6 Review the drawing in Fig. 2.87, identify the set of solid primitives, and determine the composition operations to create the represented solid model.

Problem 2.7 Create a part model for a set of six sockets as shown in Fig. 2.88d.

Problem 2.8 Bottom-up assembly modeling: manufacturers of modular robotic systems often provide the design library of part models for users to model robots. Download the part library for Vex robots from https://www.vexrobotics.com/iq/dow nloads/cad-snapcad, and build a robot similar to the one shown in Fig. 2.89, and have a motion simulation to illustrate its basic movement.

Problem 2.9 Top-bottom assembly modeling: there are thousands of rivets on a wing of commercial aircraft (see Fig. 2.90), and riveting operations heavily rely on human operators. Create a design concept of a multifunctional riveting tool which can move over the surface of wing and automate riveting processes, and create the motion simulations to demonstrate all of the required operations.

Problem 2.10 use the modular robotic system in Fig. 2.53 to build a 2D robot (open chain) and create a motion simulation model to analyze its kinematic and dynamic behaviors.

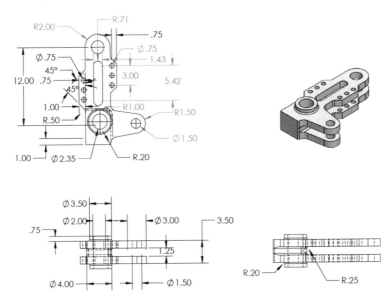

Fig. 2.86 Determine constitutive solid primitives and composition operations for solid for Problem 2.5

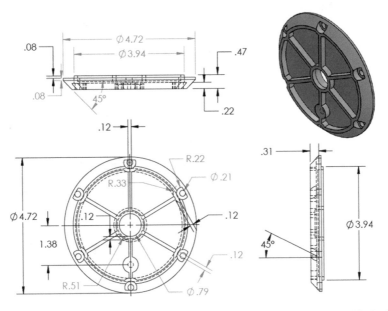

Fig. 2.87 Determine constitutive solid primitives and composition operations for solid for Problem 2.6

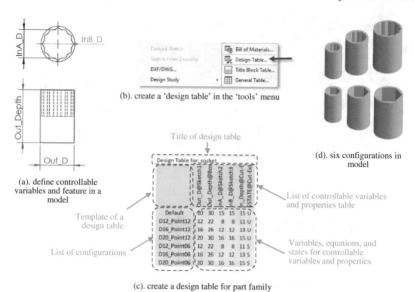

(b). create a 'design table' in the 'tools' menu

Title of design table

(a). define controllable variables and feature in a model

Template of a design table

List of configurations

(d). six configurations in model

List of controllable variables and properties table

Variables, equations, and states for controllable variables and properties

(c). create a design table for part family

Fig. 2.88 Example of creating a part model with design table

Fig. 2.89 Example of Vex robot configuration

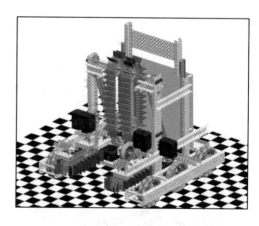

Fig. 2.90 Thousands of rivets on aircraft wing

Problem 2.11 Use the modular robotic system in Fig. 2.53 to build a 3D robot (open chain) and create a motion simulation model to analyze its kinematic and dynamic behaviors.

Problem 2.12 For a four-bar mechanism, the Grashof condition is used to justify is the driving link has a full rotation as a crank. The Grashof condition is $L + S \geq P + Q$; where L and S represent the lengths of the longest and the shortest links, and P and Q represent the lengths of other two links. (1) Create a parametric 4-bar mechanism in which the lengths of L, S, P, and Q can be adjusted to verify the Grashof condition. (2) For a 4-bar mechanism with a crank, create a motion simulation to analyze the relation of input and output angles.

References

Bi ZM (2018) Finite element analysis applications: a systematic and practical approach, 1st edn, Academic Press. ISBN-13: 978-0128099520

Bi ZM, Wang XQ (2020) Computer aided design and manufacturing, Wiley-ASME Press Serious, 1st edn. ISBN-10 1119534216

Bi ZM, Cochran D (2015) Big data analytics with applications. J Manag Anal 1(4):249–265

Chang KH (2014) Chapter 3 solid modeling, product design modelling using CAD/CAE, the Computer Aided Engineering Design Series, Academic Process, pp 125–167

Bi ZM, Yang SYT, Shen W, Wang L (2008) Reconfigurable manufacturing systems: the state of the art. Int J Prod Res 46(4):967–992

Boothroyd G (1994) Product design for manufacture and assembly. Comput Aided Des 26(7):505–520

Dassault Systems (2020) Design methods (bottom-up and top-down design). https://help.solidw orks.com/2018/English/SolidWorks/sldworks/c_Design_Methods.htm?id=4cf2b835663b4c0 ca59e0b83aa6f7dda#Pg0

Galliera J (2010) Solvers used for animation, basic motion, and motion analysis. https://forum.sol idworks.com/community/simulation/motion_studies/blog

Gui J-K, Mantyla M (1994) Functional understanding of assembly modelling. Comput Aided Des 26(6):435–451

Hooper T (2020) Top 3D CAD modeling software: the 50 best CAD Tools for ideation, rendering, prototyping and more for product engineers. https://www.pannam.com/blog/best-3d-cad-mod eling-software/

Intel (2002) Expanding Moore's law the exponential opportunity. https://www.cc.gatech.edu/com puting/nano/documents/Intel%20-%20Expanding%20Moore's%20Law.pdf

Jhanji Y (2018) Chapter 11: computer-aided design—garment designing and patternmaking. In: Automation in garment manufacturing, pp 253–290

Junk S, Kuen C (2016) Procedia CIRP 50:430–435

Lyu G, Chu X, Xue D (2017) Product modeling from knowledge, distributed computing and lifecycle perspectives: a literature review. Computer in Industry 84:1–13

Markkonen P (1999) On multi body systems simulation in product design. Doctoral thesis, Royal Institute of Technology, KTH, Stockholm

Niu N, Xu L, Bi ZM (2013) Enterprise information system architecture—analysis and evaluation. IEEE Trans Ind Inform 9(4):2147–2154

Requicha AAG (1980) Representations for rigid solids: theory, methods, and systems. ACM Comput Surv 12(4):437–464

Sanchez-Cruz H, Sossa-Azuela H, Braumann UD, Bribiesca E (2013) The Euler-Point Formula through contact surfaces of voxelized objects. J Appl Res Technol 11(1):65–78

Särestöniemi M, Pomalaza-Ráez C, Bi ZM, Kumpuniemi T, Kissi C, Sonkki M, Hämäläinen M, Iinatti J (2019) Comprehensive study on the impact of sternotomy wires on UWB WBAN channel characteristics on the human chest area. IEEE Access 7(1):74670–74682

Shalf J, Leland R (2015) Computing beyond Moor's Law. Computer 48(12):14–23

Solidworks (2018) Summary of design table parameters. Available online: https://help.solidw orks.com/2018/english/SolidWorks/sldworks/r_Summary_of_Design_Table_Parameters.htm (accessed on April 18, 2021).

Wang XV, Givehchi M, Wang L (2017) Manufacturing system on the cloud: a case study on cloud-based process planning. Procedia CIRP 63:39–45

Wikipedia (2020a) Computer aided design. https://en.wikipedia.org/wiki/Computer-aided_design

Wikipedia (2020b) Design for assembly. https://en.wikipedia.org/wiki/Design_for_assembly

Yu J, Xu L, Bi ZM, Wang C (2014) Extended interference matrices for exploded views of assembly planning. IEEE Trans Autom Sci Eng 11(1): 279–286

Zissis D, Lekkas D, Azariadis P, Papanikos P, Xidias E (2017) Collaborative CAD/CAE as a cloud service. Int J Syst Sci Oper Logist 4(4): 339–355

Chapter 3
Computer-Aided Engineering (CAE)

Abstract *Product design* involves in *modelling, simulation,* and *evaluation* of the behaviors of the product at the different stages of *its product lifecycle* (PLC). *Computer aided engineering* (CAE) uses computers to model and analyze the responses of product that is subjected to *external loads and constraints.* CAE has become essential to eliminate hidden mistakes and omissions in virtual design before the physical product is made. In this chapter, the importance of CAE in manufacturing is discussed, different CAE tools are introduced, and the focus is put on *finite element analysis* (FEA). The theoretical fundamental of FEA is introduced, the general procedure of FEA modelling and simulation is presented, and FEA is used to analyze six common types of engineering problems. The *SolidWorks Simulation* is used to illustrate how CAE is applied in analyzing and solving various engineering problems.

Keywords Computer aided engineering (CAE) · Axiomatic design theory (ADT) · Finite element analysis (FEA) · Numerical simulation · Static analysis · Modal analysis · And multidisciplinary system · Verification and validation

3.1 Introduction

Engineering concerns the design, construction, and use of structures, machines, and systems. Engineering is often multidisciplinary which relates to a broad range of particular areas such as mathematics, applied science, and types of applications such as solid mechanics, flow mechanics, aerodynamics, and electromagnetics. In *computer-aided engineering* (CAE), computers are used to model, analyze, and simulate the behaviors of products, systems, or processes for design optimization or solving various engineering problems. CAE tools are used in modeling, simulation, and design optimization of products, processes, and systems (Bahman 2018). The success of a modern enterprise depends greatly on the digitization of all of the business processes including engineering processes (Krahe et al. 2019).

Enterprises benefit from using CAE at a number of aspects. (1) CAE reduces cost and time of product development, and improves the product in its lifecycle continuously. (2) CAE generates, evaluates, and implements product design in the

Z. Bi, *Practical Guide to Digital Manufacturing,*
https://doi.org/10.1007/978-3-030-70304-2_3

minimized number of the design iterations. (3) CAE aims at the *first time right* practice by reducing the needs of physical prototypes and tests. (4) CAE uses a virtual model to evaluate the performance, reliability, and safety of products. (5) CAE can incorporate CAD, CAM, and other computer-aided tools as the holistic integration of data and process management over the product lifecycle. (6) CAE analyzes product fatigue life that can reduce the cost associated with the unanticipated failure of products or systems.

An engineering process is to convert customers' needs into a physical product or system to meet desired *functional requirements* (FRs). Following a generic procedure is helpful to ensure that an engineering process generates an ideal product or system in satisfying customers' needs. No matter that a product or system is simple or complex; the engineering process follows a generic procedure as shown in Fig. 3.1. The engineering process involves in a number of critical steps and corresponding activities to obtain an optimal solution. It begins with the identification of *customers' requirements* (CRs), and it is followed by (1) the formulation of *design problem*, (2) the definition of *solution space*, (3) *design analysis* over solution alternatives, (4) *design synthesis* for the comparison and selection of the solution alternatives, and (5) the *verification, validation* (V&V), and *implementation* of the finalized solution.

In Fig. 3.2, the case of developing an automatic fuel-recharging system (Bi et al. 2020) is used to illustrate the critical steps in applying the above engineering process.

Formulation of design problem. An engineering process aims at an optimal solution to a given engineering problem. Therefore, the first step is to formulate the design problem with clear CRs, design objectives, and design constraints. Engineers need to collect as much information as possible; in particular, the information relevant to operation environment and available resources. The engineering process for the design challenge in Fig. 3.2 can be formulated as a design problem as developing a system that is fully automated to (1) refuel gasolines for various cars; (2) make payments through the network while drivers stay in their vehicles; (3) offer the drivers with safer, cleaner, and more comfortable services than traditional self-refueling services without an extra-cost to drivers.

Definition of solution space. A solution space consists of all possible design options that may meet the given FRs and design constraints. After the design problem is formulated, engineers will determine a *conceptual design* of product or system that covers all of the feasible solutions. The conceptual design defines the boundaries of *a solution space* in terms of *the number, characteristics,* and *types* of *design variables, the ranges* and *resolutions* for the changes of design variables, and *the constraints* for the dependences of design variables. To make the complexity of a design manageable, some system design methodologies, such as axiomatic design theory (ADT), can be used to identify a set of *design variables* (DVs) for FRs (Suh 2005). To define a solution space by ADT, the system-level FRs are decomposed, so that each sub-FR can be fulfilled by a given set of DVs. Accordingly, a solution space uses a hierarchical structure to represent a set of sub-solutions and their relations.

Figure 3.3 shows the decomposition of FRs in terms of the major tasks an automatic refueling system needs to perform. The highest level *FR-0* is decomposed into the second level sub-FRs as *FR-11* for 'data collection and processing', *FR-12* for

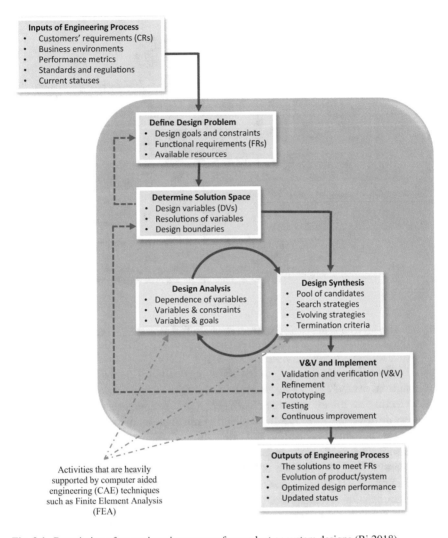

Fig. 3.1 Description of an engineering process for product or system designs (Bi 2018)

Fig. 3.2 Automatic
refueling system (Fuelmatics
2020)

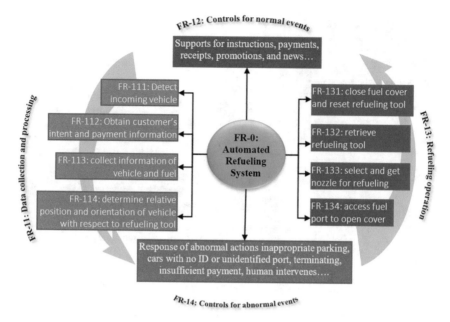

Fig. 3.3 Decomposition of system FRs of automated refueling system (Bi et al. 2020)

'controls for normal events', *FR-13* for 'refueling operation', and *FR-14* for 'controls for abnormal events'. *FR-11* is decomposed into the third level sub-FRs as *FR-111* for 'detecting an incoming vehicle', *FR-112* for 'obtaining customer's intent and payment information', *FR-113* for 'collecting information of vehicle and fuel', and *FR-114* for 'determining position and orientation of vehicle relative to the refueling tool'. Similarly, *FR-13* is further decomposed into the third level sub-eFRs as *FR-131* for 'accessing fuel port to open cover', *FR-132* for 'selecting and gripping nozzle for refueling', *FR-133* for 'retrieving refueling tool', and *FR-134* for 'closing fuel cover and resetting refueling tool'.

Figure 3.4 shows the correspondence of DVs to the FRs in Fig. 3.3. The sub-FRs at the third level under FR-11 are mapped to four DVs as *DV-111* with the options of visions, barcodes, and laser scanners, *DV-112* and DV-113 with the options of electronic chips, apps, and decentralized databases over the Internet, and DV-114 with the options of visions, barcodes, laser scanners, and controllable platforms. The sub-FRs at the third level under *FR-13* are mapped to four DVs as *DV-131, DV-132,* and *DV-114* with the options of robots, gantry systems, and specialized mechanisms equipped with multiple functional tools, and DV-133 with the options of manual inputs, *programmable logic controllers* (PLC), and IoT-based apps. The *FR-12* and *FR-14* at the second level are mapped to *DV-12* and DV-14 with the options of stand-alone systems and IoT-enabled apps.

Design synthesis. A *design space* consists of infinite or finite number of possible solutions depending on the number and types of design variables. To obtain an optimal

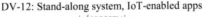

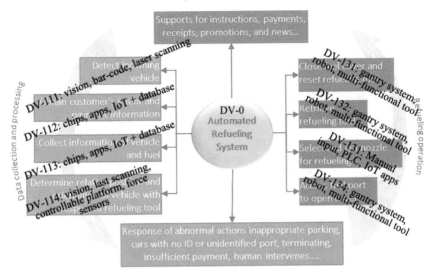

Fig. 3.4 The mapping from FRs to DVs of automated refueling system (Bi et al. 2020)

solution to the formulated problem, the performances of potential solutions in the design space must be analyzed, evaluated, and compared based on the specified optimization criteria. Therefore, *design synthesis* plays its critical role in the engineering process in (*a*) selecting an initial possible solution as a reference for evaluation and comparison; (*b*) executing search algorithms to explore better solutions, (*c*) defining terminating conditions to control the optimization process (see Fig. 3.5).

Design analysis. To implement the design synthesis in Fig. 3.5, design analysis is needed to evaluate and compare different solutions. *Design analysis* is used to model the behaviors of a product or system, and analyze a potential solution to determine (*a*) whether or not it meets all design constraints and (*b*) how well its performance is against the specified optimization criteria. The evaluation results from the design analysis are used by *design synthesis* to select better solutions. Generally, potential solutions have to be evaluated in a quantitative way. Therefore, the behaviors of a product or system have to be mathematically modeled to correspond design variables to design constraints, FRs, and evaluation criteria. Take the mechanism (*DV-131* or *DV-134* in Fig. 3.4) for opening or closing a fuel inlet as an example, performing design analysis needs the kinematic and dynamic models of the mechanism to evaluate whether or not the fuel inlet is reachable, and the inlet cover can be opened or closed without any interference within the specified period of time.

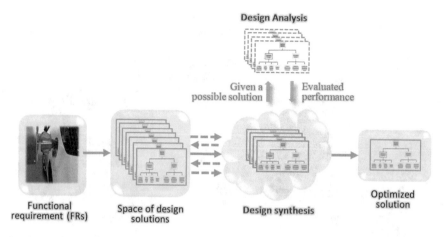

Fig. 3.5 *Design analysis* and *design synthesis* in an engineering process (Bi 2018)

Implementation of optimal solution. Once an optimal solution is determined at the conceptual design stage, the detail design is performed to implement optimal solution. Since adding the details of the design solution may affect the performance of the product or system, engineers have to refine the solution for its practical implementation. Taking the functional modules for *DV-131* or *DV-134* as the examples, joint actuators specify the ranges of motions, accelerations, and power; engineers need to re-evaluate the solution when the actuators are selected for the robot. Even the home positions of actuators will affect the workspace of the robot. The main tasks at the implementation stage include (*a*) convert the virtual model into the physical product or system, (*b*) verify and validate the performance of physical system, and refine the system whenever it is needed, and (*c*) perform the practice of *continuous improvement* (CI) in the application.

3.2 Design Analysis Methodologies

The behaviors of a product or system are governed by certain scientific principles, which can be mathematically represented by governing equations. Taking an example of a solid object subjected to external loads, the responses of the solid object to external loads can be represented by *the compatibility relations, strain–displacement relations*, and *the equations of motion*. From this perspective, design analysis begins with the understanding of the scientific and mathematical principles that govern the system behaviors.

As shown in Fig. 3.6, once a mathematic model is formulated, different engineering methods can be used to analyze the product or system. For example, design analysis methods can be the types of *graphic methods, experimental methods,* and *computational methods*. Table 3.1 gives a comparison of these analysis methods.

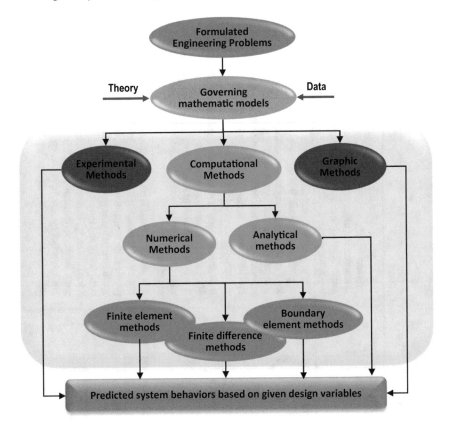

Fig. 3.6 Classification of design analysis methods

A graphic method uses the drawing tools to define the relations of inputs and outputs of system graphically. It helps to understand system behaviors visually. However, a graphic method is usually very preliminary, and it can only deal with some simple problems.

An experimental method studies the relations of inputs and outputs of system experimentally. Experimental methods are often applied in the implementation stage; appropriate *design of experiments* (DoE) will be needed to ensure that experimental results are reliable and trustable. However, an experimental method has a few of the limitations as below: (1) experiments are conducted on a physical system while it is only available at the implementation stage; fixing a design error discovered at the implementation phase involves a high cost; (2) using an experimental method means the need of additional cost on physical systems and instrumentations for measurements; (3) an experimental method uses the enumeration to investigate the impact of the changes of inputs on outputs of system. Since a system usually relates to many design variables, it is infeasible or impractical to understand the system by a large number of experiments.

Table 3.1 Comparison of different design analysis methods

Method		Description	Features
Graphic		Uses drawing tools to define the relations of inputs and outputs of system graphically	• It is easy to use, but it is only applicable to simple design problems • It requires the users with a high-level visibility and drawing capability; fixing an error takes a large amount of time for repetitive works
Experiment		Models the relations of inputs and outputs of system experimentally	• It is very reliable and can make the experiments very specific to given applications. It is applicable to the systems with any level of the complexity and uncertainty • It involves in an additional cost for setup and instrumentation; it is used only when the physical system is available, and it has its limitation in the number of experiments for practical purpose
Computational	Analytical	Obtains system outputs based on given system inputs analytically	• It leads to an explicit mathematical model to predict the trend of system in a continuous domain • It is only applicable to small-scale, simple design problems with a few of design variables and well-structured governing equations
	Numerical	FEA—integrating derivatives in the elements of the entire discretized domain FDM—approximating derivatives by finite differences in the elements of the entire discretized domain FBM—approximating the governing equations in elements of discretized boundaries of the domain	• It is used in a virtual model where no physical system is needed. It is generic to a wide scope of interdisciplinary problems in a continuous, discrete, or mixed domain. It is applicable to the engineering systems with any level of complexity and uncertainty • It leads to an approximated solution, and it needs multiple steps for the verification and validation of design solutions. The impact of system inputs on system outputs can only be studied numerically by simulation

A computational method uses computer models to simulate system behaviors. Computational methods are further classified into *analytical* methods and *numerical* methods. An analytical method uses analytical models to represent the relations of system inputs and outputs; it applies to some simple problems where system outputs can be obtained explicitly using analytical models. A numerical method uses numerical models to represent the relations of system inputs and outputs, and system outputs are obtained numerically by computer simulation. Numerical methods are applicable to both of explicit and inexplicit mathematical models. Numerical methods are widely used to analyze the systems with high complexity and uncertainties, which cannot be tackled with graphic methods, experimental methods, or analytical methods.

Numerical methods have become the default CAE tools for a number of reasons: (1) modern products or systems become so complex for other design analysis methods to evaluate their performances thoroughly; (2) in contrast to experimental methods, numerical methods are able to predict system behaviors without physical prototyping; this reduces the development cost, shortens design time, and allows to compare a large number of design options for optimization; (3) cutting-edge CAD and CAE tools are available for engineers to solve a wide scope of engineering problems without sophisticated trainings.

Many numerical methods are available, and they generally use the *divide and conquer* strategy to deal with the generality and complexity of different engineering problems. As shown in Fig. 3.6, a numerical method can be one of *finite element analysis* (FEA), *finite difference methods* (FDM), and *boundary element methods* (BEM). FEA differs from FDM in approximating the derivatives in a mathematical model. Derivatives are evaluated by the integration in FEA; while these derivatives are evaluated by finite differences in FDM. FEA differs from BEM in tackling with a continuous domain. FEA model has elements and nodes in the entire domain; while a BEM model has the elements and nodes only on the boundaries of the domain.

3.3 Numerical Simulation

As popular computational tools, numerical simulation is widely used to find approximate solutions to various engineering problems. In analyzing a system by numerical simulation, the system behaviors should firstly be formulated as a mathematic model; which is typically represented by some partial differential equations (PDEs) subjected to the specified boundary conditions. Figure 3.7 shows the dependence of a simulation model with its original engineering problem. An engineering problem is formulated as a set of design variables for inputs (I), system characteristic (S), and outputs (O); the behaviors of system can be represented by the relations of I, O, and S inexplicitly as $f\,(I, S, O) = 0$.

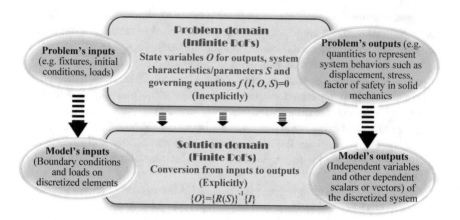

Fig. 3.7 Numerical simulation as design analysis method

The simulation model differs from its original engineering problem at the following aspects:

(1)   A simulation model discretizes the continuous domain of the original problem; in a discretized node, the behaviors on discretized nodes are used to represent the behaviors of any point in the continuous domain.

(2)   A simulation model has a limited number of degrees of freedom (DoF), each DoF corresponds to a *state variable* to be solved in simulation. The system behaviors are collectively determined by state variables on all nodes. In contrast, the original engineering problem involves infinite number of DoF since the points in a continuous domain are uncountable and infinite.

(3)   A simulation model must be solvable to find state variables *explicitly* subjected to the specified boundary conditions and loads. Solving a simulation model can be viewed as a conversion from the given boundary conditions to state variables to represent system behaviors approximately.

(4)   A simulation model is developed in such a way that all of the given parameters in the original engineering problem are defined as the boundary conditions and loads on discretized nodes and elements; all independent parameters related to system behaviors are defined as state variables or other dependent quantities.

3.4   Finite Element Analysis (FEA) and Modeling Procedure

FEA adopts the piecewise approximation where a continuous domain is discretized into *finite elements*. Each element is represented by a set of nodes; the behaviors of elements are represented by those of the nodes; a system model can then be assembled from element models. By incorporating boundary and load conditions,

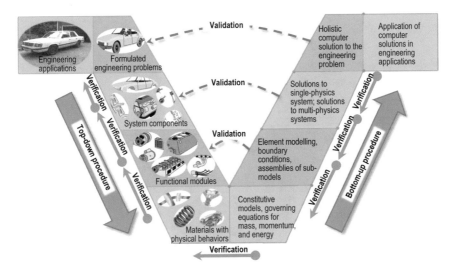

Fig. 3.8 Top-down and bottom-up procedure in FEA

the system model can be solved to obtain a numerical solution to original engineering problem. To analyze an engineering problem, FEA consists of two procedures, i.e., *the top-to-bottom procedure* and *the bottom-up procedure* as shown in Fig. 3.8.

The top-down procedure is to break down system complexity. *Firstly*, the boundaries of system are defined to clarify inputs (I), parameters (S), outputs (O), and design constraints and objectives. *Secondly*, the continuous domain is decomposed into discretized elements and nodes; multiple layers of decompositions may be performed, so that the behavior of one element and its relations with others can be defined appropriately. *Lastly*, the elemental behaviors are modeled as mathematical equations.

The bottom-up procedure is to obtain the system solution based on the sub-solutions at the element level. *Firstly*, element types and analysis types are selected to determine governing equations of elements adequately. *Secondly*, the solutions to the mathematical models are defined at the element level. *Thirdly*, the elements models are assembled into a system model based on the topological relations defined in the top-down procedure. *Finally*, the boundary conditions are applied in the system model; it is then solved to obtain the system solution. Since FEA is a technique of numerical simulation, verification and validation (V&V) that must be applied to endure the outcomes from FEA are acceptable.

To illustrate the steps in using FEA to solve an engineering problem, the idea of FEA is adopted to calculate the workspace of a robot numerically as in Example 3.1.

Example 3.1 Calculate the workspace of an Espon robot shown in Fig. 3.9a (Espon 2020).

Solution. *Firstly*, the problem of workspace calculation is formulated as a design problem with the inputs (I), outputs (O), and system parameters (S) as shown in

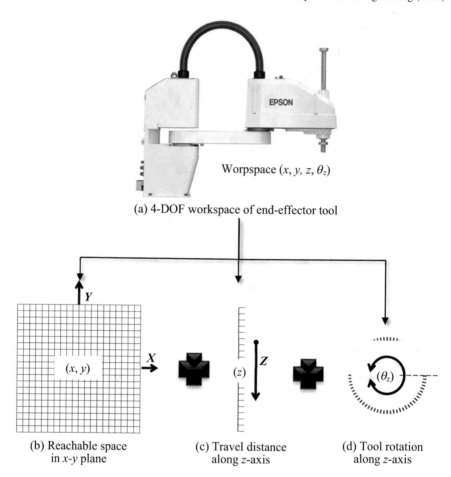

Worpspace (x, y, z, θ_z)

(a) 4-DOF workspace of end-effector tool

(b) Reachable space
in x-y plane

(c) Travel distance
along z-axis

(d) Tool rotation
along z-axis

Fig. 3.9 Decomposition of continuous domain

Table 3.2. Note that the workspace is a collection of all reachable points (x, y, z, θ_z) by the end-effector.

Secondly, the continuous domain of the workspace is discretized into three sub-domains as shown in Fig. 3.9b–d, respectively. Since the z-axis translation, z-axis

Table 3.2 Description of the formulated problem for workspace analysis of Espon robot

Inputs (*I*)	Joint	Joint 1 (θ_1)	Joint 2 (θ_2)	Joint 3 (z_3)	Joint 4 (θ_4)
	Range	±132 °	±150 °	200 mm	±360 °
	Home	0°	0°	0 mm	0°
Outputs (*O*)	*End-effector position and orientation: (x, y, z, θ_z)*				
System parameters (S)	*Name*	Arm length (L_1)		Arm length (L_2)	Z-offset
	Dimension	300 mm		300 mm	0 mm

rotation, and the motion in the x–y plane are decoupled and implemented by respective joints, the motions in these three sub-domains can be analyzed individually. The behavior of each element is represented by its central point, e.g., if the central point is reachable, the corresponding cell is reachable.

Thirdly, the kinematic model is developed to describe the relations of joint motions $(\theta_1, \theta_2, z_3, \theta_4)$ and the end-effector motion (x, y, z, θ_z). Since z_3 and θ_4 are one-to-one mapped to z and θ_z, respectively, only the mapping of (θ_1, θ_2) to (x, y) are modeled here.

In the first sub-domain, a working point (x, y) in the workspace can be calculated from given joint displacement (θ_1, θ_2) as

$$\left.\begin{array}{l} x = L_1\cos\theta_1 + L_2\cos\theta_2 \\ y = L_1\sin\theta_1 + L_2\sin\theta_2 \end{array}\right\} \tag{3.1}$$

$$A_1\cos\theta_1 + B_1\sin\theta_1 + C_1 = 0 \tag{3.2}$$

where

$$\left.\begin{array}{l} A_1 = 2xL_1 \\ B_1 = 2yL_1 \\ C_1 = L_1^2 - L_2^2 - x^2 - y^2 \end{array}\right\} \tag{3.3}$$

Let $t_1 = \tan\frac{\theta_1}{2}$, $\sin\theta_1 = \frac{2t_1}{1+t_1^2}$, and $\cos\theta_1 = \frac{1-t_1^2}{1+t_1^2}$, Equation (3.2) becomes

$$(C_1 - A_1)t_1^2 + 2B_1t_1 + (C_1 + A_1) = 0 \tag{3.4}$$

The solution of from Eq. (3.4) is found as

$$\theta_1 = 2\tan^{-1}\frac{-2B_1 \pm \sqrt{B_1^2 - C_1^2 + A_1^2}}{(C_1 - A_1)} \tag{3.5}$$

Following the similar procedure to eliminate θ_1 in Eq. (3.1) gets the solution of θ_2 as

$$\theta_2 = 2tan^{-1}\frac{-2B_2 \pm \sqrt{B_2^2 - C_2^2 + A_2^2}}{(C_2 - A_2)} \tag{3.6}$$

where

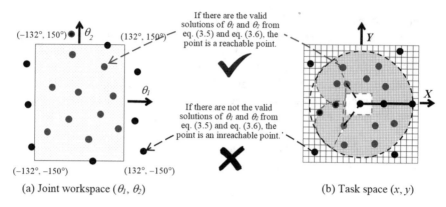

(a) Joint workspace (θ_1, θ_2) (b) Task space (x, y)

Fig. 3.10 Determine if a point in 3D space is a reachable point

$$\left.\begin{array}{l} A_2 = 2xL_2 \\ B_2 = 2yL_2 \\ C_2 = L_2^2 - L_1^2 - x^2 - y^2 \end{array}\right\} \tag{3.7}$$

Fourthly, elements are analyzed one by one to determine if their central points have corresponding inverse kinematic solutions. If an inverse kinematic solution exists in the joint space, the element is reachable and within the workspace in the task space (see Fig. 3.10).

Fifthly, the reachability of all of discretized elements is checked, and all of the reachable elements are assembled to form the workspace of the robot. Note that the workspace consists of the sub-space on the *x–y* plane, z-axis translation, and *z-axis* rotation.

Finally, the result from the workspace analysis is processed to visualize and retrieve the result. The post-processing also includes deriving some relevant quantities that help to understand the solution to the original engineering problem. Figure 3.11 visualizes the workspace of the robot with the limits of the end-effector motion along *x*-, *y*-, and *z*- axes.

The FEA modeling procedure can be generalized from the aforementioned example, and it consists of the following steps.

Step 1—Decomposition. The engineering problem is formulated to identify inputs (I), outputs (O), and characteristic parameters (S) of system. The continuous domain about S is discretized into a set of elements, and each element is represented by its nodes. The schematic of the discretization also defines element-to-element and node-to-element relations. The system behaviors are represented by the field variables on nodes, and the interpolation functions are used to describe the relation of any point and its corresponding nodes in the element.

Step 2—Develop element models. The analysis type of the FEA model is defined to reflect the physical behaviors of system. The analysis type determines governing differential equations of elements in mathematical modeling. The numerical solution

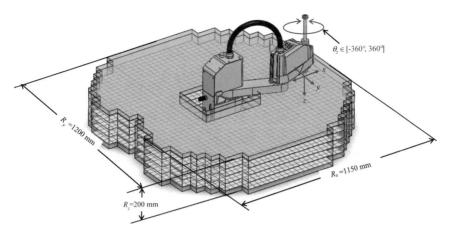

Fig. 3.11 Post-processing of workspace analysis

to the mathematic model is obtained by the approximation such as *direct methods, minimum potential energy methods*, or *weighted residual methods*.

Step 3—Assembly. The element models in *local coordinate systems* (LCSs) are transformed into the element models in a *global coordinate system* (GCS), and the transformed element models in GCS are then assembled into a system model.

Step 4—Apply boundary conditions and loads. The interactions of the system with its application environment are analyzed to define *boundary conditions* (BCs) and *loads* for the FEA model.

Step 5—Solve for primary unknowns. Sufficient BCs ensure that the assembled system model is solvable. For the majority of engineering applications, a system model refers to a set of the linear equations for the field variables on nodes. Numerous algorithms have been developed and used to solve unknown field variables in the system model.

Step 6—Calculate dependent variables. The variables used to represent the system behaviors can be classified into *independent variables* and *dependent variables*. For example, stress and strain are dependent with each other. Strains can be selected as independent variables, sequentially, the stresses are dependent, which can be determined based on the constitutive models of materials. After independent variables are solved from the system model, the post-processing can be performed to evaluate dependent variables.

3.5 FEA Theory

In FEA modeling, a virtual model of a part, process, product, or a system has to be established; the virtual model includes one or more finite volumes with specified material properties. The *divide and conquer* strategy is applied to deal with the

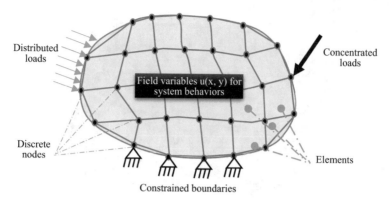

Fig. 3.12 Discretization of a continuous domain

0. As shown in Fig. 3.12, the continuous domain is decomposed into small parts called '*elements*'. Elements are connected at the interconnections, which are called as '*nodes*'. In return, each element is associated with a set of nodes, and the behavior of any material point in the element is determined by the behaviors of the nodes associated with this element. *An elemental model* is used to describe the behaviors of the nodes of the element; note that the behaviors of the nodes are represented by the field variables. *A system model* is an assembly of the elemental models of all of discretized elements in the continuous domain. By applying boundary conditions and loads to a system model, the behaviors of all nodes in the continuous domain can be determined as the solution to the system model.

3.5.1   Governing Equations of Engineering Problems

Ideally, an analytical model to an engineering problem is desired; since it leads to an exact solution. Unfortunately, an analytical solution does not always exist; even if it exists, it is mostly impractical to obtain it with a reasonable amount of time and level of the required expertise. In contrast, FEA has been a powerful CAE tool due to its capability in dealing with the complexity of various engineering problems at different aspects: (1) a continuous domain of a real-world design problem is often irregular and very complex; moreover, it is difficult to represent the boundaries of a domain accurately. (2) The solid domain may consist of a set of sub-domains with different properties, or one material with anisotropic nature. (3) The behaviors of the engineering system may involve in multiple disciplines. (4) The engineering system has the complex interactions with its application environments, which brings the challenges in defining boundary or loading conditions.

Table 3.3 shows a few examples of engineering problems with the corresponding laws of mechanics and differential equations. As long as an engineering problem is formulated as a mathematic model with specified boundary conditions, numerical simulation such as FEA can be used to find the system solution.

FEA is a generic technique to solve a mathematic model subjected to specified boundary conditions. If different engineering systems can be formulated as the same mathematic model, the same solver in the FEA implementation can be used to solve

Table 3.3 Mathematic models of common physical phenomena in engineering

Engineering applications	Laws of mechanics and differential equations
Solid mechanics where $[\sigma]^T$ and $[\varepsilon]^T$ are stress and strain vectors as $$[\sigma]^Y = \begin{bmatrix} \sigma_{xx} & \sigma_{yy} & \sigma_{zz} & \tau_{xy} & \tau_{yz} & \tau_{xz} \end{bmatrix}$$ $$[\varepsilon]^T = \begin{bmatrix} \varepsilon_{xx} & \varepsilon_{yy} & \varepsilon_{zz} & \gamma_{xy} & \gamma_{yz} & \gamma_{xz} \end{bmatrix}$$	Hooke's law $$\varepsilon_{xx} = \tfrac{1}{E}\left[\sigma_{xx} - v(\sigma_{yy} + \sigma_z)\right]$$ $$\varepsilon_{yy} = \tfrac{1}{E}\left[\sigma_{yy} - v(\sigma_{xx} + \sigma_{zz})\right]$$ $$\varepsilon_{zz} = \tfrac{1}{E}\left[\sigma_{zz} - v(\sigma_{xx} + \sigma_{yy})\right]$$ $$\gamma_{xy} = \tfrac{\tau_{xy}}{G} \quad \gamma_{yz} = \tfrac{\tau_{yz}}{G} \quad \gamma_{xz} = \tfrac{\tau_{xz}}{G}$$ where $$\varepsilon_{xx} = \tfrac{\partial u}{\partial x} \quad \varepsilon_{yy} = \tfrac{\partial v}{\partial y} \quad \varepsilon_{zz} = \tfrac{\partial w}{\partial z}$$ $$\gamma_{xy} = \tfrac{\partial u}{\partial y} + \tfrac{\partial v}{\partial x} \quad \gamma_{yz} = \tfrac{\partial v}{\partial z} + \tfrac{\partial w}{\partial y} \quad \gamma_{xz} = \tfrac{\partial u}{\partial z} + \tfrac{\partial w}{\partial x}$$ E and G are young's modulus and shear modulus, respectively
Dynamics	Lagrange's equation where $\frac{d}{dt}\left(\frac{\partial T}{\partial \dot{q}_i}\right) - \frac{\partial T}{\partial q_i} + \frac{\partial \Lambda}{\partial q_i} = Q_i \quad (i = 1, 2, 3 \cdots n)$ $t = $ time $T = $ kinetic energy of the system $q_i = $ coordinate system $\dot{q} = $ time derivate of coordinate system $\Lambda = $ potential energy of the system $Q_i = $ non-conservative forces or moments
Heat transfer problem	Fourier law of conduction $q = -k \cdot \nabla T$ Newton's law of convection (BC)$q = hA(T_s - T_f)$ Stefan-Boltzmann law of radiation (BC) $P = e\sigma A(T^4 - T_C^4)$ where $q = $ heat flux $k = $ conductivity coefficient $T = $ temperature $h = $ convection coefficient $T_s = $ temperature on surface $T_f = $ temperature in surrounding fluid $P = $ net radiated power $A = $ radiating area $\sigma = $ Stefan's constant $E = $ emissivity $T_c = $ temperature of surrounding

(continued)

Table 3.3 (continued)

Engineering applications	Laws of mechanics and differential equations
Electromagnetics E = electric field B = magnetic field D = electric displacement H = magnetic field strength ρ = charge density ε_0 = permittivity μ_0 = permeability J = current density C = speed of light P = polarization I = electric current M = magnetization	Maxwell's equations Gauss' law for electricity $\nabla \cdot (\varepsilon_0 E + P) = \rho$ Gauss's law for magnetism $\nabla \cdot B = 0$ Faraday's law of induction $\nabla \times E = \frac{\partial B}{\partial t}$ Ampere's law $\nabla \times H = J + \frac{\partial D}{\partial t}$ where $\begin{array}{l} D = (\varepsilon_0 E + P) \\ B = \mu_0 (H + M) \end{array}$ $\nabla \cdot (\bullet)$ and $\nabla \times (\bullet)$ are the *divergence* and *curl* operations of vectors, respectively
Fluid mechanics	Law of conservation of mass $\frac{\partial \rho}{\partial t} + \frac{\partial \rho u}{\partial x} + \frac{\partial \rho v}{\partial y} + \frac{\partial \rho w}{\partial z} = 0$ where t = time, ρ = density, u, v, and w are the velocities along x, y, and z axis, respectively For a steady flow: $\frac{\partial \rho}{\partial t} = 0$ From impressive flow where the density is constant: $\frac{\partial u}{\partial x} + \frac{\partial v}{\partial y} + \frac{\partial w}{\partial z} = 0$

these problems. Many engineering problems have similar governing equations, even though these problems are formulated in different disciplines. Figure 3.13 has shown a variety of engineering problems, which are governed by the well-known Poisson equations (Chandrupatla and Belegundu 2012). If an FEA tool includes a solver to the Poisson equations, it can be applied to solve all of these problems.

3.5.2 Solution by Direct Formulation

In *direct formulation*, an element model is derived from the governing equations directly. Figure 3.14 shows an example of modeling a spring element by direct formulation.

The spring element in Fig. 3.14 is composed of two nodes, which are denoted as node i and node j, respectively. The behavior of the element is represented by the displacements at two nodes U_i and U_j. U_i and U_j can be determined by the conditions

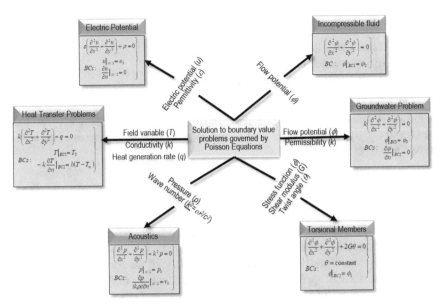

Fig. 3.13 Poisson Equations in different engineering problems

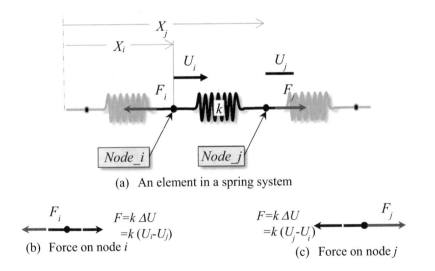

(a) An element in a spring system

(b) Force on node i $\qquad$ (c) Force on node j

Fig. 3.14 Modeling of a spring element by direct formulation

of the force balance in *the free body diagrams* (FBDs) of node i and j in Fig. 3.14b, c. Note that the behavior of a spring element is governed by the Hook's law as

$$F = k \cdot \Delta U \tag{3.8}$$

where.

> F: the force passes through the spring,
> K: the spring coefficient, and
> ΔU: the length change of the spring

In an FEA model, the behavior of an element is represented by the field variables of the associated nodes; in Fig. 3.14, the behavior of the spring element is represented by U_i and U_j. By applying the forces F_i and F_j on the nodes, the forces in the FBDs of two nodes must be balanced, i.e.,

$$\left.\begin{array}{c} k\left(U_i - U_j\right) = F_i \\ k\left(U_j - U_i\right) = F_j \end{array}\right\} \tag{3.9}$$

Equation (3.9) involves in two equations for two field variables U_i and U_j; it is not solvable until the forces F_i and F_j are known. In turn, F_i and F_j are determined by boundary conditions. For example, if the left node of the spring element is fixed and the right node is applied a force of F, the boundary conditions generate two additional equations for Eq. (3.9), i.e., $U_i = 0$, and $F_j = F$; then, the element model becomes solvable to obtain the element solution as $F_i = -F$ and $U_j = F/k$.

The above spring element model is generic and can be extended to solve a variety of simple engineering problems. When system elements respond external loads linearly, they can be modeled as equivalently as spring elements. In an equivalent spring system, the solution can be derived by mapping their physical parameters, external loads, and field variables into spring rate (k), forces (F_i and F_j), and the displacement in a spring element (U_i and U_j). Table 3.4 gives the mappings of some basic engineering problems to equivalent spring systems.

Taking an example of a resistor element in Table 3.4, its element model can be defined by mapping all of physical properties and quantities to those of a spring element, i.e.

$$\left.\begin{array}{c} \frac{1}{R} \rightarrow k \\ V \rightarrow U \\ I \rightarrow F \end{array}\right\} \tag{3.10}$$

Replacing the corresponding parameters and variables in Eq. (3.9) gets the model of a resistor element (R) as

$$\begin{bmatrix} 1/R & -1/R \\ -1/R & 1/R \end{bmatrix}\begin{bmatrix} V_i \\ V_j \end{bmatrix} = \begin{bmatrix} I_i \\ I_j \end{bmatrix} \tag{3.11}$$

Example 3.2. Find the voltage drop in each resistor of the circuit in Fig. 3.15.

Solution. Each resistor is treated as an equivalent spring element; accordingly, the circuit in Fig. 3.15a is decomposed in the equivalent spring system in Fig. 3.15b. It

Table 3.4 Similarity of a spring system with other engineering systems

Engineering problems	Equivalent terms in a spring element		
	Stiffness k	Load F	Field variable U
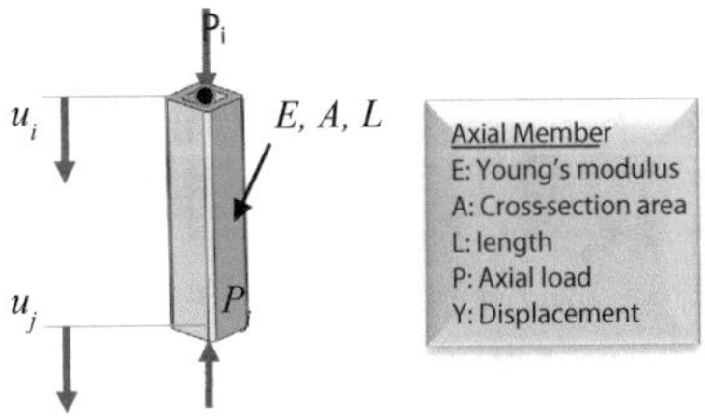P_i u_i E, A, L u_j P_j Axial Member E: Young's modulus A: Cross-section area L: length P: Axial load Y: Displacement	$\frac{EA}{L}$	P	u
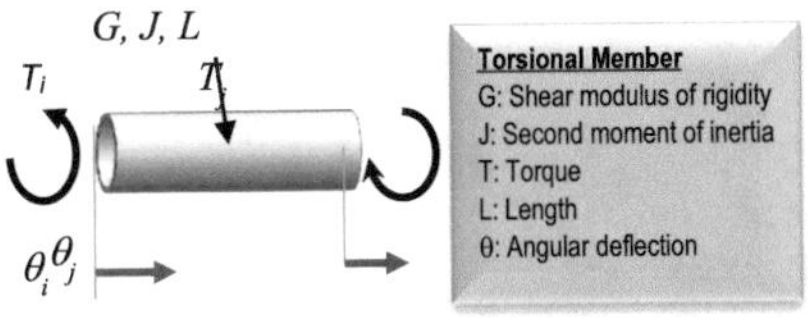G, J, L T_i T_j $\theta_i \theta_j$ Torsional Member G: Shear modulus of rigidity J: Second moment of inertia T: Torque L: Length θ: Angular deflection	$\frac{GJ}{L}$	T	θ
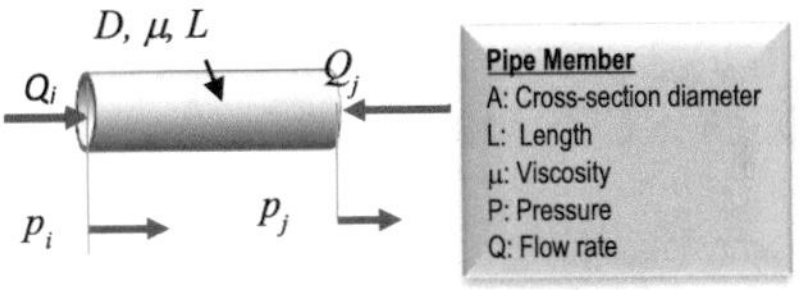D, μ, L Q_i Q_j p_i p_j Pipe Member A: Cross-section diameter L: Length μ: Viscosity P: Pressure Q: Flow rate	$\frac{\pi D^4}{128\mu}$	Q	p
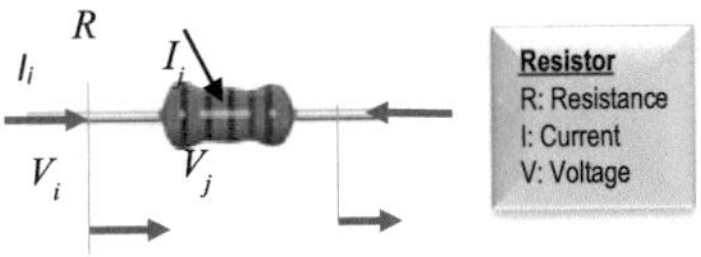R I_i I_j V_i V_j Resistor R: Resistance I: Current V: Voltage	$\frac{1}{R}$	I	V
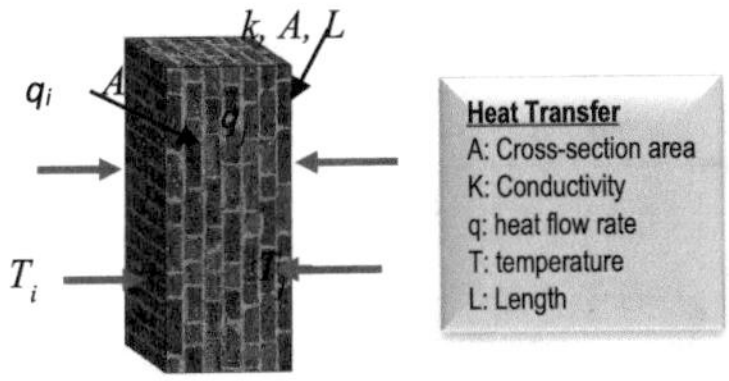k, A, L q_i A q_j T_i T_j Heat Transfer A: Cross-section area K: Conductivity q: heat flow rate T: temperature L: Length	$\frac{kA}{L}$	q	T

consists of 4 elements and 4 nodes. Table 3.5 shows the result of decompositions and corresponding element models.

Assembling the element models in Table 3.5 gets the system model as

$$\begin{bmatrix} 0.2 & -0.2 & 0 & 0 \\ -0.2 & 0.2+0.1+0.05 & -0.1-0.05 & 0 \\ 0 & -0.1-0.05 & 0.1+0.05+0.1 & -0.1 \\ 0 & 0 & -0.1 & 0.1 \end{bmatrix} \begin{bmatrix} V_1 \\ V_2 \\ V_3 \\ V_4 \end{bmatrix} = \begin{bmatrix} I_{1,1} \\ I_{2,1}+I_{2,2}+I_{2,3} \\ I_{3,2}+I_{3,3}+I_{3,4} \\ +I_{4,4} \end{bmatrix} \leftarrow \begin{bmatrix} 1 \\ 0 \\ 0 \\ -1 \end{bmatrix} \quad (3.12)$$

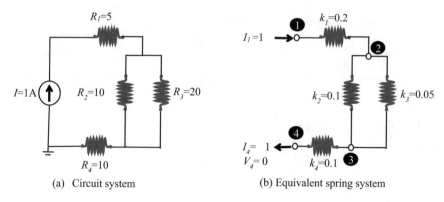

(a) Circuit system   (b) Equivalent spring system

Fig. 3.15 FEA modeling of a circuit example

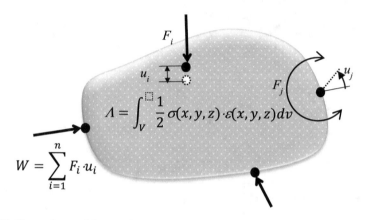

Fig. 3.16 Types of potential energy in a solid domain

Table 3.5 The decomposition of equivalent spring system

Element	Node i	Node j	ke (1/R)	Element model
1	❶	❷	0.2	$\begin{bmatrix} 0.2 & -0.2 \\ -0.2 & 0.2 \end{bmatrix} \begin{bmatrix} V_1 \\ V_2 \end{bmatrix} = \begin{bmatrix} I_{1,1} \\ I_{2,1} \end{bmatrix}$
2	❷	❸	0.1	$\begin{bmatrix} 0.1 & -0.1 \\ -0.1 & 0.1 \end{bmatrix} \begin{bmatrix} V_2 \\ V_3 \end{bmatrix} = \begin{bmatrix} I_{2,2} \\ I_{3,2} \end{bmatrix}$
3	❷	❸	0.05	$\begin{bmatrix} 0.05 & -0.05 \\ -0.05 & 0.05 \end{bmatrix} \begin{bmatrix} V_2 \\ V_3 \end{bmatrix} = \begin{bmatrix} I_{2,3} \\ I_{3,3} \end{bmatrix}$
4	❸	❹	0.1	$\begin{bmatrix} 0.1 & -0.1 \\ -0.1 & 0.1 \end{bmatrix} \begin{bmatrix} V_3 \\ V_4 \end{bmatrix} = \begin{bmatrix} I_{3,4} \\ I_{4,4} \end{bmatrix}$

The circuit system has two boundary conditions: (1) the current at nodes 1 and 4 are given as 1 A and − 1 A, respectively, and no additional current is input at node 2 and node 3; (2) node 4 is grounded as a reference ($V_4 = 0$ v). Therefore, the system model Eq. (3.12) can be simplified as

$$\begin{bmatrix} 0.2 & -0.2 & 0 & 0 \\ -0.2 & 0.35 & -0.15 & 0 \\ 0 & -0.15 & 0.25 & -0.1 \\ 0 & 0 & 0 & 0 \end{bmatrix} \begin{bmatrix} V_1 \\ V_2 \\ V_3 \\ V_4 \end{bmatrix} = \begin{bmatrix} 1 \\ 0 \\ 0 \\ 0 \end{bmatrix} \tag{3.13}$$

Solving Eq. (3.13) obtains the solution at $[V] = [21.6667, 16.6667, 10, 0]^{\mathrm{T}}$, and the voltage drops over the resistors R_1, R_2, R_3, and R_4 are 5, 6.6667, 6.6667, and 10 voltage, respectively.

3.5.3 Principle of Minimum Potential Energy

The behavior of an engineering system can be described from the perspective of *energy* and *work*. In a computational domain, energy can be transferred from one form to another, but it never vanishes. For a system in the steady condition, the principle of the minimum potential energy can be utilized to model system behaviors.

Figure 3.16 shows a solid domain subjected to a set of external loads F_i ($i = 1$, $2,....n$). The response of the solid domain is represented by (1) the displacements u_i ($i = 1, 2, ...n$) at the locations where the external loads are applied and (2) the distributions of stress $\sigma(x, y, z)$ and strain $\varepsilon(x, y, z)$ in the solid domain. The total potential energy in the solid domain is expressed as

$$\Pi = \Lambda - W = \int_V \frac{1}{2} \sigma(x, y, z) \cdot \varepsilon(x, y, z) dv - \sum_{i=1}^{n} F_i \cdot u_i \tag{3.14}$$

where Π is the total potential energy, Λ is the strain energy in solid domain, and W is total work done by external forces F_i ($i = 1, 2,....n$).

The principle of minimum potential energy states that an equilibrium system meets the conditions of minimizing the total potential energy of the system as

$$\frac{\partial \Pi}{\partial u_i} = \frac{\partial (\Lambda - W)}{\partial u_i} = 0, \quad (i = 1, 2, \ldots n) \tag{3.15}$$

Example 3.3 Use the principle of minimum potential energy to find the displacement at the free end in Fig. 3.17.

Solution. The solid domain in Fig. 3.17a is decomposed into an FEA model in Fig. 3.17b, which consists of 4 elements and 4 nodes. The details of FEA model are summarized in Table 3.6.

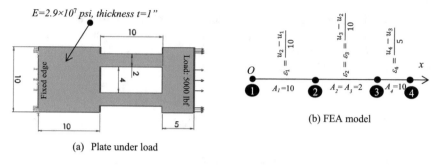

(a) Plate under load

(b) FEA model

Fig. 3.17 A two-dimensional plate subjected to axial load

Table 3.6 The decomposition of equivalent spring system

Element	Node i	Node j	ε	$A(in^2)$	Strain energy ($lbf.in$)
1	❶	❷	$\frac{u_2-u_1}{10}$	10	$\int_0^{10} \frac{EA_1}{2}\left(\frac{u_2-u_1}{10}\right)^2 dx$
2	❷	❸	$\frac{u_3-u_2}{10}$	2	$\int_0^{10} \frac{EA_2}{2}\left(\frac{u_3-u_2}{10}\right)^2 dx$
3	❷	❸	$\frac{u_3-u_2}{10}$	2	$\int_0^{10} \frac{EA_3}{2}\left(\frac{u_3-u_2}{10}\right)^2 dx$
4	❸	❹	$\frac{u_4-u_3}{5}$	10	$\int_0^{10} \frac{EA_4}{2}\left(\frac{u_4-u_3}{5}\right)^2 dx$

Using Eq. (3.14) gets the total potential energy as

$$\Pi = 1.45 \times 10^7 (u_2 - u_1)^2 + 5.8 \times 10^6 (u_3 - u_2)^2 + 5.8 \times 10^7 (u_4 - u_3)^2$$
$$- (R)u_1 - (5000)u_4 \tag{3.16}$$

Using Eqs. (3.15) and (3.16) determine the conditions for minimum potential energy as

$$\left.\begin{array}{l}
\frac{\partial \Pi}{\partial u_1} = 0 \rightarrow \qquad\qquad -2.9 \times 10^7 (u_2 - u_1) = R \\[4pt]
\frac{\partial \Pi}{\partial u_2} = 0 \rightarrow 2.9 \times 10^7 (u_2 - u_1) - 11.6 \times 10^6 (u_3 - u_2) = 0 \\[4pt]
\frac{\partial \Pi}{\partial u_3} = 0 \rightarrow 11.6 \times 10^6 (u_3 - u_2) - 11.6 \times 10^7 (u_4 - u_3) = 0 \\[4pt]
\frac{\partial \Pi}{\partial u_4} = 0 \rightarrow \qquad\qquad 11.6 \times 10^7 (u_4 - u_3) = 5000
\end{array}\right\} \tag{3.17}$$

Equation (3.17) can further be simplified as

$$2.9 \times 10^6 \begin{bmatrix} 10 & -10 & 0 & 0 \\ -10 & 14 & -4 & 0 \\ 0 & -4 & 44 & -40 \\ 0 & 0 & -40 & 40 \end{bmatrix} \begin{bmatrix} u_1 \\ u_2 \\ u_3 \\ u_4 \end{bmatrix} = \begin{bmatrix} R \\ 0 \\ 0 \\ 5000 \end{bmatrix} \tag{3.18}$$

Applying the boundary condition that the left side is fixed, i.e., $u_1 = 0$, Eq. (3.18) leads to the solution of $[u]$ as

$$[u] = \begin{bmatrix} 0.0 & 1.72 \times 10^{-4} & 6.03 \times 10^{-4} & 6.47 \times 10^{-4} \end{bmatrix}^T$$

3.5.4 Weighted Residual Methods

The direct method and the principle of the minimum potential energy are applicable to a limited number of engineering problems; they lack the flexibility to solve a broad scope of *partial differential equations* (PDEs). There is a demand to develop the methods that are more generic for PDEs. In this section, the object under an axial load in Fig. 3.18 is used as an example to illustrate the concept of approximated solution, and *the weighted residual method* is discussed to approximate the solution to a PDE. The object has a varying cross-sectional area along its length to carry the load.

Drawing the FBD of the finite segment of the solid domain identifies all of three forces and the condition of force equivalent as

$$\left. \begin{array}{l} p(x) = E \cdot b(x)(1)\frac{u(x)-u(x+\Delta x)}{x} \\ p(x + \Delta x) = E \cdot b(x + \Delta x)(1)\frac{u(x+\Delta x)-u(x)}{\Delta x} \\ p(x + \Delta x) + b(x) \cdot (1) \cdot \rho \cdot g \cdot \Delta x = p(x) \end{array} \right\} \quad (3.19)$$

Simplifying Eq. (3.19) gets

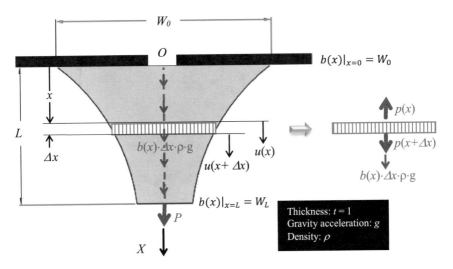

Fig. 3.18 A two-dimensional plate subjected to an axial load

$$E \cdot \{b(x + \Delta x) - b(x)\} \frac{u(x + \Delta x) - u(x)}{\Delta x} + b(x) \cdot \rho \cdot g \cdot \Delta x = 0 \qquad (3.20)$$

Taking the limit of $\Delta x \to 0$ in Eq. (3.20) gets

$$E \frac{d(b(x))}{dx} \frac{d(u(x))}{dx} + b(x) \cdot \rho \cdot g = 0 \qquad (3.21)$$

Adding the boundary conditions in Eq. (3.21) gets the mathematic model of *the boundary value problem* (BVP) as

$$\left.\begin{array}{c} \frac{d(u(x))}{dx} + f(x) = 0 \\ \frac{d(u(x))}{dx}\bigg|_{x=L} = \frac{P}{E \cdot W_L} \\ u(x)|_{x=L} = 0 \end{array}\right\} \qquad (3.22)$$

where $f(x) = b(x) \cdot \rho \cdot g / \left\{ E \cdot \frac{d(b(x))}{dx} \right\}$

The exact solution $u(x)$ of the displacement to the boundary value problem Eq. (3.22) is defined as

$$\left.\begin{array}{c} \Re(x) = \frac{d(u(x))}{dx} + f(x) \equiv 0 \quad x \in [0, L] \\ \frac{d(u(x))}{dx}\bigg|_{x=L} = \frac{P}{E \cdot W_L} \\ u(x)|_{x=L} = 0 \end{array}\right\} \qquad (3.23)$$

where $\Re(x)$ is *the residual function* over the solid domain of $x \in [0, L]$.

When an approximate solution $\bar{u}(x)$ is used in Eq. (3.23) instead of the exact solution $u(x)$, the first equation cannot be satisfied strictly. In the approximation, the conditions of a weak solution are relaxed as

$$\left.\begin{array}{c} \int_0^L v(x) \cdot \Re(x) dx = \int_0^L v(x) \cdot \left\{ \frac{d(\bar{u}(x))}{dx} + f(x) \right\} dx = 0 \\ \frac{d(\bar{u}(x))}{dx}\bigg|_{x=L} = \frac{P}{E \cdot W_L} \\ \bar{u}(x)|_{x=L} = 0 \end{array}\right\} \qquad (3.24)$$

where $\overline{\Re}(x)$ is *the residual function* for $\bar{u}(x)$ over the solid domain of $x \in [0, L]$; $v(x)$ is a test function.

For any function $u(x)$, the solution to BVP can be approximated by a part of the Taylor expansion of the function $u(x)$ as

$$\tilde{u}(x) = u(x_0) + \frac{\partial u(x_0)}{2x}(x - x_0) + \frac{1}{2!} \frac{\partial^2 u(x_0)}{\partial x^2}(x - x_0)^2 + \cdots + \frac{1}{n!} \frac{\partial^n u(x_0)}{\partial x^n}(x - x_0)^n \qquad (3.25)$$

Let $x_0 = 0$, Eq. (3.25) can be simplified as a polynomial equation as

$$\tilde{u}(x) = c_0 + c_1 x + c_2 x^2 + \ldots + c_{N-1} x^{N-1} \qquad (3.26)$$

where c_i $(i = 0, 1, 2, \ldots, N-1)$ are N constants which are determined by the weighted residual method.

Note that a total of N equations are needed to solve N unknown constants in Eq. (3.26), a few of the equations correspond to the boundary conditions of PDE, and the rest of the equations are defined by using different test functions in the expression of the residual over the domain. Table 3.7 shows some commonly used test functions.

A weighted residual method is reduced to find a set of coefficients c_i $(i = 0, 1, 2, \ldots, N-1)$ in $\bar{u}(x)$; it will be based on the conditions for a weak solution to the BVP as below.

(1) Determine the number of the polynomial terms (N). This number should be larger than the sum of the numbers of differential equation(s) and BCs in BVP, i.e.,

$$N \geq N_D + N_{BC} \qquad (3.27)$$

Taking an example of BVP in Eq. (3.23), it has one differential equation $(N_D = 1)$ and two BCs $(N_{BC} = 2)$. Therefore, the included polynomial terms of the Taylor expansion in the approximation should be equal to or larger than $N_D + N_{BC} = 3$. The more polynomial terms are used, the better approximation can be achieved but with much more computation.

Table 3.7 Common test functions in a weighted residual method

Test function	Expression	Description
Collection method	$v_i(x) = \begin{cases} 1 & x = x_i \\ 0 & x \in (a, b), x \neq x_i \end{cases}$	The test function has the unit value only at the specified point (x_i) in the domain (a, b) of x
Subdomain method	$v_i(x) = \begin{cases} 1 & x \in (x_i, x_{i+1}) \\ 0 & x \in (a, b), x \notin (x_i, x_{i+1}) \end{cases}$	The test function has the unit value only in the specified subdomain (x_i, x_{i+1}) in the domain (a, b) of x
Least-square method	$v_i(x) = \frac{\partial \Re(c_0, c_1, \ldots c_{N-1}, x)}{\partial c_i}$	The test function aims to minimize the least square residuals over the domain; the number of the test functions determines that of the constants to be determined
Galerkin method	$v_i(x) = \phi_i(x)$	One of the sub-functions in used the approximated solution $\tilde{u}(x)$ as the test function

(2) BCs in BVP are applied to define N_{BC} number of the equations, which should be satisfied by $\tilde{u}(x)$.

(3) Select $(N\text{-}N_{BC})$ number of test functions, and apply the condition Eq. (3.24) for a weak-solution. This generates $(N\text{-}N_{BC})$ number of new equations, which should be satisfied by $\tilde{u}(x)$. Combining with N_{BC} equations for boundary conditions, a system of N number of the equations are developed for N number of the constants c_i $(i = 0, 1, 2, ..., N\text{-}1)$ in $\overline{u}(x)$.

(4) Solve N equations simultaneously to obtain $c_0, c_1, ..., c_{N-1}$ for the complete solution of $\overline{u}(x)$ to VBP.

In above procedure, the number of the polynomial terms (N) and the types of test functions can be chosen arbitrarily as long as N satisfies Eq. (3.27). Different test functions lead to the different approximated solutions to BVP; but all of them have to meet the conditions for a weak solution in Eq. (3.24).

Example 3.4 Find a weak solution to the following boundary value problem:

$$\frac{d^2u}{dx^2} = 6x - \sin(x), 0 \le x \le 1$$

$$subjected\ to:$$

$$\left.\begin{array}{l} u(x)|_{x=1} = \sin(1) \\ \dfrac{du}{dx}|_{x=0} = 0 \end{array}\right\} \tag{3.28}$$

Solution. The weighted residual method is used in the steps as follows.

Step 1. Equation (3.28) has one differential equation $(N_D = 1)$ and two boundary conditions $(N_{BC} = 2)$. Therefore, the number of the polynomial terms (N) must be larger than $N_D + N_{BC} = 3$. Let N be 4, and the approximate solution is assumed as

$$\tilde{u}(x) = c_0 + c_1x + c_2x^2 + c_3x^3 \tag{3.29}$$

Equation (3.29) has unknown constants (c_0, c_1, c_2, c_3); thus, four equations about (c_0, c_1, c_2, c_3) are derived from the conditions of the weak solution.

Step 2. Equation (3.28) involves in two boundary conditions, substituting Eq. (3.29) into Eq. (3.28) gives

$$\left.\begin{array}{l} \tilde{u}(x)|_{x=0} = c_0 + c_1(1) + c_2(1)^2 + c_3(1)^3 = \sin(1) \\ \dfrac{d\tilde{u}(x)}{dx}\Big|_{x=0} = c_1 + 2c_2(0) + 3c_3(0)^2 = 0 \end{array}\right\} \tag{3.30}$$

Using Eq. (3.30) in Eq. (3.29) gets

$$\tilde{u}(x) = (\sin(1) - c_2 - c_3) + c_2x^2 + c_3x^3 \tag{3.31}$$

Equation (3.31) is used to obtain the first and second derivatives of $\tilde{u}(x)$ as

$$\left. \begin{aligned} \frac{d\tilde{u}(x)}{dx} &= 2c_2x + 3c_3x^2 \\ \frac{d^2\tilde{u}(x)}{dx^2} &= 2c_2 + 6c_3x \end{aligned} \right\}$$

(3.32)

Step 3. Equation (3.31) include two unknown constants (c_2, c_3). Therefore, two different test functions have to be selected to define another two equations by using the condition of a weak-solution (Eq. 3.23). Using Eq. (3.32) into Eq. (3.28) yields the expression of the residual as

$$\overline{\mathfrak{R}}(x) = \int_0^1 v(x) \cdot ((2c_2 + 6c_3x) - 6x + \sin(x))dx = 0$$

(3.33)

Next, different methods are used in selecting test functions $v(x)$ to evaluate residuals in Eq. (3.33).

(a). In a collocation method, the test function is selected as

$$v_i(x) = \begin{cases} 1 \ x = x_i \\ 0 \ x \in (a, b), x \neq x_i \end{cases}$$

(3.34)

where x_i is an arbitrary position in the domain of interest.

Since only two more equations are needed to solve c_2 and c_3; the condition of the weak-solution is applied twice. Let $x_1 = 1/3$ and $x_2 = 2/3$ in Eq. (3.24) and substituting it in Eq. (3.33), respectively,

$$\left. \begin{aligned} \overline{\mathfrak{R}}_1 &= \left(2c_2 + 6c_3\left(\tfrac{1}{3}\right)\right) - 6\left(\tfrac{1}{3}\right) + \sin(x) = 0 \\ \overline{\mathfrak{R}}_1 &= \left(2c_2 + 6c_3\left(\tfrac{2}{3}\right)\right) - 6\left(\tfrac{2}{3}\right) + \sin(x) = 0 \end{aligned} \right\}$$

(3.35)

Solving Eq. (3.35) yields $c_2 = -0.01801$ and $c_3 = 0.8544$. The approximated solution becomes

$$\tilde{u}(x) = 0.00508 - 0.01801x^2 + 0.8544x^3$$

(3.36)

$\overline{\mathfrak{R}}(x)$ by using the collocation method is plotted in Fig. 3.19. $\tilde{u}(x)$ in Eq. (3.36) satisfies the condition of a weak solution in sense that there is no error at a position specified by the test function in Eq. (3.34).

(b). In the sub-domain method, the test function is defined as

$$v_i(x) = \begin{cases} 1 \ x \in (x_i, x_{i+1}) \\ 0 \ x \in (a, b), x \notin (x_i, x_{i+1}) \end{cases}$$

(3.37)

where (a_i, b_i) is an arbitrary sub-domain of interest.

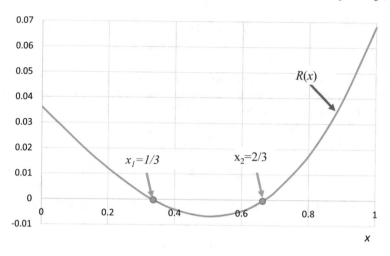

Fig. 3.19 Approximated solution using collocation methods

Two additional equations are needed to solve c_2 and c_3. Therefore, set two subdomains as $(0, 0.5)$ and $(0.5, 1.0)$, and use Eq. (3.27) in Eq. (3.3) to test the condition of a weak solution as

$$\left.\begin{array}{l} \overline{\mathfrak{R}}_1 = \int_0^{0.5} ((2c_2 + 6c_3 x) - 6x + \sin(x))dx = c_2 + \frac{3}{4}c_3 + \frac{1}{4} - \cos\left(\frac{1}{2}\right) = 0 \\ \overline{\mathfrak{R}}_2 = \int_{0.5}^{1} ((2c_2 + 6c_3 x) - 6x + \sin(x))dx = c_2 + \frac{9}{4}c_3 - \frac{9}{4} + \cos\left(\frac{1}{2}\right) - \cos(1) = 0 \end{array}\right\} \tag{3.38}$$

Solving Eq. (3.38) yields $c_2 = -0.01499$ and $c_3 = 0.85676$, and the approximated solution becomes

$$\tilde{u}(x) = 0.000299 - 0.01499x^2 + 0.85676x^3 \tag{3.39}$$

$\overline{\mathfrak{R}}(x)$ by using the subdomain method is plotted in Fig. 3.20. $\tilde{u}(x)$ in Eq. (3.39) satisfies the condition of a weak solution in the sense that there is no error in the sub-domain specified by the test function in Eq. (3.37).

(c). In a least-square method, the test function is selected as

$$v_i(x) = \frac{\partial \overline{\mathfrak{R}}(c_0, c_1, \dots c_{N-1}, x)}{\partial c_i} \tag{3.40}$$

where c_i is an elected constant to be determined. The motivation of a least-square method is to minimize the absolute value of the residual integral of the whole domain. To achieve this, c_2 and c_3 must be chosen to minimize the residual integral,

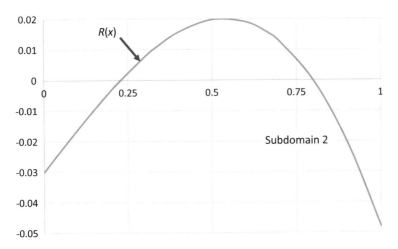

Fig. 3.20 Approximated solution using sub-domain methods

$$\tilde{R} = \int_0^1 \overline{R}^2(x)dx = 0 \tag{3.41}$$

Minimizing $\tilde{R}$ leads to the following conditions of c_2 and c_3,

$$\left.\begin{array}{l} \frac{\partial \tilde{R}}{\partial c_2} = 2\int_0^1 \frac{\partial \overline{R}}{\partial c_2}\overline{R}(x)dx = 0 \\ \frac{\partial \tilde{R}}{\partial c_3} = 2\int_0^1 \frac{\partial \overline{R}}{\partial c_3}\overline{R}(x)dx = 0 \end{array}\right\} \tag{3.42}$$

For the BVP in Eq. (3.28), the test functions are defined as

$$\left.\begin{array}{l} v_1(x) = \frac{\partial \Re_2(c_0,c_1,c_2,c_3,x)}{\partial c_2} = 2 \\ v_2(x) = \frac{\partial \overline{n}(c_0,c_1,c_2,c_3,x)}{\partial c_3} = 6x \end{array}\right\} \tag{3.43}$$

Using Eq. (3.43) for the condition of a weak solution yields

$$\left.\begin{array}{l} \overline{\Re}_1 = \int_0^1 (2)((2c_2 + 6c_3x) - 6x + \sin(x))dx = (2)(2c_2 + 3c_3 - 3 - \cos(1)) = 0 \\ \overline{\Re}_2 = \int_1^1 (6x)((2c_2 + 6c_3x) - 6x + \sin(x))dx = (6)\left(c_2 + 2c_3 - 2 + \int_0^1 x\sin(x)dx\right) = 0 \end{array}\right\} \tag{3.44}$$

Solving Eq. (3.44) yields $c_2 = -0.012462$ and $c_3 = 0.855076$, and the approximated solution becomes

$$\tilde{u}(x) = -0.00114 - 0.012462x^2 + 0.855076x^3 \tag{3.45}$$

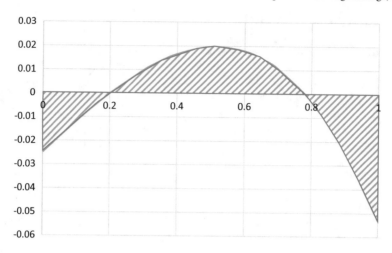

Fig. 3.21 Approximated solution using least-square methods

$\overline{\mathfrak{R}}(x)$ by using the Least square method is plotted in Fig. 3.21. $\tilde{u}(x)$ in Eq. (3.45) satisfies the condition of a weak solution in the sense that the integral of absolute residual over the whole domain is minimized.

(d). In a Galerkin method, the test functions are selected from the constitutive functions in the expression of the approximated solution $\tilde{u}(x)$ in Eq. (3.31), which includes three constitutive functions of

$$\phi_1(x) = 1, \phi_2(x) = x^2, \phi_3(x) = x^3 \tag{3.46}$$

$\phi_1(x)$ and $\phi_2(x)$ are selected as test functions to determine c_2 and c_3,

$$\left.\begin{array}{l}\overline{\mathfrak{R}}_1 = \int_0^1 \phi_1(x)((2c_2 + 6c_3x) - 6x + \sin(x))dx = (2c_2 + 3c_3 - 2 - \cos(1)) = 0 \\ \overline{R}_2 = \int_1^1 \phi_2(x)((2c_2 + 6c_3x) - 6x + \sin(x))dx = \left(\frac{2}{3}c_2 + \frac{3}{2}c_3 - \frac{3}{2} + \int_0^1 x^2\sin(x)dx\right) = 0 \end{array}\right\} \tag{3.47}$$

Solving Eq. (3.47) yields $c_2 = -0.019815$ and $c_3 = 0.8599773$, and the approximated solution becomes

$$\tilde{u}(x) = 0.00131 - 0.019815x^2 + 0.85599773x^3 \tag{3.48}$$

$\overline{\mathfrak{R}}(x)$ by using the Galerkin method is plotted in Fig. 3.22. $\tilde{u}(x)$ in Eq. (3.36) satisfies the condition of a weak solution in the sense that the integral of a weighted residual over the whole domain is minimized.

In using a weighted residual method, different test functions are defined to satisfy the condition of a weak solution. By specifying certain test function, the accuracy of the approximated solution $\tilde{u}(x)$ varies from one BVP to another. Therefore, the

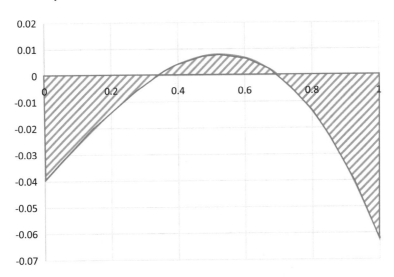

Fig. 3.22 Approximated solution using the Galerkin methods

performances of different weighted residual methods are not comparable. The accuracy of an approximation method can be verified when an exact solution is available to original BVP.

Note that the exact solution is validated by comparing the approximated solution with the exact one if it is available. The exact solution to the BVP in Eq. (3.28) can be found by the integration with the given boundary conditions as

$$u(x) = x^3 + \sin(x) - x \tag{3.49}$$

Table 3.8 and Fig. 3.23 show the comparison of the exact solution (Eq. (3.49)) with the approximated solutions (Eqs. (3.36), (3.39), (3.45), and (3.48)) over the whole domain of $x \in (0, 1)$. The comparison has shown that all of the four weight-residual methods have resulted in a weak solution $\tilde{u}(x)$ with an acceptable approximation over the whole domain.

In the practice of FEA modeling, the Galerkin method is the most commonly used one among these four weighted residual methods. In this book, the Galerkin method will be used as an example to develop element models in FEA.

3.5.5 Types of Differential Equations

The system behavior of an engineering system can be represented by a mathematical model, especially, *partial differential equations* (PDEs). Table 3.3 has shown some engineering systems with corresponding PDEs. The solution to a PDE is that to

Table 3.8 The approximations by using four different test functions

x	Exact solution $u(x)$	Approximation $\|\tilde{u}(x) - u(x)\|$			
		Collocation method	Subdomain method	Least-squares method	Galerkin method
0.00	0	0.005080985	0.000299015	0.001143015	0.005288
0.05	0.000104169	0.005038591	0.000333564	0.001171455	0.005242
0.10	0.000833417	0.004921868	0.000425572	0.001245976	0.005113
0.15	0.002813132	0.004746227	0.000557858	0.001350661	0.004918
0.20	0.006669331	0.004526454	0.000713866	0.001470218	0.004674
0.25	0.013028959	0.004276401	0.000877974	0.001590287	0.004396
0.30	0.022520207	0.004008678	0.001035802	0.00169775	0.004097
0.35	0.035772807	0.003734352	0.001174513	0.001781034	0.003789
0.40	0.053418342	0.003462642	0.001283118	0.001830414	0.003483
0.45	0.076090534	0.003200626	0.001352769	0.001838304	0.003188
0.50	0.104425539	0.002952946	0.001377054	0.001799554	0.002909
0.55	0.139062229	0.002721531	0.001352274	0.00171173	0.002649
0.60	0.180642473	0.002505311	0.001277729	0.001575393	0.002408
0.65	0.229811406	0.002299954	0.001155981	0.001394369	0.002183
0.70	0.287217687	0.002097598	0.000993122	0.001176014	0.001968
0.75	0.35351376	0.0018866	0.000799025	0.000931463	0.001753
0.08	0.429356091	0.001651294	0.000587586	0.000675874	0.001521
0.85	0.515405405	0.001371755	0.00037696	0.000428667	0.001256
0.90	0.61232691	0.001023575	0.000189785	0.000213741	0.000934
0.95	0.720790505	0.000577655	5.339E-05	5.96895E-05	0.000526
1.00	0.841470985	0	0	0	0
Maximized errors over the domain		**0.005080985**	**0.001377054**	**0.001838304**	**0.005288**

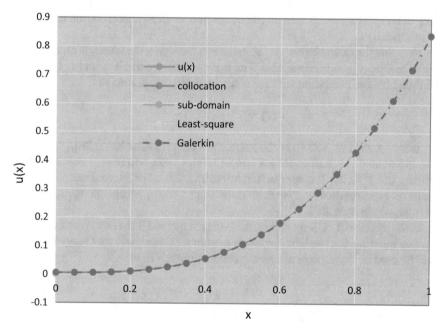

Fig. 3.23 Comparison of exact and approximated solutions

Fig. 3.24 Domain of
Dependence and Influence of
an Elliptic PDE

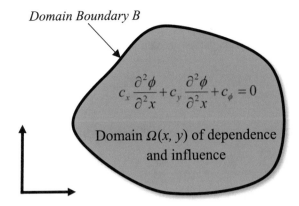

Domain Boundary B

$$c_x \frac{\partial^2 \phi}{\partial^2 x} + c_y \frac{\partial^2 \phi}{\partial^2 x} + c_\phi = 0$$

Domain $\Omega(x, y)$ of dependence
and influence

an engineering system subjected to the boundary conditions, PDEs can be *elliptic*,
parabolic, or *hyperbolic*.

In a 2D domain, an elliptic PDE only contains the derivatives with respect to
spatial variables (x, y), and no preferred path exists to seeking the solution to PDE.
Consequently, the solution at a certain point depends on that of all other points in
the physical domain. An elliptic PDE is used to represent an *equilibrium problem*
in a *closed domain* $D(x, y)$, which exists in all of the engineering fields. Figure 3.24
shows the closed solution domain $D(x, y)$ and its boundary B. An elliptic PDE is
usually an ordinary differential equation, and a typical example of an elliptic PDE is
the Laplace equation as

$$c_x \frac{\partial^2 \phi}{\partial^2 x} + c_y \frac{\partial^2 \phi}{\partial^2 y} + c_\phi = 0 \tag{3.50}$$

where $\phi(x, y)$ is a state variable such as the temperature in a heat transfer problem.
It is subjected to the boundary conditions of

$$\left. \begin{array}{l} \Gamma_1 : \phi = \phi_0 \\ \Gamma_2 : a\phi + b\frac{\partial^2 \phi}{\partial^2 n} = 0 \end{array} \right\} \tag{3.51}$$

where Γ_1 and Γ_2 correspond to essential and natural boundaries and $\frac{\partial \phi}{\partial n}$ denotes the
derivative normal to the boundary.

A parabolic PDE contains only second-order spatial derivatives; the preferred
path in seeking the solution to PDE is along the line or surface of the constant
time. At each time step, the solution at one point are dependent on that of all other
points. A *propagation problem is* an initial-value problem in an open domain. One
of independent variables, for example, of time, does not have a closed range. The
solution $f(x, y, t)$ in the domain $D(x, y, t)$ is marched forward from the initial state,
guided and modified by the given boundary conditions. A propagation problem is

governed by a parabolic or hyperbolic PDE. The majority of the propagation problems are unsteady problems. Taking an example, the following equation stands for an unsteady propagation problem,

$$\frac{\partial \phi}{\partial t} = \alpha \left(\frac{\partial^2 \phi}{\partial^2 x} + \frac{\partial^2 \phi}{\partial^2 y} \right) \qquad (3.52)$$

where $\phi (x, y)$ is a state variable and its initial condition is given as

$$\phi(x, y, t_0) = \phi_0(x, y) \;\; at \;\; t = t_0$$

The solution to a parabolic or hyperbolic PDE is propagating from the initial property distribution at time t_o. The solution is marched forward in the given time step in the time domain.

A hyperbolic PDE corresponds to *an eigenvalue problem*. The solution to a hyprtbolic PDE exists only for special values (i.e., *eigenvalues*) of system parameters. An eigenvalue problem is solved to obtain eigenvalues and the corresponding system configurations A typical hyperbolic PDE is shown as

$$\frac{\partial^2 \phi}{\partial^2 x} + \frac{\partial^2 \phi}{\partial^2 y} + \lambda \phi = 0 \qquad (3.53)$$

where $\phi (x, y)$ is a state variable and the boundary conditions are given as

$$\phi = \phi_0 \;\; on \;\; \Gamma_1$$
$$\frac{\partial \phi}{\partial n} + \alpha \phi = 0 \;\; on \;\; \Gamma_2$$

The numerical solution treats a continuous domain as a set of discrete nodes and elements. It calculates the values of the state variables on these discrete nodes. The calculation is repeated whenever a system parameter or boundary condition is changed in the model.

3.5.6 Solutions to Different PDEs

The solutions to three types of PDEs for 2D rectangle and triangle elements are discussed in the section.

3.5.6.1 Solution to Elliptic PDEs

An equilibrium problem represented by elliptic PDEs is a steady problem, and the solution at any position is influenced by these of all other points in the domain, and the solutions to all nodes in the domain are modeled and obtained simultaneously. In this section, the Galerkin method is applied to develop an element model for an elliptic PDE in Eq. (3.50), which is subjected to the boundary conditions of Eq. (3.51).

The weak-form solution to the PDE meets the condition of

$$\overline{R} = \int w(x, y)\left(c_x \frac{\partial^2 \phi}{\partial^2 x} + c_y \frac{\partial^2 \phi}{\partial^2 y} + c_\varphi - 0 \right) dx dy = 0 \qquad (3.54)$$

where $w(x, y)$ is the selected test function and c_x, c_y and c_ϕ are constants.

Equation (3.54) can be expressed by the reduced derivatives as below

$$\overline{R} = \int \left\{ c_x \{ \frac{\partial\left[w(x, y)\frac{\partial\phi}{\partial x}\right]}{\partial x} - \frac{\partial w(x, y)}{\partial x}\frac{\partial\phi}{\partial x} \} + c_y \{ \frac{\partial\left[w(x, y)\frac{\partial\phi}{\partial y}\right]}{\partial y} - \frac{\partial w(x, y)}{\partial y}\frac{\partial\phi}{\partial y} \} + c_\varphi \cdot w(x, y) \right\} dx dy \phi$$

$$= c_x \underbrace{\oint \partial\left[w(x, y)\frac{\partial\phi}{\partial x}\right]dy}_{I} - c_x \underbrace{\int \frac{\partial w(x, y)}{\partial x}\frac{\partial\phi}{\partial x}dx dy}_{II} + c_y \underbrace{\oint \partial\left[w(x, y)\frac{\partial\phi}{\partial y}\right]dx}_{III}$$

$$- c_y \underbrace{\int \frac{\partial w(x, y)}{\partial y}\frac{\partial\phi}{\partial y}dx dy}_{IV} + \underbrace{\int w(x, y) \cdot c_\phi dx dy}_{V} \qquad (3.55)$$

where the residual caused by approximation consists of five parts (i.e., *I*, *II*, *III*, *IV*, and *V*). The parts of *I* and *III* are the integrals along the boundary of the 2D domain. The parts of *II*, *IV* and *V* are the integrals in the domain.

Figure 3.25 shows a heat transfer domain, which is governed by an elliptic PDE. The continuous domain has its two types of boundaries, i.e., *essential boundary* Γ_1

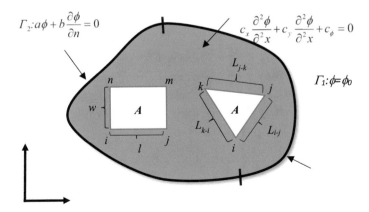

Fig. 3.25 Rectangle and triangle elements in a heat transfer domain

and *natural boundary* Γ_2. In addition, rectangle and triangular elements are shown as the examples of discretized elements and nodes.

To solve Eq. (3.55) for an elemental model, this section limits the discussions on the parts of *II*, *IV* and *V*, since the parts of *I* and *III* are the integrals over the boundary surfaces rather than the others in the solid domain.

For a rectangle element in Fig. 3.25, a state variable at a given position (x, y) can be determined by the values at four nodes $(\phi_i, \phi_j, \phi_m, \phi_n)$ using the shape functions as

$$\phi = [S]\{\varphi\} = S_i(x, y) \cdot \phi_i + S_j(x, y) \cdot \phi_j + S_m(x, y) \cdot \phi_m + S_n(x, y) \cdot \phi_n \tag{3.56}$$

where $\phi(x, y)$ is the state variable at an arbitrary position (x, y) in the element, ϕ_i, ϕ_j, ϕ_m, and ϕ_n are the values of the state variable at nodes, and the shape functions in a rectangle elements are

$$\left. \begin{array}{l} S_i(x, y) = \left(\frac{l-x}{l}\right)\left(\frac{w-y}{w}\right) \\ S_j(x, y) = \frac{x}{l}\left(\frac{w-y}{w}\right) \\ S_m(x, y) = \frac{x}{l}\frac{y}{w} \\ S_n(x, y) = \left(\frac{l-x}{l}\right)\frac{y}{w} \end{array} \right\} \tag{3.57}$$

where l and w are the length and width of the rectangle element.

Since the Galerkin method is used, the shape functions $S_i(x, y)$, $S_j(x, y)$, $S_m(x, y)$, and $S_n(x, y)$ in Eq. (3.56) are used as the test functions, respectively. Substituting the shape functions into Eq. (3.55) obtain the parts of *II*, *IV*, and *V* as below.

Part II for the rectangle element is found as

$$c_x \int \frac{\partial [S]^T}{\partial x} \frac{\partial \phi}{\partial x} dx dy = c_x \int \frac{\partial S]^T}{\partial x} \frac{d[S][\phi])}{\partial x} dx dy = [K_x][\phi] \tag{3.58}$$

where

$$[K_x] = c_x \begin{bmatrix} \int \frac{\partial S_i}{\partial x} \cdot \frac{\partial S_i}{\partial x} dx dy & \int \frac{\partial S_i}{\partial x} \cdot \frac{\partial S_j}{\partial x} dx dy & \int \frac{\partial S_i}{\partial x} \cdot \frac{\partial S_m}{\partial x} dx dy & \int \frac{\partial S_i}{\partial x} \cdot \frac{\partial S_n}{\partial x} dx dy \\ \int \frac{\partial S_i}{\partial x} \cdot \frac{\partial S_j}{\partial x} dx dy & \int \frac{\partial S_j}{\partial x} \cdot \frac{\partial S_j}{\partial x} dx dy & \int \frac{\partial S_j}{\partial x} \cdot \frac{\partial S_m}{\partial x} dx dy & \int \frac{\partial S_j}{\partial x} \cdot \frac{\partial S_n}{\partial x} dx dy \\ \int \frac{\partial S_m}{\partial x} \cdot \frac{\partial S_i}{\partial x} dx dy & \int \frac{\partial S_m}{\partial x} \cdot \frac{\partial S_j}{\partial x} dx dy & \int \frac{\partial S_m}{\partial x} \cdot \frac{\partial S_m}{\partial x} dx dy & \int \frac{\partial S_m}{\partial x} \cdot \frac{\partial S_n}{\partial x} dx dy \\ \int \frac{\partial S_n}{\partial x} \cdot \frac{\partial S_i}{\partial x} dx dy & \int \frac{\partial S_n}{\partial x} \cdot \frac{\partial S_j}{\partial x} dx dy & \int \frac{\partial S_n}{\partial x} \cdot \frac{\partial S_m}{\partial x} dx dy & \int \frac{\partial S_n}{\partial x} \cdot \frac{\partial S_n}{\partial x} dx dy \end{bmatrix}$$

$$= \frac{c_x w}{6l} \begin{bmatrix} 2 & -2 & -1 & 1 \\ -2 & 2 & 1 & -1 \\ -1 & 1 & 2 & -2 \\ 1 & -1 & -2 & 2 \end{bmatrix}$$

and $[\boldsymbol{\phi}]^T = \begin{bmatrix} \phi_i & \phi_j & \phi_m & \phi_n \end{bmatrix}$

Part IV for the rectangle element is found as

$$c_y \int \frac{\partial [S]^T}{\partial y} \frac{\partial \phi}{\partial y} dxdy = c_y \int \frac{\partial [S]^T}{\partial y} \frac{\partial ([S][\boldsymbol{\phi}])}{\partial y} dxdy = \begin{bmatrix} K_y \end{bmatrix}[\boldsymbol{\phi}] \qquad (3.59)$$

where

$$\begin{aligned}
\begin{bmatrix} K_y \end{bmatrix} &= c_y
\begin{bmatrix}
\int \frac{\partial S_i}{\partial y} \cdot \frac{\partial S_i}{\partial y} dxdy & \int \frac{\partial S_i}{\partial y} \cdot \frac{\partial S_j}{\partial y} dxdy & \int \frac{\partial S_i}{\partial y} \cdot \frac{\partial S_m}{\partial y} dxdy & \int \frac{\partial S_i}{\partial y} \cdot \frac{\partial S_n}{\partial y} dxdy \\
\int \frac{\partial S_i}{\partial y} \cdot \frac{\partial S_j}{\partial y} dxdy & \int \frac{\partial S_j}{\partial y} \cdot \frac{\partial S_j}{\partial y} dxdy & \int \frac{\partial S_j}{\partial y} \cdot \frac{\partial S_m}{\partial y} dxdy & \int \frac{\partial S_j}{\partial y} \cdot \frac{\partial S_n}{\partial y} dxdy \\
\int \frac{\partial S_m}{\partial y} \cdot \frac{\partial S_i}{\partial y} dxdy & \int \frac{\partial S_m}{\partial y} \cdot \frac{\partial S_j}{\partial y} dxdy & \int \frac{\partial S_m}{\partial y} \cdot \frac{\partial S_m}{\partial y} dxdy & \int \frac{\partial S_m}{\partial y} \cdot \frac{\partial S_n}{\partial y} dxdy \\
\int \frac{\partial S_n}{\partial y} \cdot \frac{\partial S_i}{\partial y} dxdy & \int \frac{\partial S_n}{\partial y} \cdot \frac{\partial S_j}{\partial y} dxdy & \int \frac{\partial S_n}{\partial y} \cdot \frac{\partial S_m}{\partial y} dxdy & \int \frac{\partial S_n}{\partial y} \cdot \frac{\partial S_n}{\partial y} dxdy
\end{bmatrix} \\
&= \frac{c_y l}{6w}
\begin{bmatrix}
2 & 1 & -1 & -2 \\
1 & 2 & -2 & -1 \\
-1 & -2 & 2 & 1 \\
-2 & -1 & 1 & 2
\end{bmatrix}
\end{aligned}$$

Part V for the rectangle element is found as

$$[\boldsymbol{Q}] = \int [S]^T c_\phi dxdy = c_\phi
\begin{bmatrix}
\int S_i dxdy \\
\int S_j dxdy \\
\int S_m dxdy \\
\int S_n dxdy
\end{bmatrix}
= \frac{c_\phi (lw)}{4}
\begin{bmatrix}
1 \\
1 \\
1 \\
1
\end{bmatrix} \qquad (3.60)$$

Substituting Eqs. (3.58)–(3.60) into Eq. (3.54) gets the rectangle element model with no boundary edge (no Part I or III) as

$$\left([K_x] + \begin{bmatrix} K_y \end{bmatrix} \right)\{\boldsymbol{\phi}\} = [\boldsymbol{Q}] \qquad (3.61)$$

where $[K_x]$ and $[K_y]$ are obtained in Eqs. (3.58) and (3.59), and the load vector $[\boldsymbol{Q}]$ is given in eq. (3.60).

Similarly, for a triangle element in Fig. 3.25, a state variable at a given position (x, y) can be determined by the values at four nodes (ϕ_i, ϕ_j, ϕ_k) using the shape functions as

$$\phi = [S]\{\boldsymbol{\phi}\} = S_i(x, y)\phi_i + S_{ij}(x, y)\phi_j + S_m(x, y)\phi_m + S_n(x, y)\phi_n \qquad (3.62)$$

where $\phi(x, y)$ is the state variable at an arbitrary position (x, y) in the triangle element, and ϕ_i, ϕ_j and ϕ_k are the values of state variables at nodes. The shape functions in a triangular element are given as (Bi 2018)

$$\left.\begin{array}{l} S_i(x, y) = \frac{|\Delta_i|}{|\Delta|} = \frac{\alpha_i + \beta_i \cdot x + \delta_i \cdot y}{|\Delta|} \\ S_j(x, y) = \frac{|\Delta_j|}{|\Delta|} = \frac{\alpha_j + \beta_j \cdot x + \delta_j \cdot y}{|\Delta|} \\ S_k(x, y) = \frac{|\Delta_k|}{|\Delta|} = \frac{\alpha_k + \beta_k \cdot x + \delta_k \cdot y}{|\Delta|} \end{array}\right\} \tag{3.63}$$

where the constants $\boldsymbol{\alpha}, \boldsymbol{\beta}, \boldsymbol{\delta}, |\Delta|$ are determined by the coordinates of three nodes (x_i, y_i), (x_j, y_j), and (x_k, y_k) as

$$\left.\begin{array}{l} |\Delta| = \begin{vmatrix} 1 & x_i & y_i \\ 1 & x_j & y_j \\ 1 & x_k & y_k \end{vmatrix} \\ \alpha_i = x_j y_k - x_k y_j, \alpha_j = x_k y_i - x_i y_k, \alpha_k = x_i y_j - x_j y_i \\ \beta_i = y_j - y_k, \beta_j = y_k - y_i, \beta_k = y_i - y_j \\ \delta_i = x_k - x_j, \delta_j = x_i - x_k, \delta_k = x_j - x_i \end{array}\right\} \tag{3.64}$$

In using the Galerkin method, the shape functions in Eq. (3.62) are used as the test functions. Using these functions in Eq. (3.55) obtain the parts of II, IV, and V as below.

Part II for the triangle element is found as

$$c_x \int \frac{\partial [S]^T}{\partial x} \frac{\partial \phi}{\partial x} dx dy = c_x \int \frac{\partial [S]^T}{\partial x} \frac{\partial [S][\phi])}{\partial x} dx dy = [K_x][\phi] \tag{3.65}$$

where

$$[K_x] = c_x \begin{bmatrix} \int \frac{\partial S_i}{\partial x} \cdot \frac{\partial S_i}{\partial x} dx dy & \int \frac{\partial S_i}{\partial x} \cdot \frac{\partial S_j}{\partial x} dx dy & \int \frac{\partial S_i}{\partial x} \cdot \frac{\partial S_k}{\partial x} dx dy \\ \int \frac{\partial S_i}{\partial x} \cdot \frac{\partial S_j}{\partial x} dx dy & \int \frac{\partial S_j}{\partial x} \cdot \frac{\partial S_j}{\partial x} dx dy & \int \frac{\partial S_j}{\partial x} \cdot \frac{\partial S_k}{\partial x} dx dy \\ \int \frac{\partial S_k}{\partial x} \cdot \frac{\partial S_i}{\partial x} dx dy & \int \frac{\partial S_k}{\partial x} \cdot \frac{\partial S_j}{\partial x} dx dy & \int \frac{\partial S_k}{\partial x} \cdot \frac{\partial S_k}{\partial x} dx dy \end{bmatrix}$$

$$= \frac{c_x}{4A} \begin{bmatrix} \beta_i^2 & \beta_i \beta_j & \beta_i \beta_k \\ \beta_j \beta_i & \beta_j^2 & \beta_j \beta_k \\ \beta_k \beta_i & \beta_k \beta_j & \beta_k^2 \end{bmatrix}$$

and $[\boldsymbol{\phi}]^T = \begin{bmatrix} \phi_i & \phi_j & \phi_k \end{bmatrix}$

Part IV for the triangle element is found as

$$c_y \int \frac{\partial [S]^T}{\partial y} \frac{\partial \phi}{\partial y} dx dy = c_y \int \frac{\partial S]^T}{\partial y} \frac{\partial [S][\phi])}{\partial y} dx dy = [K_y][\phi] \tag{3.66}$$

where

$$[K_y] = c_y \begin{bmatrix} \int \frac{\partial S_i}{\partial y} \cdot \frac{\partial S_i}{\partial y} dxdy & \int \frac{\partial S_i}{\partial y} \cdot \frac{\partial S_j}{\partial y} dxdy & \int \frac{\partial S_i}{\partial y} \cdot \frac{\partial S_k}{\partial y} dxdy \\ \int \frac{\partial S_i}{\partial y} \cdot \frac{\partial S_j}{\partial y} dxdy & \int \frac{\partial S_j}{\partial y} \cdot \frac{\partial S_j}{\partial y} dxdy & \int \frac{\partial S_j}{\partial y} \cdot \frac{\partial S_k}{\partial y} dxdy \\ \int \frac{\partial S_k}{\partial y} \cdot \frac{\partial S_i}{\partial y} dxdy & \int \frac{\partial S_k}{\partial y} \cdot \frac{\partial S_j}{\partial y} dxdy & \int \frac{\partial S_k}{\partial x} \cdot \frac{\partial S_k}{\partial y} dxdy \end{bmatrix}$$

$$= \frac{c_y}{4A} \begin{bmatrix} \delta_i^2 & \delta_i \delta_j & \delta_i \delta_k \\ \delta_j \delta_i & \delta_j^2 & \delta_j \delta_k \\ \delta_k \delta_i & \delta_k \delta_j & \delta_k^2 \end{bmatrix}$$

Part V for the triangle element becomes

$$[Q] = \int [S]^T c_\phi dxdy = c_\phi \begin{bmatrix} \int S_i dxdy \\ \int S_j dxdy \\ \int S_k dxdy \end{bmatrix} = \frac{c_\phi A}{3} \begin{bmatrix} 1 \\ 1 \\ 1 \end{bmatrix} \tag{3.67}$$

Substituting Eqs. (3.65)–(3.67) into Eq. (3.54) gets the triangle element model with no boundary edge (no Part I or III) as

$$([K_x] + [K_y])\{\phi\} = [Q] \tag{3.68}$$

where $[K_x]$ and $[K_y]$ are defined in Eqs. (3.65) and (3.66) and the load vector $[Q]$ is defined in Eq. (3.67).

3.5.6.2 Solution to Parabolic PDEs

A parabolic PDE is used to model a unsteady propagate problem. The solution in the continuous domain is evolved with respect to time, and the initial condition affects the time-dependent solution. In the following, the procedure of element modeling is illustrated by using the parabolic PDE as

$$c_x \frac{\partial^2 \phi}{\partial^2 x} + c_y \frac{\partial^2 \phi}{\partial^2 x} + c_\phi = \alpha \frac{\partial \phi}{\partial t} \tag{3.69}$$

where $\phi(x, y)$ is the state variable in the element,
c_x, c_y, c_ϕ, α are the constants, and the initial conditions are

$$\phi(x, y, t_0) = \phi_0(x, y) \ \ at \ \ t = t_0$$

The solution to Eq. (3.69) is approximated by satisfying the condition of $\overline{R} = 0$ where the residual $\overline{R}$ is expressed as

$$\overline{R} = \int \left\{ c_x \{ \frac{\partial \left[w(x, y) \frac{\partial \phi}{\partial x} \right]}{\partial x} - \frac{\partial w(x, y)}{\partial x} \frac{\partial \phi}{\partial x} \} + c_y \{ \frac{\partial \left[w(x, y) \frac{\partial \phi}{\partial y} \right]}{\partial y} - \frac{\partial w(x, y)}{\partial y} \frac{\partial \phi}{\partial y} \} + c_\phi \cdot w(x, y) - \alpha \cdot w(x, y) \frac{\partial \phi}{\partial t} \right\} dxdy$$

$$
\begin{aligned}
= &\underbrace{c_x \oint \partial\left[w(x,y)\frac{\partial\phi}{\partial x}\right]dy}_{I} - \underbrace{c_x \int \frac{\partial w(x,y)}{\partial x}\frac{\partial\phi}{\partial x}dxdy}_{II} + \underbrace{c_y \oint \partial\left[w(x,y)\frac{\partial\phi}{\partial y}\right]dx}_{III} \\
&- \underbrace{c_y \int \frac{\partial w(x,y)}{\partial y}\frac{\partial\phi}{\partial y}dxdy}_{IV} + \underbrace{\int w(x,y)\cdot c_\phi dxdy}_{V} - \underbrace{\int \alpha w(x,y)\cdot\frac{\partial\phi}{\partial t}dxdy}_{VI}
\end{aligned}
\tag{3.70}
$$

Equation (3.70) includes the parts of *I*, *II*, *III*, *IV*, and *V* which are the same to Eq. (3.55) and discussed in the previous section. Only part *VI* is new.

To evaluate part *VI*, let $\frac{\partial\phi(x,y)}{\partial t}$ be a function of the state variable $\phi(x,y)$ which can be evaluated by taking the derivative of the interpolation in the element as

$$
\frac{\partial\phi(x,y)}{\partial t} = [S]\{\dot{\phi}\}
\tag{3.71}
$$

Therefore, part *VI* for a rectangle element can be found as

$$
\int [S]^T [S]\{\dot{\phi}\}dxdy = [C]\{\dot{\phi}\}
\tag{3.72}
$$

where

$$
\begin{aligned}
[C] = \alpha &\begin{bmatrix}
\int S_i^2 dxdy & \int S_i S_j dxdy & \int S_i S_m dxdy & \int S_i S_n dxdy \\
\int S_j S_i dxdy & \int S_j^2 dxdy & \int S_j S_m dxdy & \int S_j S_n dxdy \\
\int S_m S_i dxdy & \int S_m S_j dxdy & \int S_m^2 dxdy & \int S_m S_n dxdy \\
\int S_n S_i dxdy & \int S_n S_j dxdy & \int S_n S_m dxdy & \int S_n^2 dxdy
\end{bmatrix} \\
&= \frac{(lw)\alpha}{18}\begin{bmatrix}
2 & 1 & 1 & 2 \\
1 & 2 & 2 & 1 \\
1 & 2 & 2 & 1 \\
2 & 1 & 1 & 2
\end{bmatrix}
\end{aligned}
$$

$[\dot{\varphi}]$ is the vector for the derivative of state variable with respect to time.

Merging Eqs. (3.61) and (3.72) yields the rectangle element model to Eq. (3.69) as

$$
\left([K_x] + [K_y]\right)\{\phi\} + [C]\{\dot{\phi}\} = [Q]
\tag{3.73}
$$

where $[K_x]$, $[K_y]$ and $[Q]$ are given in Eq. (3.61) and $[C]$ is given in Eq. (3.72) for the rectangle element.

Similarly, $[C]$ in part *VI* for a triangle element can be found as

$$
[C] = \alpha \begin{bmatrix}
\int S_i^2 dxdy & \int S_i S_j dxdy & \int S_i S_k dxdy \\
\int S_j S_i dxdy & \int S_j^2 dxdy & \int S_k S_j dxdy \\
\int S_k S_i dxdy & \int S_k S_j dxdy & \int S_k^2 dxdy
\end{bmatrix} = \frac{A\alpha}{12}\begin{bmatrix}
2 & 1 & 1 \\
1 & 2 & 1 \\
1 & 1 & 2
\end{bmatrix}
\tag{3.74}
$$

Merging Eqs. (3.68) and (3.74) yields the triangle element model to Eq. (3.69) as

$$\left([K_x] + [K_y]\right)\{\phi\} + [C]\{\dot{\phi}\} = [Q]$$ (3.75)

where $[K_x]$, $[K_y]$ and $[Q]$ are given in Eq. (3.61) and $[C]$ is given in Eq. (3.74) for the triangle element.

Note that $\{\dot{\phi}\}$ depends on $\{\phi\}$ and changes over time, the time-dependent solutions are obtained iteratively based on the given initial condition. For example, *the forward difference method* can be used in Eq. (3.73) or Eq. (3.75) to evaluate $\{\dot{\phi}\}$ at time step t based on $\{\phi\}$. Once $\{\dot{\phi}\}$ at time t is obtained, $\{\phi\}$ at the next time step $t + \Delta t$ can be found as

$$\{\phi\}_{t+\Delta t} = \{\phi\}_t + \{\dot{\phi}\}_t \Delta t$$ (3.76)

Once $\{\phi\}_{t+\Delta t}$ is known, $\{\dot{\phi}\}_t$ can be obtained from two time steps as

$$\{\dot{\phi}\}_t = \frac{1}{\Delta t}\left\{\{\phi\}_{t+\Delta t} - \{\phi\}_t\right\}$$ (3.77)

Substituting Eq. (3.77) back to Eq. (3.73) or Eq. (3.75) finds

$$\frac{[C]}{\Delta t}\left(\{\phi\}_{t+\Delta t} - \{\phi\}_t\right) = [Q] - \left([K_x] + [K_y]\right)\{\phi\}_t$$ (3.78)

Rearranging Eq. (3.78) leads to an explicit element model without the term of $\{\dot{\phi}\}_t$ as

$$\frac{[C]}{\Delta t}\{\phi\}_{t+\Delta t} = [Q] + \left(\frac{[C]}{\Delta t} - \left([K_x] + [K_y]\right)\right)\{\phi\}_t$$ (3.79)

In Eq. (3.79), the time step Δt must be set appropriately to the convergence and stability.

3.5.6.3 Solution to Hyperbolic PDEs

A hyperbolic PDE can be used to model *an eigenvalue problem,* which has the solution only for specific values (i.e., *eigenvalues*). In a 2D domain, the Helmholtz equation in Fig. 3.26 is used as an example to model an eigenvalue problem.

The hyperbolic PDE for such an eigenvalue problem is

$$\frac{\partial^2 \phi}{\partial^2 x} + \frac{\partial^2 \phi}{\partial^2 y} + \lambda\phi = 0$$ (3.80)

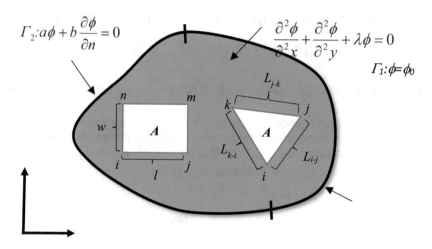

Fig. 3.26 Helmholtz equation for an eigenvalue problem

where $\phi(x, y)$ is the state variable and the boundary conditions are given as

$$\left.\begin{array}{l} \Gamma_1 : \phi = \phi_0 \\ \Gamma_2 : a\phi + b\frac{\partial \phi}{\partial n} = 0 \end{array}\right\} \tag{3.81}$$

The solution to Eq. (3.80) is approximated by satisfying the condition of $\overline{R} = 0$ where the residual $\overline{R}$ is expressed as

$$\overline{R} = \int \left\{ \left\{ \frac{\partial\left[w(x, y)\frac{\partial\phi}{\partial x}\right]}{\partial x} - \frac{\partial w(x, y)}{\partial x}\frac{\partial\phi}{\partial x} \right\} + \left\{ \frac{\partial\left[w(x, y)\frac{\partial\phi}{\partial y}\right]}{\partial y} - \frac{\partial w(x, y)}{\partial y}\frac{\partial\phi}{\partial y} \right\} + w(x, y)\lambda\phi \right\} dxdy$$

$$= \underbrace{\oint \partial\left[w(x, y)\frac{\partial\phi}{\partial x}\right]dy}_{I} - \underbrace{\int \frac{\partial w(x, y)}{\partial x}\frac{\partial\phi}{\partial x}dxdy}_{II} + \underbrace{\oint \partial\left[w(x, y)\frac{\partial\phi}{\partial y}\right]dx}_{III}$$

$$- \underbrace{\int \frac{\partial w(x, y)}{\partial y}\frac{\partial\phi}{\partial y}dxdy}_{IV} + \underbrace{\int w(x, y)\cdot\lambda\phi dxdy}_{V} \tag{3.82}$$

where $w(x, y)$ is a selected test function; part I and III are the integrals along the boundary, part II, IV, and V are three integrals over the continuous domain. The discussion on the integrals along the boundary (part I and III) will be left out here; interested readers can find the details in other literatures (Bi 2018). Here, the integrals over the domain are discussed in modeling rectangle and triangle elements.

In using the Galekin method, the shape functions are used as the test functions in evaluating the residual in Eq. (3.82), and part II, IV, and V are obtained accordingly.

For part II in the rectangle element,

$$\int \frac{\partial [S]^T}{\partial x} \frac{\partial \phi}{\partial x} dxdy = \int \frac{\partial [S]^T}{\partial x} \frac{\partial ([S][\phi])}{\partial x} dxdy = [K_x][\phi] \qquad (3.83)$$

where

$$[K_x] = \begin{bmatrix} \int \frac{\partial S_i}{\partial x} \cdot \frac{\partial S_i}{\partial x} dxdy & \int \frac{\partial S_i}{\partial x} \cdot \frac{\partial S_j}{\partial x} dxdy & \int \frac{\partial S_i}{\partial x} \cdot \frac{\partial S_m}{\partial x} dxdy & \int \frac{\partial S_i}{\partial x} \cdot \frac{\partial S_n}{\partial x} dxdy \\ \int \frac{\partial S_i}{\partial x} \cdot \frac{\partial S_i}{\partial x} dxdy & \int \frac{\partial S_j}{\partial x} \cdot \frac{\partial S_j}{\partial x} dxdy & \int \frac{\partial S_j}{\partial x} \cdot \frac{\partial S_m}{\partial x} dxdy & \int \frac{\partial S_j}{\partial x} \cdot \frac{\partial S_n}{\partial x} dxdy \\ \int \frac{\partial S_m}{\partial x} \cdot \frac{\partial S_i}{\partial x} dxdy & \int \frac{\partial S_m}{\partial x} \cdot \frac{\partial S_j}{\partial x} dxdy & \int \frac{\partial S_m}{\partial x} \cdot \frac{\partial S_m}{\partial x} dxdy & \int \frac{\partial S_m}{\partial x} \cdot \frac{\partial S_n}{\partial x} dxdy \\ \int \frac{\partial S_n}{\partial x} \cdot \frac{\partial S_i}{\partial x} dxdy & \int \frac{\partial S_n}{\partial x} \cdot \frac{\partial S_j}{\partial x} dxdy & \int \frac{\partial S_n}{\partial x} \cdot \frac{\partial S_m}{\partial x} dxdy & \int \frac{\partial S_n}{\partial x} \cdot \frac{\partial S_n}{\partial x} dxdy \end{bmatrix}$$

$$= \frac{w}{6l} \begin{bmatrix} 2 & -2 & -1 & 1 \\ -2 & 2 & 1 & -1 \\ -1 & 1 & 2 & -2 \\ 1 & -1 & -2 & 2 \end{bmatrix}$$

and $[\varphi]^T = [\phi_i \ \phi_j \ \phi_m \ \phi_n]$.

For part IV in the rectangle element,

$$\int \frac{\partial [S]^T}{\partial y} \frac{\partial \phi}{\partial y} dxdy = \int \frac{\partial [S]^T}{\partial y} \frac{\partial ([S][\phi])}{\partial y} dxdy = [K_y][\phi] \qquad (3.84)$$

where

$$[K_y] = \begin{bmatrix} \int \frac{\partial S_i}{\partial y} \cdot \frac{\partial S_i}{\partial y} dxdy & \int \frac{\partial S_i}{\partial y} \cdot \frac{\partial S_j}{\partial y} dxdy & \int \frac{\partial S_i}{\partial y} \cdot \frac{\partial S_m}{\partial y} dxdy & \int \frac{\partial S_i}{\partial y} \cdot \frac{\partial S_n}{\partial y} dxdy \\ \int \frac{\partial S_i}{\partial y} \cdot \frac{\partial S_i}{\partial y} dxdy & \int \frac{\partial S_j}{\partial y} \cdot \frac{\partial S_j}{\partial y} dxdy & \int \frac{\partial S_j}{\partial y} \cdot \frac{\partial S_m}{\partial y} dxdy & \int \frac{\partial S_j}{\partial y} \cdot \frac{\partial S_n}{\partial y} dxdy \\ \int \frac{\partial S_m}{\partial y} \cdot \frac{\partial S_i}{\partial y} dxdy & \int \frac{\partial S_m}{\partial y} \cdot \frac{\partial S_j}{\partial y} dxdy & \int \frac{\partial S_m}{\partial y} \cdot \frac{\partial S_m}{\partial y} dxdy & \int \frac{\partial S_m}{\partial y} \cdot \frac{\partial S_n}{\partial y} dxdy \\ \int \frac{\partial S_n}{\partial y} \cdot \frac{\partial S_i}{\partial y} dxdy & \int \frac{\partial S_n}{\partial y} \cdot \frac{\partial S_j}{\partial y} dxdy & \int \frac{\partial S_n}{\partial y} \cdot \frac{\partial S_m}{\partial y} dxdy & \int \frac{\partial S_n}{\partial y} \cdot \frac{\partial S_n}{\partial y} dxdy \end{bmatrix}$$

$$= \frac{l}{6w} \begin{bmatrix} 2 & 1 & -1 & -2 \\ 1 & 2 & -2 & -1 \\ -1 & -2 & 2 & 1 \\ -2 & -1 & 1 & 2 \end{bmatrix}$$

For Part V in the rectangle element,

$$\int [S]^T \lambda [S][\phi] dxdy = [M][\phi] \qquad (3.85)$$

where

$$[M] = \begin{bmatrix} \int S_i^2 dxdy & \int S_i S_j dxdy & \int S_i S_m dxdy & \int S_i S_n dxdy \\ \int S_j S_i dxdy & \int S_j^2 dxdy & \int S_j S_m dxdy & \int S_j S_n dxdy \\ \int S_m S_i dxdy & \int S_m S_j dxdy & \int S_m^2 dxdy & \int S_m S_n dxdy \\ \int S_n S_i dxdy & \int S_n S_j dxdy & \int S_n S_m dxdy & \int S_n^2 dxdy \end{bmatrix} = \frac{(lw)}{18} \begin{bmatrix} 2 & 1 & 1 & 2 \\ 1 & 2 & 2 & 1 \\ 1 & 2 & 2 & 1 \\ 2 & 1 & 1 & 2 \end{bmatrix}$$

Substituting Eqs. (3.84–3.86) into Eq. (3.82) yields the element model as

$$\{([K_x] + [K_y] - \lambda[M])\}\{\phi\} = 0 \tag{3.86}$$

where $[K_x]$, $[K_y]$ and $[M]$ are given in Eqs. (3.84–3.86).

A specific solution λ_i, (or an eigenvalue) to Eq. (3.86) must ensure that the determinate of $\{([K_x] + [K_y] - [M])\}$ is 0, and the corresponding ϕ_i is an eigenvector.

The similar procedure is applied to a triangle element to obtain the triangle element model in the same format of Eq. (3.86) but with different $[K_x]$, $[K_y]$ and $[M]$ as below

$$[K_x] = \begin{bmatrix} \int \frac{\partial S_i}{\partial x} \cdot \frac{\partial S_i}{\partial x} dxdy & \int \frac{\partial S_i}{\partial x} \cdot \frac{\partial S_j}{\partial x} dxdy & \int \frac{\partial S_i}{\partial x} \cdot \frac{\partial S_k}{\partial x} dxdy \\ \int \frac{\partial S_i}{\partial x} \cdot \frac{\partial S_i}{\partial x} dxdy & \int \frac{\partial S_j}{\partial x} \cdot \frac{\partial S_j}{\partial x} dxdy & \int \frac{\partial S_j}{\partial x} \cdot \frac{\partial S_k}{\partial x} dxdy \\ \int \frac{\partial S_k}{\partial x} \cdot \frac{\partial S_i}{\partial x} dxdy & \int \frac{\partial S_k}{\partial x} \cdot \frac{\partial S_j}{\partial x} dxdy & \int \frac{\partial S_k}{\partial x} \cdot \frac{\partial S_k}{\partial x} dxdy \end{bmatrix}$$
$$= \frac{1}{4A} \begin{bmatrix} \beta_i^2 & \beta_i\beta_j & \beta_i\beta_k \\ \beta_j\beta_i & \beta_j^2 & \beta_j\beta_k \\ \beta_k\beta_i & \beta_k\beta_j & \beta_k^2 \end{bmatrix} \tag{3.87}$$

$$[K_y] = \begin{bmatrix} \int \frac{\partial S_i}{\partial y} \cdot \frac{\partial S_i}{\partial y} dxdy & \int \frac{\partial S_i}{\partial y} \cdot \frac{\partial S_j}{\partial y} dxdy & \int \frac{\partial S_i}{\partial y} \cdot \frac{\partial S_k}{\partial y} dxdy \\ \int \frac{\partial S_i}{\partial y} \cdot \frac{\partial S_j}{\partial y} dxdy & \int \frac{\partial S_j}{\partial y} \cdot \frac{\partial S_j}{\partial y} dxdy & \int \frac{\partial S_j}{\partial y} \cdot \frac{\partial S_k}{\partial y} dxdy \\ \int \frac{\partial S_k}{\partial y} \cdot \frac{\partial S_i}{\partial y} dxdy & \int \frac{\partial S_k}{\partial y} \cdot \frac{\partial S_j}{\partial y} dxdy & \int \frac{\partial S_k}{\partial x} \cdot \frac{\partial S_k}{\partial y} dxdy \end{bmatrix}$$
$$= \frac{1}{4A} \begin{bmatrix} \delta_i^2 & \delta_i\delta_j & \delta_i\delta_k \\ \delta_j\delta_i & \delta_j^2 & \delta_j\delta_k \\ \delta_k\delta_i & \delta_k\delta_j & \delta_k^2 \end{bmatrix} \tag{3.88}$$

$$[M] = \begin{bmatrix} \int S_i^2 dxdy & \int S_i S_j dxdy & \int S_i S_k dxdy \\ \int S_j S_i dxdy & \int S_j^2 dxdy & \int S_k S_j dxdy \\ \int S_k S_i dxdy & \int S_k S_j dxdy & \int S_k^2 dxdy \end{bmatrix} = \frac{A}{12} \begin{bmatrix} 2 & 1 & 1 \\ 1 & 2 & 1 \\ 1 & 1 & 2 \end{bmatrix} \tag{3.89}$$

3.5.7 System Model and Solution

3.5.7.1 Build System Model

In an FEA model, the number of independent variables to be determined is called *degrees of freedom* (DoF). The size of a system model is determined by the DoF of the model as

$$[K]_{N \times N}\{U\}_{N \times 1} = \{F\}_{N \times 1} \tag{3.90}$$

where N is the degrees of freedom; $\{U\}$ is the vector of N independent variables; $\{F\}$ is the vector of N loads over the corresponding independent variables; $[K]$ is a $N \times N$ relational matrix of $\{U\}$ and $\{F\}$.

The system model is assembled from element models. Assume that the FEA model has N_e elements, and each element involves in ND_i independent variables $(nd_{i,1}, nd_{i,2}, \ldots, nd_{i,NDi})$. The system model is assembled as

$$[K]_{N \times N} = \sum_{i=1}^{i=N_e} [A_i]^T [K_i][A_i] \tag{3.91}$$

where $[K_i]$ is the $ND_i \times ND_i$ relational matrix of the ND_i–th element, $[A_i]^T$ is the transpose of $[A_i]$, and $[A_i]$ is a $ND_i \times N$ assistive matrix defined as

$$[A_i] = \begin{matrix} nd_{i,1} \quad nd_{i,2} \quad nd_{i,nd_i} \\ \begin{bmatrix} \cdots\cdots\cdots\cdots\cdots\cdots \\ \cdots\ 1\ \cdots\ 0\ \cdots\ 0\ \cdots \\ \cdots\cdots\cdots\cdots\cdots\cdots \\ \cdots\ 0\ \cdots\ 1\ \cdots\ 0\ \cdots \\ \cdots\cdots\cdots\cdots\cdots\cdots \\ \cdots\ 0\ \cdots\ 0\ \cdots\ 1\ \cdots \end{bmatrix} \end{matrix} \tag{3.92}$$

where all other entities in $[A_i]$ are zeros.

Example 3.5 Figure 3.27a shows a FEA model for a 2D heat transfer problem. The model is to find the temperature distribution, and the FEA model consists of 10 nodes for three triangular elements and three rectangle elements. Build a system model based on the given element models as follows:

The stiffness matrices $[K_i]$ for elements are found as

Triangle elements (1), (4) and (6) :   Rectangle elements (2), (3) and (4) :

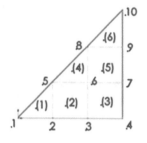

Elements	ND_i	$nd_{i,1}$	$nd_{i,2}$	$nd_{i,3}$	$nd_{i,4}$
(1)	3	1	2	5	
(2)	4	2	3	6	5
(3)	4	3	4	7	6
(4)	3	5	6	8	
(5)	4	6	7	9	8
(6)	3	8	9	10	

Fig. 3.27 **a** Elements and nodes **b** Relations of nodes and elements

$$[\mathbf{K}]^{(e)} = \begin{bmatrix} 0.5 & -0.5 & 0 \\ -0.5 & 1 & -0.5 \\ 0 & -0.5 & 0.5 \end{bmatrix} \quad [\mathbf{K}]^{(e)} = \begin{bmatrix} 4 & -1 & -2 & -1 \\ -1 & 4 & -1 & -2 \\ -2 & -1 & 4 & -1 \\ -1 & -2 & -1 & 4 \end{bmatrix}$$

Solution. Each node in the FEA element has one DoF, and the model has 10 nodes. Therefore, the system model has $N = 10$ DoF, and $[K]$ for the system has a size of 10×10.

The system model is assembled from element models. The model as $N_e = 6$ elements, the involved DoF in these elements are described in Fig. 3.27b.

To use Eq. (3.91) to assemble system model, $[A_i]$ ($i = 1, 2, ..., 6$) are determined based on Fig. 3.27b as

$$[A_1] = \begin{bmatrix} 1 & 0 & 0 & 0 & 0 & 0 & 0 & 0 & 0 & 0 \\ 0 & 1 & 0 & 0 & 0 & 0 & 0 & 0 & 0 & 0 \\ 0 & 0 & 0 & 0 & 1 & 0 & 0 & 0 & 0 & 0 \end{bmatrix}$$

$$[A_2] = \begin{bmatrix} 0 & 1 & 0 & 0 & 0 & 0 & 0 & 0 & 0 & 0 \\ 0 & 0 & 1 & 0 & 0 & 0 & 0 & 0 & 0 & 0 \\ 0 & 0 & 0 & 0 & 0 & 1 & 0 & 0 & 0 & 0 \\ 0 & 0 & 0 & 0 & 1 & 0 & 0 & 0 & 0 & 0 \end{bmatrix}$$

$$[A_3] = \begin{bmatrix} 0 & 0 & 1 & 0 & 0 & 0 & 0 & 0 & 0 & 0 \\ 0 & 0 & 0 & 1 & 0 & 0 & 0 & 0 & 0 & 0 \\ 0 & 0 & 0 & 0 & 0 & 0 & 1 & 0 & 0 & 0 \\ 0 & 0 & 0 & 0 & 0 & 1 & 0 & 0 & 0 & 0 \end{bmatrix}$$

$$[A_4] = \begin{bmatrix} 0 & 0 & 0 & 0 & 1 & 0 & 0 & 0 & 0 & 0 \\ 0 & 0 & 0 & 0 & 0 & 1 & 0 & 0 & 0 & 0 \\ 0 & 0 & 0 & 0 & 0 & 0 & 0 & 1 & 0 & 0 \end{bmatrix}$$

$$[A_5] = \begin{bmatrix} 0\,0\,0\,0\,0\,1\,0\,0\,0\,0 \\ 0\,0\,0\,0\,0\,0\,1\,0\,0\,0 \\ 0\,0\,0\,0\,0\,0\,0\,0\,1\,0 \\ 0\,0\,0\,0\,0\,1\,0\,1\,0\,0 \end{bmatrix}$$

$$[A_6] = \begin{bmatrix} 0\,0\,0\,0\,0\,0\,0\,1\,0\,0 \\ 0\,0\,0\,0\,0\,0\,0\,0\,1\,0 \\ 0\,0\,0\,0\,0\,0\,0\,0\,0\,1 \end{bmatrix}$$

Substituting the above assistive matrices and given stiffness matrices in Eq. (3.91) yield the stiffness matrix of the system as

$$[K]_{10\times10} = \sum_{i=1}^{i=6} [A_i]^T [K_i][A_i] = \begin{bmatrix} 0.5 & -0.5 & 0 & 0 & 0 & 0 & 0 & 0 & 0 & 0 \\ -0.5 & 0.5 & -1 & 0 & -1.5 & -2 & 0 & 0 & 0 & 0 \\ 0 & -1 & 8 & -1 & -2 & -2 & -2 & 0 & 0 & 0 \\ 0 & 0 & -1 & 4 & 0 & -2 & -1 & 0 & 0 & 0 \\ 0 & -1.5 & -2 & 0 & 5 & -1.5 & 0 & 0 & 0 & 0 \\ 0 & -2 & -2 & -2 & -1.5 & 13 & -2 & -1.5 & -2 & 0 \\ 0 & 0 & -2 & -1 & 0 & -2 & 8 & -2 & -1 & 0 \\ 0 & 0 & 0 & 0 & 0 & -1.5 & -2 & 5 & -2 & 0 \\ 0 & 0 & 0 & 0 & 0 & -2 & -1 & -1.5 & 5 & -0.5 \\ 0 & 0 & 0 & 0 & 0 & 0 & 0 & 0 & -1 & 0.5 \end{bmatrix}$$

$$(3.93)$$

3.5.7.2   Solution to Steady Problems

When FEA is used to model a steady engineering problem, the system model becomes a set of linear equations in Eq. (3.90), and the solution to the steady problem is found as

$$\{U\} = [K]^{-1}\{F\} \tag{3.94}$$

where $\{F\}$ is the vector of loads, $\{U\}$ is the vector of state variables, and $[K]$ is the matrix for the relation of loads and state variables, such as the stiffness matrix in static analysis. $[K]^{-1}$ is an inverse matrix $[K]$.

Note that Eq. (3.94) can be *linear* or *nonlinear*. A *linear system model* have all constant entities in $[K]$ and $\{F\}$; otherwise, the system model is nonlinear. To solve a nonlinear system model, an iterative procedure is applied to solve $\{U\}$. At each iteration, the system model is simplified as a linear system model for $\{U\}$. The iterative procedure will be terminated until all of the constraints are satisfied.

Since an FEA model for a real-world engineering problem involves in thousands of unknown variables, an effective algorithm to find the solution to Eq. (3.94) is critical.

Solvers to a linear system model can be classified into *direct* or *iterative* methods. A direct method, such as Gaussian elimination or LU decomposition, obtains the solution in a definite number of steps, and the computation can be predicted based on the size of $[K]$. The computation of an iterative method is hard to be estimated since the number of iterations to find a converged solution depends on the characteristic of $[K]$. Theoretically, a direct method can find an exact solution to a system of linear equations. However, it suffers a number of significant disadvantages in its implementation. For example,

(1)   A direct method provides an open-loop solution. It implies that the solution does not have the mechanism to compensate round-off errors in numerical calculation; however, computation errors are accumulated rapidly in the solving process.

(2)   The implementation of a direct method is very intrigue if all of the special cases are considered. For example, a special treatment is needed when the coefficient at the pivoting position becomes zero in the Gaussian elimination.

(3)   A direct method requires a large amount of computation when the number of unknown variables increase. In addition, it lacks the flexibility to reduce the computation for the system model with a sparse matrix.

An iterative method is an ideal alternative since it provides a closed-loop solution by iterations. The accuracy of the solution is evaluated at every iteration, and the iterative process will not be terminated until the solution reaches the required accuracy.

To solve the system model iteratively, rewrite $[K]\{U\} = \{F\}$ as $([K_L] + [K_D] + [K_U])\{U\} = \{F\}$ where the original full $[K]$ is decomposed into three sub-matrices, i.e., *lower*, *diagonal*, and *upper* sub-matrices as

$$
[K_L] = \begin{bmatrix} 0 & 0 & \cdots & 0 \\ k_{2,1} & \ddots & \cdots & \cdots \\ \cdots & \ddots & \ddots & \cdots \\ k_{N,1} & \cdots & k_{N,N-1} & 0 \end{bmatrix}, \ [K_D] = \begin{bmatrix} k_{1,1} & 0 & \cdots & 0 \\ 0 & \ddots & \cdots & \cdots \\ \cdots & \ddots & \ddots & \cdots \\ 0 & \cdots & 0 & k_{N,N} \end{bmatrix},
$$

$$
[K_U] = \begin{bmatrix} 0 & k_{1,2} & \cdots & k_{1,N} \\ 0 & \ddots & \cdots & \cdots \\ \cdots & \ddots & \ddots & k_{2,1} \\ 0 & \cdots & 0 & 0 \end{bmatrix}
$$
(3.95)

Accordingly, the system solution Eq. (3.94) can be written as

$$
\{U\} = [K_D]^{-1}\{\{F\} - ([K_L] + [K_U])\{U\}\}
$$
(3.96)

where $[K_L]$, $[K_D]$, and $[K_U]$ are lower, diagonal, and upper matrices shown in Eq. (3.95).

Equation (3.97) can further be written in an iterative form from *the k*-th step to *the k + 1*-th step as

$$\{U\}^{(k+1)} = [K_D]^{-1}\{\{F\} - ([K_L] + [K_U])\{U\}^{(k)}\} \tag{3.97}$$

where $\{U\}^{(k)}$ and $\{U\}^{(k+1)}$ are the iterated solutions at the k-th step to the k + 1-th steps, respectively.

The interactive process (i.e., Eq. (3.97)) will continue until the termination condition is satisfied,

$$\|\{U\}^{(k+1)} - \{U\}^{(k)}\| < \varepsilon \tag{3.98}$$

where ε is the specified accuracy of system solution.

The above method is referred as the *Jacobi' method*. Assumed that the diagonal entries in $[K]$ are nonzero, the scalar version of Eq. (3.99) is

$$U_i^{(k+1)} = \frac{1}{k_{ii}}\left[F_i - \sum_{j=1; j\neq i}^{N} k_{ij}U_j^{(k)}\right], \ (i = 1, 2, \ldots N) \tag{3.99}$$

Note that an iterative method can guarantee the convergence of the iterative process under the conditions of

$$|k_{ii}| > \sum_{j=1; j\neq i}^{N} |k_{ij}| \ for \ i = 1, 2, \ldots N \tag{3.100}$$

And

$$\||K_D|^{-1}(K_L + K_U)\| = _{i=1,2,\ldots N}^{max}|k_{ij}/k_{ii}| < 1 \tag{3.101}$$

The conditions in Eqs. (3.100) and (3.101) ensure that the converged solution will be guaranteed. Not all of FEA models can strictly meet these conditions, but they can still be solved by an iterative method effectively.

Example 3.6 Use the iterative method to solve $[K]\{U\} = \{F\}$ where.

$$[K] = \begin{bmatrix} 2 & -1 & 0 & 0 & 0 \\ -1 & 2 & -1 & 0 & 0 \\ 0 & -1 & 2 & -1 & 0 \\ 0 & 0 & -1 & 2 & -1 \\ 0 & 0 & 0 & -1 & 2 \end{bmatrix} \text{ and } \{F\} = \begin{Bmatrix} 5 \\ 2 \\ 2 \\ 2 \\ 5 \end{Bmatrix}$$

Solution. Let $\{U\}^{(0)} = [1\ 1\ 1\ 1\ 1]$ and the terminate tolerance $\varepsilon = 0.0001$.

Using Eq. (3.99) to execute the iterative process, the process to obtain the final solutions is.

Step	$\{x_i\}^{(k)}$					Convergence
0	[1	1	1	1	1]	
1	[3.0000	2.0000	2.0000	2.0000	3.0000]	3.5000
10	[6.5762	8.6270	9.1523	8.6270	6.5762]	0.7750
20	[7.6621	10.4369	11.3242	10.4369	7.6621]	0.1839
30	[7.9198	10.8664	11.8396	10.8664	7.9198]	0.0436
40	[7.9810	10.9683	11.9619	10.9683	7.9810]	0.0104
50	[7.9955	10.9925	11.9910	10.9925	7.9955]	0.0025
60	[7.9989	10.9982	11.9979	10.9982	7.9989]	5.8325e-04
70	[7.9997	10.9996	11.9995	10.9996	7.9997]	1.3841e-04
73	[7.9998	10.9997	11.9997	10.9997	7.9998]	8.9898e-05

3.5.7.3   Solution to Eigenvalue Problems

To understand the response of an engineering system to dynamic loads, eigenvalues must be found so that the stead ability of system under the stimulations of different frequencies can be evaluated. *An eigenvalue problem* aims to find the eigenvalues and eigenvectors for the given system model.

Let a system model of an eigenvalue problem be

$$[K]\{u\} - \lambda[M]\{u\} = 0 \qquad (3.102)$$

An eigenvalue λ_i should satisfy the condition for a valid solution of the system model,

$$det([K]\{u\} - \lambda[M]) = 0 \qquad (3.103)$$

where $det(\bullet)$ is the determinant of the matrix $(\bullet)$.

Equation (3.103) is commonly referred as *the characteristic equation* in an eigenvalue problem of Eq. (3.102). λ_i that satisfies Eq. (3.103) is called as an eigenvalue ($i = 1, 2, ...N$), and an eigenvector u_i corresponding to λ_i satisfies the condition of

$$([K] - \lambda_i[M])\{u\} = 0 \qquad (3.104)$$

If a system model with a few DoF, eigenvalues to Eq. (3.102) can be obtained by solving the polynomial equation about λ, which is defined based on Eq. (3.103). For a system model with a large number of DoF, the following numerical method is used to find a set of the most significant eigenvalues.

Assume that the eigenvalue problem has N DoF, i.e., the matrix $[K]$ or $[M]$ has the size of $N \times N$, a numerical method aims to find M number of the lowest eigenvalues and the corresponding eigenvectors, and these eigenvalues are found iteratively in the following procedure:

(1)   select M non-zero vectors for a $N \times M$ modal matrix $[X_i]$, where each of its columns corresponds to one vector; set the iteration index $i = 1$,

(2) solve the system of linear equations $[K]\{Y_i\} = [M]\{X_i\}$ to get $N \times M$ modal matrix $\{Y_i\}$,

(3) build a new eigenvalue problem with a reduced size of $M \times M$ as

$$[K]_i^*\{u\}^* - \lambda[M]_i^*\{u\}^* = 0$$

where $[K]_i^* = \{Y_i\}^T[K]\{Y_i\}$ and $[M]_i^* = \{Y_i\}^T[M]\{Y_i\}$

(4) solve the new eigenvalue problem for M eigenvalues and eigenvalues as

$$\lambda_m^{(i)} = diag\left[\lambda_1^{(i)}, \lambda_2^{(i)}, \cdots \lambda_m^{(i)}\right]$$
$$[u^*] = \left[u_1^{(i)}, u_2^{(i)}, \cdots u_m^{(i)}\right]$$

(5) update the iteration as $i \leftarrow i + 1$, $[X_i] \leftarrow [X_{i+1}] = normc\ ([Y_i][u^*]$, where $normc(\bullet)$ is a function to make every vector (column) in a matrix as a unit vector;

(6) check the terminating condition as

$$max\left(\left|\frac{\lambda_1^{(i)} - \lambda_1^{(i-1)}}{\lambda_1^{(i)}}\right|, \left|\frac{\lambda_2^{(i)} - \lambda_2^{(i-1)}}{\lambda_2^{(i)}}\right|, \cdots \left|\frac{\lambda_m^{(i)} - \lambda_m^{(i-1)}}{\lambda_m^{(i)}}\right|\right) \le \varepsilon$$

If it is satisfied, the iteration is terminated. Otherwise, return to step 2) for the continuation.

Example 3.7 Find the first two natural frequencies of the following system model with the given accuracy ε of 0.001:

$$[M]\{\ddot{u}\} + [K]\{u\} = 0 \qquad (3.105)$$

where

$$[M] = \begin{bmatrix} 20 & 5 & 0 \\ 5 & 10 & 5 \\ 0 & 5 & 10 \end{bmatrix}, [K] = \begin{bmatrix} 2 & -1 & 0 \\ -1 & 2 & -1 \\ 0 & -1 & 1 \end{bmatrix}$$

Solution. Equation (3.105) can be turned into an eigenvalue problem with three DoF ($N = 3$) as.

$$[K]\{u\} - \lambda_i[M]\{u\} = 0 \text{ where } \lambda_i = \omega_i^2 \qquad (3.106)$$

Since the first two natural frequencies ($M = 2$) of the system model is interested, let assume that an initial vector as $\{u\}_{2\times3}$, i.e.,

$$[u]_{23} = [u_1 \ u_2] = \begin{bmatrix} 1 & 0 \\ 0 & 1 \\ 0 & 0 \end{bmatrix}$$

Let $i = 1$ and $[X_1] = [u]_{23}$ to initialize the iterative process.

Iteration 1. Solving the equation set $[K][Y_i] = [M][X_i]$ yields

$$[Y_i] = \begin{bmatrix} 25 & 20 \\ 30 & 35 \\ 30 & 40 \end{bmatrix}$$

The reduced eigenvalue problem at this iteration becomes

$$[K]^*[u]^* - \lambda[M]^*[u]^* = 0$$

where

$$[K]^* = [Y_i]^T[K][Y_i] = \begin{bmatrix} 650 & 575 \\ 575 & 650 \end{bmatrix}$$

$$[M]^* = [Y_i]^T[M][Y_i] = 1.0 \times 10^4 \begin{bmatrix} 4.7 & 5.1125 \\ 5.1125 & 5.725 \end{bmatrix}$$

The eigenvalues and eigenvectors of the above reduced problem can be found as $\lambda_1 = 0.1052$, $\lambda_2 = 0.0113$, and $[u]^* = \begin{bmatrix} 0.7455 & 0.0450 \\ -0.6665 & 0.9990 \end{bmatrix}$.

Next, $[X_i]$ can be updated as

$$[X_i] = normc([Y_i][u]^*) = normc\left(\begin{bmatrix} 5.3075 & 21.1040 \\ -0.9625 & 36.3137 \\ -4.295 & 41.3087 \end{bmatrix}\right) = \begin{bmatrix} 0.7697 & 0.3582 \\ -0.1396 & 0.6164 \\ -0.6229 & 0.7012 \end{bmatrix}$$

Iteration 2. Solving the equation set $[K][Y_i] = [M][X_i]$ with the updated $[X_i]$ yields

$$[Y_i] = \begin{bmatrix} 7.1085 & 31.8025 \\ -0.4800 & 53.3581 \\ -7.4069 & 63.4522 \end{bmatrix}$$

The reduced eigenvalue problem at this iteration becomes

$$[K]^*[u]^* - \lambda[M]^*[u]^* = 0$$

where

$$[K]^* = [Y_i]^T[K][Y_i] = \begin{bmatrix} 650 & 575 \\ 575 & 650 \end{bmatrix}$$

$$[M]^* = [Y_i]^T[M][Y_i] = 1.0 \times 10^4 \begin{bmatrix} 4.7 & 5.1125 \\ 5.1125 & 5.725 \end{bmatrix}$$

The eigenvalues and eigenvectors of the above reduced problem can be found as $\lambda_1 = 0.1052$, $\lambda_2 = 0.0113$, and $[u]^* = \begin{bmatrix} 1.000 & -0.0069 \\ 0.0054 & 1.0000 \end{bmatrix}$.

Next, $[X_i]$ can be updated as

$$[X_i] = normc([Y_i][u]^*) = normc\left(\begin{bmatrix} 7.2799 & 31.7526 \\ -0.1923 & 53.3601 \\ -7.0647 & 63.5019 \end{bmatrix} \right) = \begin{bmatrix} 0.7175 & 0.3575 \\ -0.0190 & 0.6008 \\ -0.6963 & 0.7150 \end{bmatrix}$$

Check the convergence: $\max\left(\left| \frac{\lambda_1^{i+1} - \lambda_1^{i+1}}{\lambda_1^{i+1}} \right|, \left| \frac{\lambda_1^{i+1} - \lambda_1^{i+1}}{\lambda_1^{i+1}} \right| \right) = 0.05 > \varepsilon$; therefore, the iteration has to be continued.

Iteration 3. Solving the equation set $[K][Y_i] = [M][X_i]$ with the updated $[X_i]$ yields

$$[Y_i] = \begin{bmatrix} 7.1139 & 31.6789 \\ -0.0274 & 53.2034 \\ -7.0852 & 63.3574 \end{bmatrix}$$

The reduced eigenvalue problem at this iteration becomes

$$[K]^*[u]^* - \lambda[M]^*[u]^* = 0$$

where

$$[K]^* = [Y_i]^T[K][Y_i] = 1.0 \times 10^3 \begin{bmatrix} 0.1514 & -0.0000 \\ -0.0000 & 1.5700 \end{bmatrix}$$

$$[M]^* = [Y_i]^T[M][Y_i] = 1.0 \times 10^5 \begin{bmatrix} 0.0151 & -0.0000 \\ -0.0000 & 1.3908 \end{bmatrix}$$

The eigenvalues and eigenvectors of the above reduced problem can be found as $\lambda_1 = 0.1001$, $\lambda_2 = 0.0113$, and $[u]^* = \begin{bmatrix} 1.000 & -0.000 \\ 0.000 & 1.0000 \end{bmatrix}$.

Next, $[X_i]$ can be updated as

$$[X_i] = normc([Y_i][u]^*) = normc\left(\begin{bmatrix} 7.1143 & 31.6788 \\ -0.0267 & 53.2034 \\ -7.0844 & 63.3575 \end{bmatrix}\right) = \begin{bmatrix} 0.7086 & 0.3576 \\ -0.0027 & 0.6006 \\ -0.7056 & 0.7152 \end{bmatrix}$$

Check the convergence: $max\left(\left|\frac{\lambda_1^{i+1}-\lambda_1^i}{\lambda_1^{i+1}}\right|, \left|\frac{\lambda_2^{i+1}-\lambda_2^i}{\lambda_2^{i+1}}\right|\right) < \varepsilon$. The convergence condition is satisfied.

Finally, the first two natural frequencies are found as $\omega_1 = 0.3164$ and $\omega_2 = 0.1063$.

3.5.7.4 Solution to Transient Problems

In a transient problem, the dynamic response of a system in a transient period is of interest. Other than the spatial domain, the time domain has to be discretized. The solution to a transient problem can be treated as a set of sub-solutions in a series of time steps from the initial time to the terminate time. The solving procedure is *step by step* where the sub-solution in the next step is updated based on those in precedent steps. While some special consideration have to be taken in determining time steps and updating system models, the sub-solution at each step is similar to an equilibrium problem. In this section, a transient heat transfer problem is used as an example to illustrate the solving procedure for a transient problem.

In a transient heat transfer problem, the temperature distribution in a body varies over time; the transient response is critical to some manufacturing processes such as heat treatments. To find the solution to a transient heat transfer problem, the time domain is discretized into steps, while the time step is critical to both of the stability of the iterative process and accuracy of sub-solutions: if the time step is too small, a spurious oscillation may occur which leads to meaningless results. If the time step is too large, the solutions may not be converged since the temperature gradients are evaluated appropriately.

To discretize the time domain, Biot number ($B_i = h\Delta x/k$) and Fourier number ($F_0 = \Delta t/(\Delta x)^2$) are used to determine a reasonable time step, where h and k are convective and conductive coefficients, respectively; α is the thermal diffusivity; Δx denotes the mean length of an element in one spatial direction and Δt is the time step.

When the Biot number Bi is less than 1 ($Bi < 1$), the time step should be determined to meet the condition of convergence of $0.1 \leq F_0 \leq 0.5$; therefore,

$$\Delta t = \frac{(\Delta x)^2 F_0}{\alpha} \tag{3.107}$$

When the Biot number Bi is larger than 1 ($Bi > 1$), the time step should be determined by considering both of the Biot and Fourier numbers as $(Fo)(Bi) = b$,

$$\Delta t = \frac{(\Delta x)k_{solid}}{\alpha h}b = \frac{(\Delta x)\rho c}{h}b \tag{3.108}$$

where $b = F_0 \cdot B_i$; ρ and c are the density and specific heat of materials.

Assume that a transient heat transfer problem is formulated as the following differential equations (Bi 2018),

$$[C] \cdot \left\{ \frac{\partial \mathbf{T}}{\partial t} \right\} + [\mathbf{K}] \cdot \{\mathbf{T}\} = \{\mathbf{Q}\} \tag{3.109}$$

where $[C]$ and $[K]$ are the heat storage and heat transfer matrices, respectively; $\{Q\}$ is the vector of the heat loads; $T\}$ and $\{\frac{\partial \mathbf{T}}{\partial t}\}$ are the vectors of temperatures and temperature gradients, respectively.

Discretizing the time domain by setting the time step of Δt gets the iteration equation at step i as

$$\{T\}^{(i+1)}\{T\}^{(i)} + \Delta t \{\dot{T}\}^{(i)} + \frac{\Delta t^2}{2} \{\ddot{T}\}^{(i)} \tag{3.110}$$

The flexibility of Eq. (3.110) can be improved by adjusting the weight in the front of $\{\ddot{T}\}^{(i)}$ by the Euler parameter $\theta \in (0, 1)$ as

$$\{T\}^{(i+1)}\{T\}^{(i)} + \Delta t \{\dot{T}\}^{(i)} + \theta \Delta t^2 \{\ddot{T}\}^{(i)} \tag{3.111}$$

A different value of the Euler parameter corresponds to a different scheme of the iteration, i.e.,

$$\theta = \begin{cases} 0 \ forward\ difference(Euler) \\ 1/2 \ Crank - Nicolson\ difference \\ 1 \ Backward\ difference \end{cases} \tag{3.112}$$

In Eq. (3.111), the second-order of temperature can be approximated by the first-order derivative at two time steps as

$$\{\ddot{T}\}^{(i)} = \frac{\{\dot{T}\}^{(i+1)} - \{\dot{T}\}^{(i)}}{t} \tag{3.113}$$

Substituting Eq. (3.113) into (3.111) yields

$$\{T\}^{(i+1)}\{T\}^{(i)} + (1 - \theta)\Delta t \{\dot{T}\}^{(i)} + \theta \Delta t \{\dot{T}\}^{(i+1)} \tag{3.114}$$

Equation (3.114) shows that $\{\dot{T}\}^{(i+1)}$ depends on $\{T\}^{(i+1)}$ as

$$\{\dot{T}\}^{(i+1)} = \frac{(\{T\}^{(i+1)} - \{T\}^{(i)})}{\theta \Delta t} - \left(\frac{1}{\theta} - 1\right)\{\dot{T}\}^{(i)} \tag{3.115}$$

Since both of $\{T\}^{(i)}$ and $\{\dot{T}\}^{(i)}$ are known at the previous time step i; using Eq. (3.115) in Eq. (3.114) gets the explicit equation of iteration as

$$\left(\frac{[C]}{\theta\Delta t}+[K]\right)\{T\}^{(i+1)}=\{Q\}+\frac{[C]}{\theta\Delta t}\{T\}^{(i)}+\left(\frac{1}{\theta}-1\right)[C]\{\dot{T}\}^{(i)} \qquad (3.116)$$

where $[K_m]=\frac{[C]}{\theta\Delta t}+[K]$ is the modified heat transfer matrix, and.

$\{Q_m\}=\{Q\}+\frac{[C]}{\theta\Delta t}\{T\}^{(i)}+\left(\frac{1}{\theta}-1\right)[C]\{\dot{T}\}^{(i)}$ is the modified vector of heat loads.

Example 3.8 A 2-in thick plate of 1060 alloy is uniformly heated to 300 °C. It is air-quenched in the forced air. Assume that the temperature of the plate surface is immediately set to 25 °C. The materials density $\rho = 0.0975437$ lb/in³, the conductivity $k = 0.002675$ Btu/(s·in·°C), and the specific heat $c = 0.214961$ Btu/(lb·°C). Determine the cooling curve of temperature at the neutral plane of the plate with respect to time in 5 s.

Solution. Figure 3.28 shows a 1D transient heat transfer element described in the *local coordinate system* (LCS). It consists of node i and node j. The state variables at two nodes are the temperatures (T_i, T_j) and temperature gradient $(\dot{T}_i, \dot{T}_j)$. The elemental behavior is governed by

$$k\frac{\partial^2 T}{\partial x^2}-c\frac{\partial T}{\partial t}+\dot{q}=0 \qquad (3.117)$$

where k is conductivity, ρ is the density, and c is the specific heat.

The element model is developed based on Eq. (3.117) as follows. Let the shape functions of an element be $S_i(x)=1-x/L$ and $S_j(x)=x/L$, using the shape functions as the test function in the Galerkin method leads to the weak-form solution Eq. (3.117) as (Bi 2018),

$$[K_m]\left\{\begin{array}{c}T_i\\T_j\end{array}\right\}^{(p+1)}=\{Q_m\} \qquad (3.118)$$

Fig. 3.28 1D transient heat transfer element

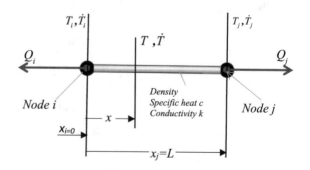

where p is the step index and $[K_M]$ is the modified conductive matrix as

$$[K_M] = \left(\frac{\rho c AL}{6\theta \Delta t} \begin{bmatrix} 2 & 1 \\ 1 & 2 \end{bmatrix} + \frac{kA}{L} \begin{bmatrix} 1 & -1 \\ -1 & 1 \end{bmatrix} \right) \tag{3.119}$$

$[Q_M]$ is the modified load vector determined by the state of the precedent step (i.e., step i) as

$$\{Q_m\} = \left\{ \begin{matrix} Q_i \\ Q_j \end{matrix} \right\} + \frac{\rho c AL}{6\theta \Delta t} \begin{bmatrix} 2 & 1 \\ 1 & 2 \end{bmatrix} \left\{ \begin{matrix} T_i \\ T_j \end{matrix} \right\}^{(p)} + \left(\frac{1}{\theta} - 1 \right) \frac{\rho c AL}{6} \begin{bmatrix} 2 & 1 \\ 1 & 2 \end{bmatrix} \left\{ \begin{matrix} \dot{T}_i \\ \dot{T}_j \end{matrix} \right\}^{(p)} \tag{3.120}$$

After the temperature at $(p+1)$ have been found, the temperature gradients can be derived by .

$$\left\{ \begin{matrix} \dot{T}_i \\ \dot{T}_j \end{matrix} \right\}^{(p+1)} = \frac{1}{\theta \Delta t} \left\{ \begin{matrix} T_i^{(p+1)} - T_i^{(p)} \\ T_j^{(p+1)} - T_j^{(i)} \end{matrix} \right\} - \left(\frac{1}{\theta} - 1 \right) \left\{ \begin{matrix} \dot{T}_i \\ \dot{T}_j \end{matrix} \right\}^{(p)} \tag{3.121}$$

Since the plate is symmetric about its neutral axis, a 1D FEA model is developed for half of the plate thickness. Accordingly, Fig. 3.29 shows a decomposition which leads to an FEA model with 5 elements and 6 nodes. Every element has the length of 0.2-in.

Let the Fourier number be $F_0 = 0.5$, and the step time Δt is found from Eq. (3.107) as

$$\Delta t = \frac{(\Delta x)^2 F_0}{\alpha} = \frac{(0.2)^2 (0.5)}{(0.002675/(0.0975437 * 0.214961))} = 0.1568 (s)$$

If the history of the temperature change in 5 s is concerned, the total number N of the time steps is

$$N = \frac{(5)}{\Delta t} = 32 (steps)$$

Fig. 3.29 Nodes and elements in 1D transient heat transfer problem

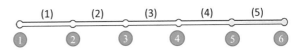

$$T_1^{(0)} = T_2^{(0)} = T_3^{(0)} = T_4^{(0)} = T_5^{(0)} = T_6^{(0)} = 300°C$$
$$\dot{T}_1^{(0)} = \dot{T}_2^{(0)} = \dot{T}_3^{(0)} = \dot{T}_4^{(0)} = \dot{T}_5^{(0)} = \dot{T}_6^{(0)} = 0$$
$$T_6^{(p)} = 25°C \quad p > 0$$

Let the Euler parameter $\theta = 0.5$ in Eq. (3.121) where the Crank-Nicolson difference is used in the approximation. At each time step p from 1 to N, the system model can be assembled from 5 element models expressed in Eq. (3.118) as

$$
[K]^{(G)} \left\{ \begin{array}{c} T_1 \\ T_2 \\ T_3 \\ T_4 \\ T_5 \\ T_6 \end{array} \right\}^{(p+1)} = \{Q\}^{(p)} = \left\{ \begin{array}{c} 2c_1 T_1^{(p)} + c_1 T_2^{(p)} + 2c_2 \dot{T}_1^{(p)} + c_2 \dot{T}_2^{(p)} \\ c_1 T_1^{(p)} + 4c_1 T_2^{(p)} + c_2 \dot{T}_1^{(p)} + 4c_2 \dot{T}_2^{(p)} + c_1 T_3^{(p)} + c_2 \dot{T}_3^{(p)} \\ c_1 T_2^{(p)} + 4c_1 T_3^{(p)} + c_2 \dot{T}_2^{(p)} + 4c_2 \dot{T}_3^{(p)} + c_1 T_4^{(p)} + c_2 \dot{T}_4^{(p)} \\ c_1 T_3^{(p)} + 4c_1 T_4^{(p)} + c_2 \dot{T}_3^{(p)} + 4c_2 \dot{T}_4^{(p)} + c_1 T_5^{(p)} + c_2 \dot{T}_5^{(p)} \\ c_1 T_4^{(p)} + 4c_1 T_5^{(p)} + c_2 \dot{T}_4^{(p)} + 4c_2 \dot{T}_5^{(p)} + c_1 T_6^{(p)} + c_2 \dot{T}_6^{(p)} \\ c_1 T_5^{(p)} + 2c_1 T_6^{(p)} + c_2 \dot{T}_5^{(p)} + 2c_2 \dot{T}_6^{(p)} \end{array} \right\}
$$

$$(3.122)$$

where

$$
[K]^{(G)} = \begin{bmatrix} 0.0312 & -0.0045 & 0 & 0 & 0 & 0 \\ -0.0045 & 0.0624 & -0.0045 & 0 & 0 & 0 \\ 0 & -0.0045 & 0.0624 & -0.0045 & 0 & 0 \\ 0 & 0 & -0.0045 & 0.0624 & -0.0045 & 0 \\ 0 & 0 & 0 & -0.0045 & 0.0624 & -0.0045 \\ 0 & 0 & 0 & 0 & -0.0045 & 0.0312 \end{bmatrix}
$$

$$(3.123)$$

c_1 and c_2 are two constants as below

$$
\left. \begin{array}{l} c_1 = \frac{\rho c A L}{6 \theta \Delta t} = 0.0089 \\ c_2 = \left(1 - \frac{1}{\theta}\right) \frac{\rho c A L}{6} = -0.00069893 \end{array} \right\}
$$

$$(3.124)$$

At each step p, after $\{T\}^{(p+1)}$ is found from Eq. (3.118), the temperature gradients on nodes can be updated based on Eq. (3.121) as

$$
\left\{ \begin{array}{c} \dot{T}_1 \\ \dot{T}_2 \\ \dot{T}_3 \\ \dot{T}_4 \\ \dot{T}_5 \\ \dot{T}_6 \end{array} \right\}^{(p+1)} = \left\{ \begin{array}{c} \frac{1}{\theta \Delta t}\left(T_1^{(p+1)} - T_1^{(p)}\right) - \left(\frac{1}{\theta} - 1\right)\dot{T}_1^{(p)} \\ \frac{1}{\theta \Delta t}\left(T_2^{(p+1)} - T_2^{(p)}\right) - \left(\frac{1}{\theta} - 1\right)\dot{T}_2^{(p)} \\ \frac{1}{\theta \Delta t}\left(T_3^{(p+1)} - T_3^{(p)}\right) - \left(\frac{1}{\theta} - 1\right)\dot{T}_3^{(p)} \\ \frac{1}{\theta \Delta t}\left(T_4^{(p+1)} - T_4^{(p)}\right) - \left(\frac{1}{\theta} - 1\right)\dot{T}_4^{(p)} \\ \frac{1}{\theta \Delta t}\left(T_5^{(p+1)} - T_5^{(p)}\right) - \left(\frac{1}{\theta} - 1\right)\dot{T}_5^{(p)} \\ \frac{1}{\theta \Delta t}\left(25 - T_6^{(p+1)}\right) - \left(\frac{1}{\theta} - 1\right)\dot{T}_6^{(p)} \end{array} \right\}
$$

$$(3.125)$$

Note that the calculation for the temperature gradient of the boundary node 6 is different since the temperature at this node is fixed as 25 °C. Accordingly, when the calculation result at step p is moved to next step $p + 1$, the temperature at node 6 must be set back to 25 °C as the boundary constraint. The temperature changes on nodes with respect to time are obtained in Table. 3.9.

The plots of temperature on node 1 (i.e., the center of plate) are illustrated in Figs. 3.30 and 3.31 respectively.

Table 3.9 The History of nodal temperatures in 5 s

Step		Temperature (°C)					
No.	Time (sec)	Node 1	Node 2	Node 3	Node 4	Node 5	Node 6
0	0	300	300	300	300	300	25
1	0.1568	299.9945	299.9618	299.4711	292.6341	197.4066	25
2	0.3136	299.8771	299.3362	293.5781	248.8355	120.4615	25
3	0.4704	298.8845	295.6335	274.0357	212.8518	138.0203	25
4	0.6272	294.7467	286.1748	255.8324	204.9803	127.9079	25
5	0.784	285.8694	275.4648	245.3924	192.5891	122.1926	25
6	0.9408	275.5284	265.5057	234.3123	183.8132	115.9687	25
7	1.0976	265.4401	255.0242	224.6841	175.3348	110.9712	25
8	1.2544	255.091	245.0755	215.2958	167.9028	106.3968	25
9	1.4112	245.1051	235.2671	206.5492	160.9382	102.3165	25
10	1.568	235.3315	225.8797	198.1785	154.4945	98.5393	25
11	1.7248	225.9358	216.8209	190.2477	148.4187	95.0359	25
12	1.8816	216.8846	208.1518	182.6812	142.6915	91.7441	25
13	2.0384	208.2126	199.8444	175.4773	137.2591	88.6407	25
14	2.1952	199.9055	191.9035	168.6055	132.101	85.7005	25
15	2.352	191.9622	184.3126	162.0528	127.1924	82.9092	25
16	2.5088	184.3696	177.0622	155.8008	122.5178	80.254	25
17	2.6656	177.1169	170.1378	149.8358	118.0622	77.7256	25
18	2.8224	170.1904	163.5267	144.1437	113.8137	75.3161	25
19	2.9792	163.577	157.2152	138.7118	109.7612	73.0187	25
20	3.136	157.2633	151.1904	133.5279	105.895	70.8274	25
21	3.2928	151.2364	145.4396	128.5806	102.206	68.737	25
22	3.4496	145.4835	139.9505	123.859	98.6858	66.7424	25
23	3.6064	139.9925	134.7113	119.3527	95.3265	64.8391	25
24	3.7632	134.7514	129.7108	115.0518	92.1205	63.0228	25
25	3.92	129.7491	124.9382	110.9471	89.0608	61.2894	25
26	4.0768	124.9747	120.383	107.0294	86.1406	59.6352	25
27	4.2336	120.4178	116.0353	103.2904	83.3536	58.0563	25
28	4.3904	116.0686	111.8859	99.7218	80.6937	56.5495	25
29	4.5472	111.9177	107.9255	96.3159	78.1551	55.1114	25
30	4.704	107.9559	104.1457	93.0652	75.7322	53.7389	25
31	4.8608	104.1746	100.5382	89.9627	73.4197	52.4289	25
32	5.0176	100.5658	97.095	87.0016	71.2127	51.1787	25

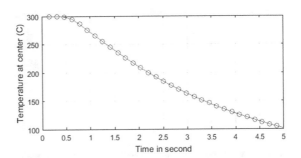

Fig. 3.30 Temperature at Center of Plate with respect to Time

Fig. 3.31 Temperature
Gradient at Center of Plate
with respect to Time

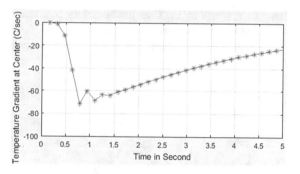

3.6 CAE Implementation—SolidWorks Simulation

Numerous commercial and open-source computer-aided engineering packages have
been developed based on the aforementioned FEA theory. In the following section,
the Solidworks Simulation is used as an example to illustrate the capabilities of
commercial CAE tools for pre-processing, processing, and post-processing of FEA,
which are discussed in Sect. 3.4. Figure 3.32 shows the Graphic User Interfaces and
main functional modules for FEA modeling.

3.6.1 Computational Domain

FEA aims to investigate the system response in a continuous domain subjected to
boundary conditions. A continuous domain can be in *one-dimensional* (1-D), *two-
dimensional* (2-D), or *three-dimensional* (3-D). FEA begins with the construction of
solid models as *a computational domain*. A solid model with the geometric information of vertices, edges, and boundary surfaces is sufficient to FEA; however, it
is desirable to use an advanced solid model that includes parametrized sketches,
dimensions, and features. A parametric model simplifies the changes and supports
the system integration. Modern CAE systems are integrated with their solid modeling
tools to create native solid models; in addition, CAE tools can use solid models from
other computer-aided platforms directly. For example, SolidWorks Simulation is
integrated in the solid modeling environment, and allow to import and export solid
models in over 30 formats of solid models as shown in Fig. 3.33.

3.6.2 Materials Library

In an FEA model, material properties are defined in a computational domain. Material
properties are classified into five groups as (1) *physical properties* such as density

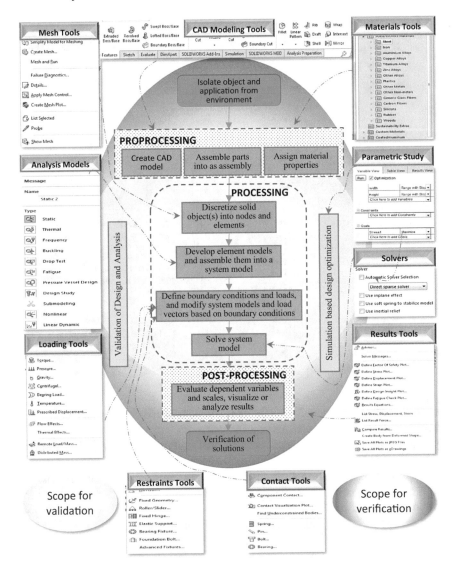

Fig. 3.32 Solidworks Simulation for computer-aided engineering (CAE)

and melting temperature, (2) *mechanical properties* such as the modulus of elasticity, Poisson's ratio, yield strength, and hardness, (3) *thermal properties* such as thermal conductivity, specific heat, and the coefficient of thermal expansion, (4) *electric properties* such as resistivity and di-electricity, and (5) *acoustic properties* such as compression wave velocity, and shear wave velocity, and bar velocity. Different analysis types require the information of different material properties. For example, the modulus of elasticity, the Poison's ratio, yield strength, and the density must be

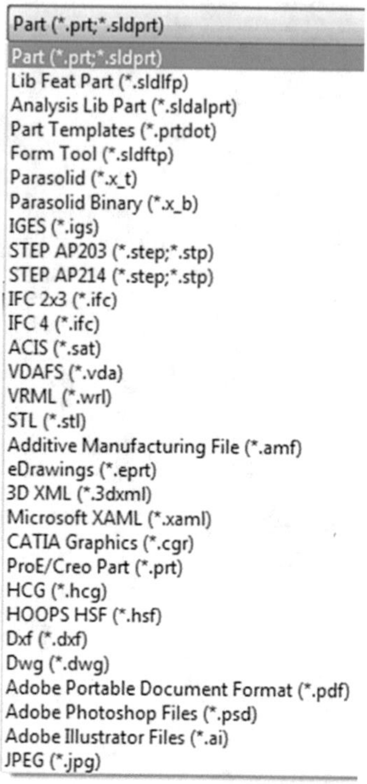

Fig. 3.33 Compatible file
formats in Solidworks

Part (*.prt;*.sldprt)
Part (*.prt;*.sldprt)
Lib Feat Part (*.sldlfp)
Analysis Lib Part (*.sldalprt)
Part Templates (*.prtdot)
Form Tool (*.sldftp)
Parasolid (*.x_t)
Parasolid Binary (*.x_b)
IGES (*.igs)
STEP AP203 (*.step;*.stp)
STEP AP214 (*.step;*.stp)
IFC 2x3 (*.ifc)
IFC 4 (*.ifc)
ACIS (*.sat)
VDAFS (*.vda)
VRML (*.wrl)
STL (*.stl)
Additive Manufacturing File (*.amf)
eDrawings (*.eprt)
3D XML (*.3dxml)
Microsoft XAML (*.xaml)
CATIA Graphics (*.cgr)
ProE/Creo Part (*.prt)
HCG (*.hcg)
HOOPS HSF (*.hsf)
Dxf (*.dxf)
Dwg (*.dwg)
Adobe Portable Document Format (*.pdf)
Adobe Photoshop Files (*.psd)
Adobe Illustrator Files (*.ai)
JPEG (*.jpg)

defined in a static and modal analysis. The specific heat and the thermal conductivity have to be defined in heat transferring analysis.

Engineering systems mostly use conventional materials such as metals and alloys, plastics, and construction materials. The properties of those materials are well studied and documented. A CAE tool usually provides *a material library* for conventional industrial materials; it also provides the interface and template for users to define custom materials for solids.

In the SolidWorks Simulation, the material library has the levels of *Library*, *Category*, and *Materials*, and custom materials are defined in a *Custom Material Library* as shown in Fig. 3.34. To create a new material, the designer begins with the construction of a custom material library, follows with the creation of a new material category in this custom library, and finally, defines a new material in the new custom category.

3.6.3 Meshing Tools

The continuous domain in an original engineering problem has infinite degrees of freedom (DoF). The continuous domain is discretized as a discrete FEA model

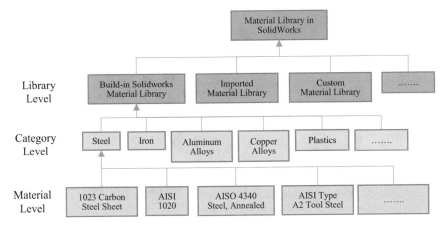

Fig. 3.34 Structure of material library in Solidworks Simulation

with a finite number of *nodes* and *elements* by *meshing*. The SolidWorks Simulation supports *automatic meshing* that is capable of (1) estimating the element sizes based on the volume, surface area, and other geometric attributes of a computational domain, and (2) generating elements and nodes based on element sizes and mesh controls. The scale of a system model increases with an increase of elements and nodes, and the system solution can be generally improved when the numbers of elements and nodes are increased. To reduce the approximation errors at some areas of interest, *local mesh controls* can be applied to refine element sizes on the selected features such as edges, faces, and solids.

An CAE tool supports many element types for different analyses. Figure 3.35 shows three element types with different degrees of freedom. For a *bulky* object, 3D solid elements are suitable. For *thin* objects, 2D shell elements can be used to simplify the FEA model. For sparse structures built from trusses and beams, 1D truss or beam elements are appropriate.

When an assembly model is analyzed, manual intervenes are often required to (1) ensure there is no interference of solids, and (2) define mesh controls to achieve

(a). 1-D elements   (b). 2-D elements   (c). 3-D elements

Fig. 3.35 1D, 2D and 3D elements

Fig. 3.36 Detecting and removing interferences in an assembly model

compatible meshes when they are needed. *Interface detection* in the SolidWorks can be used to identify and eliminate an interference of two solids. Figure 3.36 shows the interface of the interface detection tool to detect the interferences in an object group. When an interference is detected, the assembly model must be revised to eliminate the mutual intrusions of solids.

As assembly model involves in some boundary surfaces where the parts make contacts with each other. If no relative motion is allowed on a boundary surface, *a bonded contact* can be defined for this boundary surface. A bonded contact means that the nodes on the contact surface of the respective parts have the same displacements in system modeling. The mesh at a contact surface can be either *a compatible mesh* or *an incompatible mesh* as shown in Fig. 3.37. In a compatible mesh, the nodes at the contacts are shared by two solids; in an incompatible mesh, the nodes on two solids do not have one-to-one correspondences. Since nodes in an incompatible mesh

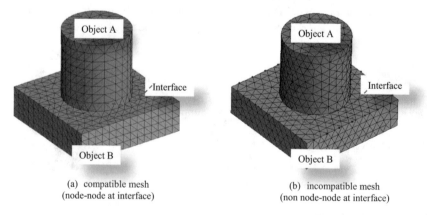

Fig. 3.37 Nodes in compatible and incompatible meshes

are restrained by displacement equations, errors in numerical calculation may lead to an artificial issue of the stress concentration at contacts. Therefore, a compatible mesh usually achieves better accuracy than an incompatible mesh. For an assembly model, designers should refine the meshing parameters to obtain a compatible mesh whenever it is feasible.

For a complex model, it is common that an automatic meshing process is not successful on the first try. When the meshing process fails, the *Failure Diagnostics* can be utilized to diagnose the problems, and an iterative meshing process can be used to modify element sizes, apply mesh controls on the features of interest, and change the settings at boundary surfaces until the mesh in the whole computational domain is successfully generated.

To improve the solution, the *automatic mesh refinement* tool can be used. It supports the mesh refinement in two ways, i.e., *h-adaptive meshing* and *p-adaptive meshing*. H-adaptive meshing is to refine the mesh by increasing the number of elements and reducing element sizes, and p-adaptive meshing is to use the high-order elements without increasing the number of elements. Both of h-adaptive and p-adaptive methods can be performed to meet the expected meshing accuracy automatically.

3.6.4 Analysis Types

The solution to an FEA model is developed for a given mathematic model; therefore, two FEA models would be different if their governing equations are different. By defining *the analysis type* of an FEA model, the SolidWorks Simulation provides the solutions to a variety of engineering problems. The analysis types supported by the Solidworks Simulation are listed in Table 3.10.

3.6.5 Boundary Conditions

To find a particular solution to differential equations, boundary conditions (BCs) must be given in an FEA model. Taking an example of static analysis, BCs are classified into *fixtures* and *loads*. The *PropertyManager* in the SolidWorks Simulation provides an interface to define BCs on vertices, edges, faces, features, and bodies. *Restrained* DoF can be *zero* or a non-zero fixed displacement. Note that BCs are essential to some analysis types such as static analysis, modal analysis, and dynamic analysis. Table 3.11 provides a list of common restraints in the SolidWorks Simulation.

Table 3.12 gives the options to define a load for a static analysis in the SolidWorks Simulation.

Table 3.10 Analysis types supported by SolidWorks Simulation

Analysis Type	Description
Static analysis	Static analysis is used to analyze stress, strain, and deflection of solids subjected to static loads. The materials are assumed to behave in their elastic ranges. A static failure occurs when the stress exceeds the strength of materials. Static analysis can not represent the behaviors of solids when the stress exceeds the material strength.
Nonlinear Analysis	Nonlinear analysis mainly concerns the nonlinearities when 1) both of elastic and plastic deformations occur to solids and 2) the material properties are nonlinear. The first type of nonlinearity occurs when a solid involves in plastic deformation; which is also referred to large displacement. The second type of nonlinearity is represented by a nonlinear stress-strain curve.
Modal Analysis	Modal analysis is to characterize an engineering system by determining the natural frequencies of system. With a nature frequency, an external excitation may lead to amplified system responses. Modal analysis generates 1) a list of natural frequencies and 2) the mode shapes corresponding to these frequencies.
Dynamic Analysis	Dynamic analysis takes into consideration of the dynamic loads such as impact forces or fluctuated loads. *The modal superposition method* can be used to analyze the system responses; the overall system response is obtained by adding the responses to individual loads. The types of dynamic loads in the Solidworks include 1) time-dependent acceleration or load, 2) a fluctuated load or acceleration, and (3) non-deterministic inputs such as random vibration expressed by a power spectrum density (PSD) curve.
Thermal Analysis	Thermal analysis is to analyze the heat storage, conduction, radiation, and convection in a computation domain. The solution to a heat transfer problem was developed based on the energy conservation.
Flow Simulation	Flow simulation aims to analyze an air and fluid flow that passes around or through solid objects. Flow simulation can be integrated with the thermal analysis, so that the heat transfer behaviors such as conduction, convection and radiation between flow and solids can be analyzed simultaneously.
Fatigue Analysis	Fatigue analysis is used to analyze the fatigue damage over time when the solids are subjected to dynamic loads; a fatigue analysis is built upon a static analysis in which the stress distribution of the solids is evaluated based on given nominal loads.
	Fatigue analysis evaluates an accumulated damage on solids caused by periodical loads. The damage increases with the number of loading cycles. The strength of material is characterized by the S-N curve for the relation of fatigue strength and the number of cycles. Once the dynamic loads are provided, the fatigue analysis uses the S-N curve to predict the fatigue life and the safety factor of solids subjected to given dynamic loads.
Drop Test	Drop test is to calculate stresses, strains, and deformations as well as changes over a transient time when an object experiences an impact force. In the drop test, defining the materials as *elasto-plastic* will allow the CAE system to take into account of energy lost in the dynamic simulation.

3.6.6 Solvers to FEA Models

In numerical simulation, an engineering problem is formulated as an FEA model, and the solution to an FEA model is the state variables that satisfy the mathematic equations and constraints in the model. As shown in Fig. 3.38, mathematic equations can be classified from the perspectives of *probability*, *linearity*, *time-dependence*, *explicitness*, and *continuity*.

- **Deterministic versus probabilistic**: the design variables in a deterministic model are uniquely determined without uncertainty. The deterministic model leads to certain results of design variables when the initial conditions are given. A

Table 3.11 BCs for static analysis

Restraints	Description
Fixed Geometry	A fixed geometry specifies the restrained DoF based on element types. Three translational displacements are fixed for a solid or truss element; both of translational and rotational displacements are fixed for a shell or beam element.
Immovable	An immovable geometry fixes all of translational motions for any element type. An immovable BC can be applied to nodes, edges, faces and bodies. For a solid element, an immovable geometry is equivalent to a fixed geometry.
Roller/Sliding	A roller/sliding condition refers to the case in which nodes are free to move over a contact face freely; the contact face is allowed to be shrunk or expanded subjected to the loading conditions; however, the displacement in the normal direction of the plane is restrained.
Fixed Hinge	A fixed hinge condition is defined for a cylindrical surface where the nodes are free to rotate about its rotational axis.
Symmetry	When the geometry and BCs are both symmetric about a reference plane, the computation domain can be simplified as one portion of the whole model. The symmetry condition is used to define the constraints over the reference plane. On a reference plane for the symmetry of a solid mesh, the translation normal to the reference plane is restrained. On a reference plane for the symmetry of a shell mesh, the translation normal to the reference plane and two rotations are restrained.

Table 3.12 Load types in static analysis

Loads	Description
Pressure	Pressure is a line or surface load over a unit length or area. Pressure can be uniform or varying along an edge or over a surface.
Force	Force is a generic term; it can be in forms of torque, moment, or force. Force is defined as a net load, which is a uniformly distributed over the selected entity such as faces, edges, and vertices.
Gravity	Gravity is a body load, which is represented as a linear acceleration of masses. Gravity load has to be defined when the self-weights affect the performance of an engineering system.
Centrifugal	Centrifugal is another type of body load when an object experiences a constant rotation. The CAE tool calculates the centrifugal load based on the specified angular velocity and the mass of materials.
Remote loads and restraints	To analyze an assembly model, insignificant objects should be excluded from the computational domain, and the loads and restraints on those objects should be moved to the modelled objects. The equivalent loads, constraints, and masses are defined as *remote loads*, *remote displacements*, and *remote masses*.

probabilistic model involves in the uncertainties where the design variables are distributed around their mean values with certain probability.

- **Linear versus nonlinear**: in a linear model, all of the relations among design variables are linear; otherwise, it is a nonlinear model. The common approach to solve a nonlinear model is the linearization, in which a nonlinear model is linearized into a set of linear models.

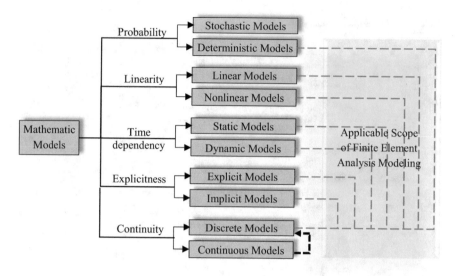

Fig. 3.38 Types of mathematic models

- **Static versus dynamic**: a static model is steady in the sense that the corresponding engineering system is in equilibrium state, and the design variables are time-invariants, and a dynamic model treats the time as another dimension of design problems; it takes into account of time-dependent changes of design variables.
- **Explicit versus implicit**: an explicit model can be solved in a series of the steps explicitly while an implicit model have to be solved iteratively.
- **Discrete versus continuous**: the computational domain in a discrete model consists of discrete elements (for example, truss elements in a truss structure) and that of a continuous model corresponding to a continuous domain.

As shown in Fig. 3.38, FEA modelling can be applied to solve all types of the above mathematical models except a probabilistic model.

The solutions to three types of differential equations have been discussed in Sect. 3.5.6 in detail, and the corresponding engineering problems are called as *equilibrium problems*, *eigenvalue problems*, and *propagation problems*, respectively. An equilibrium problem concerns the system states subjected to static, quasi-static, or repetitive loads. An eigenvalue problem is an extension of an equilibrium problem whose solutions are characterized by identifying a set of unique system configurations such as resonances and bulking. A propagation problem concerns the changes of state variables over time. The Solidworks Simulation provides four generic solvers to the above three types of engineering problems: *Auto*, *FFEPlus*, *Direct Sparse*, and *Large Problem Direct Sparse*.

Auto solver authorizes the software to select an appropriate solver based on the statistics of an FEA model automatically.

Fast Finite Elements (FFEPlus) uses an implicit method to find the solution iteratively. It is very effective when an FEA model involves in large DoF (typically, a model over 100, 000 DoF). However, FFEPLus may be ineffective when (1) the mesh is incompatible and a local bonded contact is not covered by a global bonded contact; (2) a modal analysis which involves in external forces or gravity; (3) a linear dynamic study which involves in base excitation; (4) a computational domain which involves in varying moduli of elasticity; (5) a nonlinear analysis.

Direct Sparse uses an analytic method to solve a system model. '*Sparse*' refers to the sparsity (zeroes) of a matrix; it shows the relations of design variables and loads. Direct Sparse is applicable to an FEA model with a small number of DoF and nonlinear analysis with better accuracy. However, it requires large memory; therefore, the limits of DoFs for a linear analysis, nonlinear analysis, and heat transfer problem are 100,000, 50,000, and 500,000 DoF, respectively.

Large Problem Direct Sparse (LPDS) is an enhanced version of the Direct Sparse solver fora large FEA models. Intensive computation should be handled by parallel computation. A LPDS solver is used when the Direct Sparse solver is disabled due to the limit of random access memory (RAM). LPDS should be the last resolution to an FEA model.

3.6.7 Post-processing

Numerical simulation to an FEA model usually generates a large amount of data; the post-processing tool helps users to classify, visualize, and export simulation data. The post-processing in the Solidwork Simulation allows users to (1) visualize the distributions and contours of state variables, (2) animate system responses from the perspectives of deformations, vibrational modes, and contact behaviours and others, (3) view trajectories, traces, paths in transit or dynamic model, (4) define sectional views to look into state variables internally, and (5) use a probe tool to examine a state variable or any dependent variable at selected points, edges, faces, or components.

3.6.8 Design Optimization

FEA modeling supports the simulation-based optimization. Any system optimization requires analyzing and comparing different system solutions, and FEA modeling is an ideal and generic tool to analyze system solutions in system optimization (Bi and Zhang 2001). Many CAE tools, such as the SolidWorks Simulation, provide *Design Study* for simulation-based optimization. A *Design Study* is defined to (1) discretize the design variables of interest, (2) define a set of design options by combining a set of design variables, (3) specify the optimization criteria based on the simulation data, (4) run the simulations on all of design options, and (5) compare the performances of design options and finalize the design option with the best performance.

3.7 CAE Applications

This section discusses the applications of CAE in different types of engineering problems.

3.7.1 Structural Analysis

A mechanical system responds to different types of loads such as forces, pressures, and heats. In engineering design, *a structural analysis* serves for multiple purposes such as (1) modeling and analyzing the system response to given loads, (2) evaluating stress distributions to identify critical areas subjected to the given loading conditions, and (3) determining if an engineering design is safe by comparing the maximum stress in the computational domain and the strength of material.

Figure 3.39 describes an engineering problem in structural analysis. The computational domain is the solid with a finite volume. The loads applied in the solid are classified as (1) *volume force* such as the weight by gravity, (2) *surface force* such as a drag force by surrounding airflow, and (3) *concentrated load* such as a point load on body. A structural analysis is for an equilibrium problem where the solution at any position is affected by those of all other positions in solids. At a position with an infinitesimal volume, the state variable is the stress, which is described as

$$\boldsymbol{\sigma} = \begin{bmatrix} \sigma_x & \sigma_y & \sigma_z & \tau_{xy} & \tau_{xz} & \tau_{yz} \end{bmatrix}^T \qquad (3.126)$$

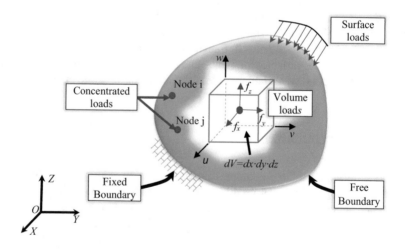

Fig. 3.39 Engineering problem in structural analysis

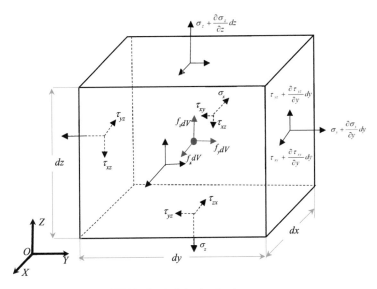

Fig. 3.40 Free body diagram (FBD) of an infinitesimal volume

where σ_x, σ_y, and σ_z are normal stresses along X, Y, and Z; τ_{xy}, τ_{xz}, τ_{yz} are the shear stresses over YZ, XZ, and XY planes, respectively.

Figure 3.40 shows the free-body diagram (FBD) of an infinitesimal volume ($dx \times dy \times dz$). The force-balancing equation for the solid can be defined in any of X-, Y-, and Z-axis.

Taking an example of the force balancing equation along Y-axis, we have

$$\left.\begin{aligned}\sum F_y = \left(\sigma_y + \frac{\partial \sigma_y}{\partial y}dy\right)dxdz - \sigma_y dxdz + \left(\tau_{xy} + \frac{\partial \tau_{xy}}{\partial x}dx\right)dydz - \tau_{xy}dxdz \\ + \left(\tau_{zy} + \frac{\partial \tau_{zy}}{\partial z}dz\right)dxdy - \tau_{zy}dxdy - f_y dxdydz = 0\end{aligned}\right\} \tag{3.127}$$

Equation (3.127) can be simplified as

$$\frac{\partial \tau_{xy}}{\partial x} + \frac{\partial \sigma_y}{\partial y} + \frac{\partial \tau_{yz}}{\partial z} + f_y = 0 \tag{3.128}$$

Repeating the process for the force balancing equations along X- and Z-axes gets

$$\left.\begin{aligned}\frac{\partial \sigma_x}{\partial x} + \frac{\partial \tau_{xy}}{\partial y} + \frac{\partial \tau_{xz}}{\partial z} + f_x = 0 \\ \frac{\partial \tau_{xy}}{\partial x} + \frac{\partial \sigma_y}{\partial y} + \frac{\partial \tau_{yz}}{\partial z} + f_y = 0 \\ \frac{\partial \tau_{xz}}{\partial z} + \frac{\partial \tau_{yz}}{\partial y} + \frac{\partial \sigma_z}{\partial z} + f_z = 0\end{aligned}\right\} \tag{3.129}$$

To model the system response in terms of the deformation of the solid, the strain at the infinitesimal volume is described as $\boldsymbol{\varepsilon} = \begin{bmatrix} \varepsilon_x & \varepsilon_y & \varepsilon_z & \gamma_{xy} & \gamma_{zy} & \gamma_{xz} \end{bmatrix}^T$ and the constitutive relations to the stress state is defined as

$$
\left.
\begin{aligned}
\varepsilon_x &= \tfrac{1}{2}\left(\sigma_x - v\left(\sigma_y + \sigma_z\right)\right) \\
\varepsilon_y &= \tfrac{1}{2}\left(\sigma_y - v\left(\sigma_y + \sigma_z\right)\right) \\
\varepsilon_z &= \tfrac{1}{2}\left(\sigma_z - v\left(\sigma_x + \sigma_y\right)\right) \\
\gamma_{xy} &= \tfrac{\tau_{xy}}{G}, \gamma_{yz} = \tfrac{\tau_{yz}}{G}, \gamma_{xz} = \tfrac{\tau_{xz}}{G}
\end{aligned}
\right\}
\tag{3.130}
$$

where E is the Young's modulus, v is the Poisson's ratio, and G is the shear modulus, which relates to the Young's modulus as

$$
G = \frac{E}{2(1 + v)}
\tag{3.131}
$$

The stress and strain relation in Eq. (3.130) can be converted to calculate stress state based on given strain state as

$$
\{\sigma\} = [D]\{\varepsilon\}
\tag{3.132}
$$

where $[D]$ is the matrix for the stress and strain relation in the 3D space,

$$
[D] = \frac{E}{(1 + v)(1 - 2v)}
\begin{bmatrix}
1 - v & v & v & 0 & 0 & 0 \\
v & 1 - v & v & 0 & 0 & 0 \\
v & 0 & 1 - v & 0 & 0 & 0 \\
0 & 0 & 0 & \tfrac{1}{2} - v & 0 & 0 \\
0 & 0 & 0 & 0 & \tfrac{1}{2} - v & 0 \\
0 & 0 & 0 & 0 & 0 & \tfrac{1}{2} - v
\end{bmatrix}
\tag{3.133}
$$

Note the strain state $\{\boldsymbol{\varepsilon}\}$ can be determined based on the displacement $\{\boldsymbol{u}\} = \begin{Bmatrix} u_x, u_y, u_z \end{Bmatrix}^T$ at position (x, y, z) as

$$
\{\boldsymbol{\varepsilon}\} =
\begin{Bmatrix}
\varepsilon_x \\
\varepsilon_y \\
\varepsilon_z \\
\gamma_{xy} \\
\gamma_{yz} \\
\gamma_{xz}
\end{Bmatrix}
=
\begin{Bmatrix}
\frac{\partial u_x}{\partial x} \\
\frac{\partial u_y}{\partial y} \\
\frac{\partial u_z}{\partial z} \\
\frac{1}{2}\left(\frac{\partial u_x}{\partial y} + \frac{\partial u_y}{\partial x}\right) \\
\frac{1}{2}\left(\frac{\partial u_y}{\partial z} + \frac{\partial u_z}{\partial y}\right) \\
\frac{1}{2}\left(\frac{\partial u_x}{\partial z} + \frac{\partial u_z}{\partial x}\right)
\end{Bmatrix}
\tag{3.134}
$$

Let $\{U\}$ be the vector of state variables, $\{S(x, y, z)\}$ be the vector of the shape functions in an element, the displacement $\{\boldsymbol{u}\}$ in the position (x, y, z) can be interpolated as

$${u} = [S(x, y, z)]{U}$$ (3.135)

where the size of $[S(x, y, z)]$ depends on (1) the number of nodes and (2) the degrees on freedom on each node.

Substituting Eq. (3.135) into Eq. (3.14) obtains the relation of the strain state ${\varepsilon}$ and the vector of state variables ${U}$ as

$${\varepsilon} = [B]{U}$$ (3.136)

where $[B]$ is the matrix for the relation of strain state and state variables in an element of the solid.

In structural analysis, *the principle of the minimized potential energy* can be used to obtain element models about the displacements on nodes. Assume that external force ${F}$ is applied on nodes, and the potential energy of an element can be evaluated as

$$\Pi = \Lambda - \sum_{i=1}^{n} F_i U_i$$ (3.137)

where Π is the potential energy, Λ is the strain energy, n is DoF of the element, F_i is an external load applied on the *i-th* DoF, and U_i is the displacement on the *i-th* DoF.

Example 3.9 A wall hook has a base size of $30 \times 30 \times 5$ mm. The hook as a diameter of $\phi 5$ mm, and its center is offset from the top surface of base by 7.5 mm. The wall hook is printed with ABS plastics, and the yield strength is given as $S_y = 35$ MPa. Assume the safety factor is $n_d = 2.0$, determine the maximum load the hook can carry without a failure.

Solution. Using the SolidWorks Simulation, the wall hook is modeled as shown in Fig. 3.41a. The corresponding FEA model is defined in Fig. 3.41b; it is assumed that the hook is fixed on wall (e.g., the base plate), and the nominal force $F_{nominal} = 10$ N is applied on hook. The simulation of the FEA model results in the stress and displacement distribution, which are shown in Fig. 3.42a, b, respectively.

The maximum von Mises stress σ_{max} is 6.585 MPa subjected to a nominal force $F_{nominal} = 10$ N. Since the FEA model is a linear model, the stress is proportional to the load, and the maximum allowable load for a safety factor of $n_d = 2.0$ can be determined as

$$F_{allowable} = \frac{F_{nominal}}{n_d} \frac{S_y}{\sigma_{max}} = \left(\frac{10}{2}\right)\left(\frac{35 10^6}{6.585 10^6}\right) = 26.58(N)$$

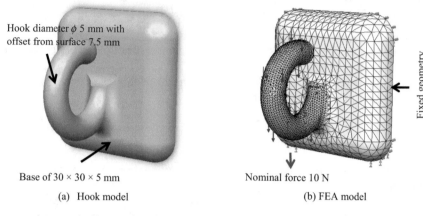

(a) Hook model

(b) FEA model

Fig. 3.41   FEA model of wall hook in Example 3.9

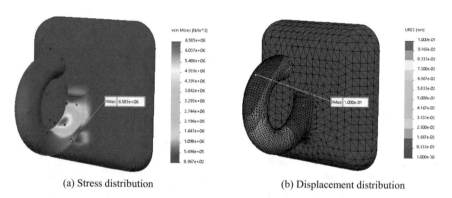

(a) Stress distribution

(b) Displacement distribution

Fig. 3.42   Simulation result of the FEA model in Example 3.9

3.7.2   Modal Analysis

Modal analysis is to address an eigenvalue problem in which a set of the natural frequencies are determined in an engineering system. As shown in Fig. 3.43, modal analysis is critical to many engineering systems subjected to dynamic loads; since the frequencies of external excitations should be away from natural frequencies except that the systems such as vibrators utilize self-vibration to implement the system functions. The mathematic models and solutions of eigenvalue problems have been discussed in Sect. 3.5.6. Here, the CAE application in modal analysis is discussed.

Example 3.10 A manufacturer produces an integrated ventilating system for customers, and Fig. 3.44 shows one of its products with two blowers. In mounting

(a). Wind turbine   (b). Construction   (c). machine tool

Fig. 3.43 Modal analysis for example engineering systems

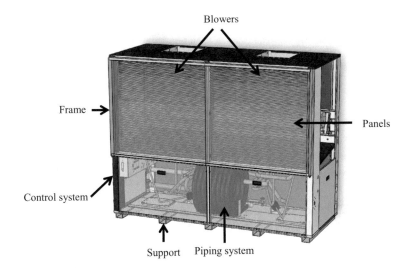

Fig. 3.44 Example of integrated ventilating system with two blowers

the blowers, a misalignment of the transmission plane of a blower will cause an excitation to the system. Conduct a modal analysis to confirm that two blowers operate at the frequencies that are below to the natural frequencies of the integrated system.

Solution. The SolidWorks Simulation is used to create an FEA model for *frequencies analysis* of the ventilating system. *Firstly*, the assembly model is imported as shown in Fig. 3.45a, and it is simplified to (1) suppress insignificant components such as the control system, the piping system, and panels, (2) the interfaces at the contacts of two components are checked to eliminate possible interferences for appropriate contact conditions, (3) the material properties are defined for all of the inclusive components. *Secondly*, a modal analysis is defined, the conditions of contact surfaces are specified, the support components are defined as fixed geometries, the applied controls are defined for some critical parts or areas, and the mesh is created for the assembly model as shown in Fig. 3.45b. *Finally*, the number of the natural frequencies of

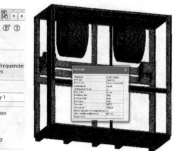

(a) Simplified assembly model   (b) Meshed model

Fig. 3.45 Example of integrated ventilating system with two blowers

Table 3.13 The first six natural frequencies of the integrated ventilating system

Mode No	Frequency (Rad/s)	Frequency (Hertz)	Period (s)
1	73.049	11.626	0.086013
2	91.969	14.637	0.068318
3	136.22	21.679	0.046127
4	144.25	22.958	0.043557
5	207.12	32.964	0.030336
6	256.1	40.76	0.024534

interest is specified, the FEA model is solved, the list of calculated natural frequencies is exported in Table 3.13, and the corresponding modal shapes are shown in Fig. 3.46.

Since the excitations are from the blower motors that are operated at the speed of 1750 *resolutions per minute* (RPM). Therefore, the excitations have the frequency of 29.167 (Hz) or 183.25 (rad/s). This operating frequency lies between the fourth (22.958 Hz) and the fifth (32.964 Hz) natural frequencies of the system in Table 3.13. It implies that the mounting misalignment may lead to an amplified vibration in operation and a premature failure of some fastening parts such as brackets and screws.

3.7.3 Heat Transfer Systems

Three primary physical quantities in thermodynamics are *temperature, energy,* and *entropy,* and there four thermodynamic laws that describe their relations:

- **The zeroth law**: when the systems *A* and *B* are in an equilibrium with the third system *C*. System *A* is in an equilibrium with system *B*; the concept of *temperature* is defined to describe a thermal equilibrium.

(a)

(b)

(c)

(d)

(e)

(f)

Fig. 3.46 The first six mode shapes from simulation

- **The first law**: when the energy is transferred into or out of a system in the form of *work*, *heat*, or *matter*, the energy in the system changes based on the conservation of energy. Energy can be neither created nor destroyed, but it can be transferred from one matter to another. Making a perpetual motion machine is impossible.
- **The second law**: in a natural thermodynamic process, the interaction of thermodynamic systems increase the summed entropies of systems.
- **The third law**: the system *entropy* approaches a constant value when the temperature in the system approaches absolute zero. The system entropy at absolute zero is close to zero except for non-crystalline solids; the system entropy equals to the logarithm of the product of the quantum ground states.

Heat transfer refers to an exchange of heat energy, which is governed by the first and second thermodynamic laws. According to the first thermodynamic law, the internal energy of system changes when heat transfer occurs. According to the second thermodynamic law, heat is transferred from the matter with a high temperature to the matter with a low temperature. Analyzing heat transfer behaviors is very important to many engineering systems due to a number of reasons: (1) an engineering system is usually to transfer the motion and/or energy in one format to another or from one component to another. (2) The properties of any materials are closely related to temperature and heat. (3) Many products and manufacturing processes are the implementation of heat transfer processes. Figure 3.47 shows three examples of the engineering systems whose heat transfer behaviors must be analyzed in designing and operating the systems.

Heat can be transferred in one of three basic modes: *conduction, convection,* and *radiation.* In *heat conduction,* the thermal energy is transferred in the materials by the electronic movements and the collisions of microscopic particles such as atoms and molecules. As shown in Fig. 3.48, the Fourier's law can be used to represent a heat transfer behavior as

(a). Heat treatment

(b). Renewable energy

(c). Air conditioning

Fig. 3.47 Example of heat transfer systems

Fig. 3.48 Heat transfer and generation in a continuous domain

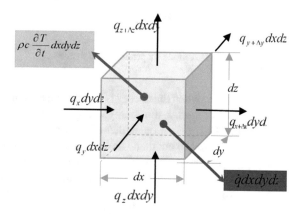

$$
\left.\begin{array}{l}
q_x = \dot{q}_x A_x = -k_x A_x \frac{\partial T}{\partial x} \\
q_y = \dot{q}_y A_y = -k_y A_y \frac{\partial T}{\partial y} \\
q_z = \dot{q}_z A_z = -k_z A_z \frac{\partial T}{\partial z}
\end{array}\right\} \tag{3.138}
$$

where q_x, q_y, and q_z are heat transfer rates along x-, y-, and z-axes;

$\dot{q}_x$, $\dot{q}_y$, and $\dot{q}_z$ are heat fluxes along x-, y-, and z-axes;

k_x, k_y, and k_z are the conductivity coefficients along x-, y-, and z-axes;

$\frac{\partial T}{\partial x}$, $\frac{\partial T}{\partial y}$, and $\frac{\partial T}{\partial z}$ are the temperature gradients along x-, y-, and z-axes;

A_x, A_y, and A_z are the effective areas of conductions perpendicular to x-, $-y$-, and z-axes, respectively.

In *heat convection*, the heat is transferred by the fluid motion over the surface of an object. Heat convection occurs when the surface temperature is different from the surrounding environment, which is governed by the Newton's law as,

$$
q_h = hA(T_s - T_f) \tag{3.139}
$$

where q_h is the rate of heat convection over the surface;

A is the effective area of heat convection;

T_s is the temperature on surface;

T_f is the fluid temperature.

In *heat radiation*, the heat is transferred in the form of electromagnetic waves. The thermal energy is converted into the electromagnetic energy and emitted as the radiation over surfaces. Heat radiation is governed by the Stefan-Boltzmann law as,

$$
q_r = \sigma \varepsilon A(T_s^4 - T_O^4) \tag{3.140}
$$

where σ is the Stefan-Boltzmann constant;

ε is the coefficient of surface emission;

A is the effective area of heat radiation;

T_0 is the absolute temperature of the environment;

T_s is the absolute temperature over the surface;

q_r is the radiant heat flow over the surface.

In an infinitesimal volume ($dxdydz$), the flows of heat transfers are shown in Fig. 3.48. The heat may be conducted along x, y, and z axes. The amount of the thermal energy in the volume is determined by (1) heat conduction along three axes, (2) the temperature change in the volume, and (3) the generated energy. heat for temperature changes of the mass in unit, and (4) the heat generated from any source in unit. The corresponding amount of thermal energy for each heat transfer mode can be calculated as below.

The energy E_{in} that is flowed into the volume by conduction is,

$$E_{in} = q_x + q_y + q_z = -k_x dydz \frac{\partial T}{\partial x} - k_y dxdz \frac{\partial T}{\partial y} - k_z dxdy \frac{\partial T}{\partial z} \qquad (3.141)$$

where q_x, q_y, and q_z are given in Eq. (3.138) and $\frac{\partial T}{\partial x}$, $\frac{\partial T}{\partial y}$, and $\frac{\partial T}{\partial z}$ are the temperature gradients along x, y, and z axes.

The energy E_{out} that is flowed out of the volume by conduction is,

$$E_{out} = q_{x+dx} + q_{y+dy} + q_{z+dz}$$

$$= q_x + q_y + q_z - k_x dxdydz \frac{\partial \left(\frac{\partial T}{\partial x} \right)}{\partial x} - k_y dxdydz \frac{\partial \left(\frac{\partial T}{\partial y} \right)}{\partial y} - k_z dxdydz \frac{\partial \left(\frac{\partial T}{\partial z} \right)}{\partial z}$$

$$(3.142)$$

where $\frac{\partial \left(\frac{\partial T}{\partial x} \right)}{\partial x}, \frac{\partial \left(\frac{\partial T}{\partial y} \right)}{\partial y}$, and $\frac{\partial \left(\frac{\partial T}{\partial z} \right)}{\partial z}$ are the second derivatives of temperature along x, y, z axes, respectively.

The amount of energy E_g generated by the internal source is,

$$E_g = \dot{q} dxdydz \qquad (3.143)$$

where $\dot{q}$ is the rate of heat generated in unit.

The amount of the energy E_s due to the temperature change is,

$$E_s = \rho c \frac{\partial T}{\partial t} dxdydz \qquad (3.144)$$

where ρ and c are the density and specific heat of materials, and $\frac{\partial T}{\partial t}$ is the temperature gradient over time.

According to the first thermodynamic law, the thermal energy in the volume must be conserved; therefore, we have,

$$E_{in} - E_{out} + E_g = E_s \qquad (3.145)$$

Substituting Eqs. (3.141)–(3.144) into Eq. (3.145) yields the governing equations of heat transfer behaviors as,

$$\frac{\partial T}{\partial x} \left(k_x \frac{\partial T}{\partial x} \right) + \frac{\partial T}{\partial y} \left(k_y \frac{\partial T}{\partial y} \right) + \frac{\partial T}{\partial z} \left(k_z \frac{\partial T}{\partial z} \right) - c \frac{\partial T}{\partial t} + \dot{q} = 0 \qquad (3.146)$$

When $\dot{q} = 0$ and $\frac{\partial T}{\partial t} = 0$, Eq. (3.146) becomes a differential equation for an equilibrium problem; when $\neq \frac{T}{t} 0$, Eq. (3.147) becomes a differential equation for a propagating problem. The FEA solutions to both types of differential equations have been discussed in Sect. 3.5.6.

Note that Eq. (3.146) describes the heat transfers within a given volume; however, the heat convection and radiation may occur on the boundary surfaces. Therefore, the heat convection and radiation should be modelled as the boundary conditions. Figure 3.49 shows five types of the boundary conditions in a heat transfer problem: (*a*) a fixed temperature, (*b*) a given heat flux, (*c*) a coefficient for convection, (*d*) the parameters for radiation, and (*e*) an insulated condition when no heat transfer occurs.

Example 3.11 A heatsink model of a central processing unit (CPU) is shown in Fig. 3.50. Assume that the heatsink uses copper materials, the heat transfer coefficient under the forced air is $h = 15$ W/(m^2·K), and the ambient temperature of the operating environment is given as $T_0 = 25$ °C. Find (1) the maximum temperature in CPU

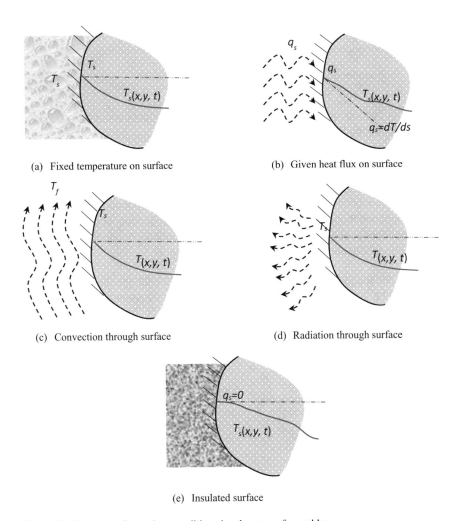

(a) Fixed temperature on surface   (b) Given heat flux on surface

(c) Convection through surface   (d) Radiation through surface

(e) Insulated surface

Fig. 3.49 Five types of boundary conditions in a heat transfer problem

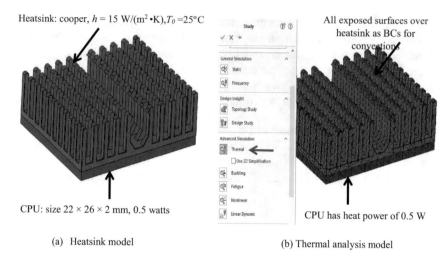

(a) Heatsink model   (b) Thermal analysis model

Fig. 3.50 Heat transfer model in Example 3.11

and (2) the maximum heat power the heatsink can dissipate when the maximum temperature can be below $T_{max} = 60\ °C$.

Solution. The heat transfer model for the heatsink has been defined in Fig. 3.50b. Two boundary conditions are (1) the convection condition on all of the exposed surfaces of heatsink, and (2) the heat power defined in the body of CPU. The materials of CPU is set as silicon. The mesh is created using the default settings. Running the simulation yields the results of Fig. 3.51a, b for the distributions of the temperature and temperature gradients, respectively. It has been found the maximum temperature at the center of CPU is $T_{max} = 30.95\ °C$.

The above thermal analysis model is used to create a design study as shown in Fig. 3.52. The heat power (P) from CPU is set as the simulation variable, and the analysis range is set as $P \in [0.5, 3]$ with a step of 0.25 (W). The result of the design

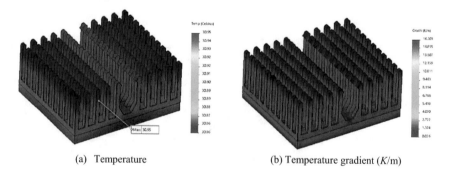

(a) Temperature   (b) Temperature gradient (K/m)

Fig. 3.51 Simulation result of the heat transfer model in Example 3.11

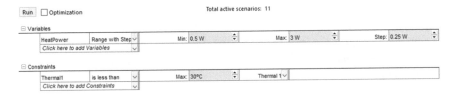

Fig. 3.52 Design study for different level of heat power from CPU

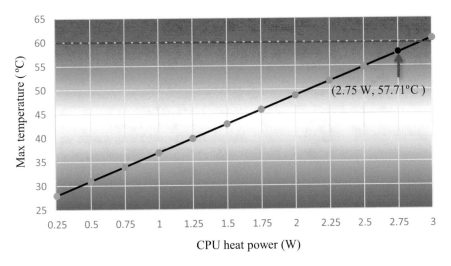

Fig. 3.53 Max temperature (°C) versus heat power (W)

study in Fig. 3.53 shows that the heatsink reach its maximum temperature $T_{max} = 57.71\ °C < 60\ °C$ when the heat power of CPU $P = 2.75\ (W)$.

3.7.4 Transient Heat Transfer Problems

The above thermal analysis is for an equilibrium-engineering problem; the steady solution is found when the input and output heat powers in solids are balanced. However, the changes of state variables at the transient stages are more critical in some manufacturing processes such as heat treatments and injection molding processes. An FEA model for a transient problem requires discretizing the time domain; the solution at each step is approximated as an equilibrium problem whose boundary conditions are updated based on the simulation results at previous steps. In this section, the heat treatment of an aluminum part is used as an example to show how a CAE tool such as SolidWorks Simulation can be used to solve a transient problem.

Example 3.12 The par geometry is shown in Fig. 3.54a; it is made of 6061-T6 aluminum alloy, the initial temperature of the heat treatment process is $T_0 = 532\ °C$. The ambient temperature is $T_a = 30\ °C$ and the coefficient of heat convection is $h = 60\ W/(m^2 \cdot K)$, find out the cycle time of the heat treatment process for the parts to reach a temperature of $T_f = 35\ °C$.

Solution. The SolidWorks Simulation is used to define a transient thermal analysis model shown in Fig. 3.55. The duration of simulation is set as 300 s, and the step for the discretization over the time domain is 10 s. Figure 3.56 shows that the FEA model

Density ρ=2.7x10³ kg/m³
Mass m=0.210391 kg
Volume V=7.79x10⁻⁵ m³
Surface area S=6.51815 x10⁻² m²
Specific Heat C_p=1005.3 W/kg ·K at (30 °C)

(a) part geometry   (b) part properties

Fig. 3.54 Geometry and properties of part in Example 3.12

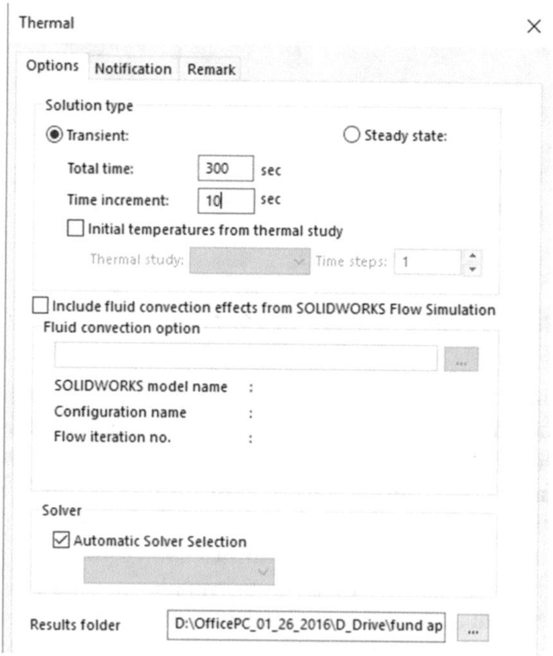

Fig. 3.55 Define a transient thermal analysis in SolidWorks Simulation

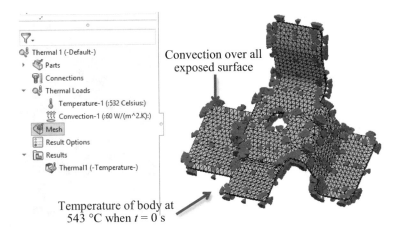

Convection over all
exposed surface

Temperature of body at
543 °C when $t = 0$ s

Fig. 3.56 Boundary conditions (BCs) of transient thermal analysis model

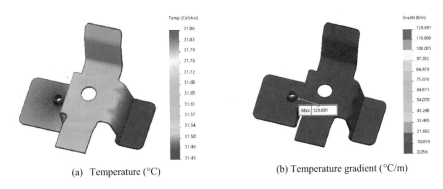

(a) Temperature (°C)

(b) Temperature gradient (°C/m)

Fig. 3.57 The distributions of temperature and temperature gradients at 300 s

includes two boundary conditions: (1) the initial temperature in the solid is $T_0 = 532°$C at $t = 0$ (s) and (2) the convections over all exposed surfaces have $h = 60$ W/(m^2·K). Figure 3.57 shows the distribution of temperature and temperature gradient when $t = 300$ s. Figure 3.57 shows the cooling curves of the part in heat treatment. It has been found that the maximum temperature of the part $T_{max} = 34.77$ °C < 35 °C at the time step of $t = 240$ s. Therefore, the cycle time of the heat treatment process is 240 s (Fig. 3.58).

3.7.5 Fluid Mechanics

Fluid mechanics is a branch of continuous mechanics. Fluid mechanics studies the behaviors of fluid systems subjected to static and dynamics loads. The matter in a

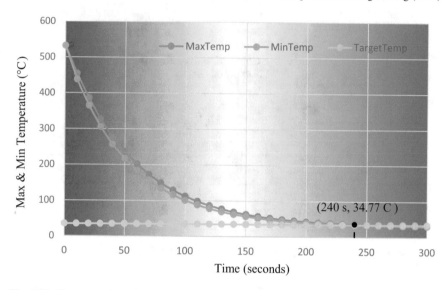

Fig. 3.58 The max, min, and target temperatures over time in heat treatment

fluid system should be modelled as continuous masses rather than discrete particles; since the matter such as fluid or gas does not hold a shape like a solid (Bar-Meir 2008). In addition, the density of the matter may vary with time and position. Figure 3.59 shows continuous mechanics includes the branches of *solid mechanics, fluid structure interaction*, and fluid mechanics. Fluid mechanics can be further classified into *fluid statics*, and *fluid dynamics*. Fluid systems can also be classified from the perspective of applications such as internal/external flows, single/multi-phase flows, and laminar/turbulent flows, pipe flows, and groundwater flows. In this section, FEA modelling of fluid systems is discussed.

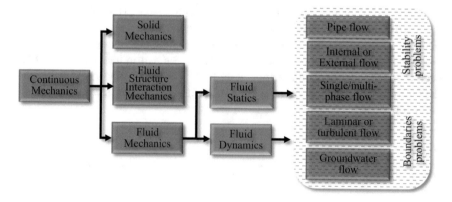

Fig. 3.59 Examples of engineering problems in fluid mechanics

Solid and fluid respond shear stresses differently. A fluid deforms continuously subjected to shear stresses, and it does not return to its original state after the deformation. In contrast, a solid exhibits a fine deformation that does not change with time. A fluid moves or flows subjected to shear stresses. A fluid can be a *liquid* or *gas*.

The earliest study on computer simulation of fluid systems was for potential flows in the 60 s. A *potential flow* was modelled as a velocity field of incompressible fluids; it was later expanded to compressible fluids (Ritchmeyer and Morton 1967). A *fluid flow* transfers *mass*, *momentum* and *energy*, and its motion is governed by the Navier–Stokes equations for mass, momentum and energy conservations. The numerical simulation to the Navier–Stokes equations was implemented in the 70's, and nowadays, CAE has been widely used to analyze fluid systems in heat, ventilation and air-conditioning (HVAC) and manufacturing industry.

3.7.5.1   Constitutive Models

A fluid flow is different from solids in sense that a fluid flow can sustain a hydrostatic stress but not shear stress. Therefore, the behavior of a fluid flow is not represented by the displacement like a solid; it is represented by the flow velocity (v) in a three dimensional space as,

$$v(x, y, z) = u(x, y, z) \cdot \boldsymbol{i} + u(x, y, z) \cdot \boldsymbol{j} + w(x, y, z) \cdot \boldsymbol{k} \qquad (3.147)$$

where (x, y, z) is an arbitrary position in the 3-D space; $v(x, y, z)$ is the vector of the fluid flow velocity; and $u(x, y, z)$, $v(x, y, z)$, and $w(x, y, z)$ are the components of the velocity vector along x-, y- and z- axes, respectively.

With the given velocity, the strain rate $\varepsilon_{ij}(i, j = 1, 2, 3)$ in the fluid flow can be derived as,

$$\left.\begin{matrix} \dot{\varepsilon}_{11} = \frac{\partial u}{\partial x} \\ \dot{\varepsilon}_{22} = \frac{\partial v}{\partial y} \\ \dot{\varepsilon}_{33} = \frac{\partial w}{\partial z} \end{matrix}\right\} \ and \ \left.\begin{matrix} \dot{\varepsilon}_{12} = \frac{1}{2}\left(\frac{\partial u}{\partial y} + \frac{\partial v}{\partial x}\right) \\ \dot{\varepsilon}_{23} = \frac{1}{2}\left(\frac{\partial v}{\partial z} + \frac{\partial w}{\partial y}\right) \\ \dot{\varepsilon}_{13} = \frac{1}{2}\left(\frac{\partial u}{\partial z} + \frac{\partial w}{\partial x}\right) \end{matrix}\right\} \qquad (3.148)$$

where u, v, and w are the components of the velocity vector given in Eq. (3.147).

Equation (3.148) can be further written in the matrix form as,

$$\dot{\varepsilon}^T = \begin{bmatrix} \dot{\varepsilon}_{11} & \dot{\varepsilon}_{22} & \dot{\varepsilon}_{33} & 2\dot{\varepsilon}_{12} & 2\dot{\varepsilon}_{23} & 2\dot{\varepsilon}_{13} \end{bmatrix}^T = [S][v] \qquad (3.149)$$

where $[v] = \begin{bmatrix} u(x, y, z) & v(x, y, z) & z(x, y, z) \end{bmatrix}^T$ is given in Eq. (3.147).

the subscripts 1, 2, and 3 represent x, y, and z axes, respectively, and

$[S]$ is the operator matrix for the strain rate given by Eq. (3.148),

$$[S] = \begin{bmatrix} \frac{\partial}{\partial x} & 0 & 0 \\ 0 & \frac{\partial}{\partial y} & 0 \\ 0 & 0 & \frac{\partial}{\partial z} \\ \frac{1}{2}\frac{\partial}{\partial y} & \frac{1}{2}\frac{\partial}{\partial x} & 0 \\ 0 & \frac{1}{2}\frac{\partial}{\partial z} & \frac{1}{2}\frac{\partial}{\partial y} \\ \frac{1}{2}\frac{\partial}{\partial z} & 0 & \frac{1}{2}\frac{\partial}{\partial x} \end{bmatrix} \tag{3.150}$$

Once the strain rate is obtained, the stress σ_{ij} can be derived accordingly. However, two constants should be defined to calculate the stress from the strain rate. *The first one* is called the deviatoric stress τ_{ij}, which relates to the deviatoric strain rate as,

$$\tau_{ij} = 2\mu\left(\dot{\varepsilon}_{ij} - \delta_{ij}\frac{\dot{\varepsilon}_{kk}}{3}\right) \tag{3.151}$$

where the term $\left(\dot{\varepsilon}_{ij} - \delta_{ij}\frac{\dot{\varepsilon}_{kk}}{3}\right)$ is called the deviatoric strain rate,

δ_{ij} is the Kronecker delta constant,

the subscripts $i, j, k = 1, 2, 3$, and 1, 2, and 3 represent x, y, and z axes, respectively, and

μ is the coefficient for the viscosity.

The second constant is the pressure in a fluid flow, which depends on the strain rate. The pressure is calculated from the volumetric strain rate as,

$$p = -\kappa\dot{\varepsilon}_{ii} + p_0 \tag{3.152}$$

where κ is the coefficient of volumetric viscosity similar to the bulk modulus for an elastic solid; p_0 is the initial pressure.

he constitutive model for the stress–strain relation is derived from Eqs. (3.149)–(3.152) as,

$$\sigma_{ij} = \tau_{ij} - \delta_{ij}p = 2\mu\dot{\varepsilon}_{ij} + \delta_{ij}\left[\left(\kappa - \frac{2}{3}\mu\right)\dot{\varepsilon}_{ii} + p_0\right] \tag{3.153}$$

Equation (3.153) can be further simplified since the volumetric viscosity can be ignorable. Therefore, the constitutive model of the fluid flow is written as,

$$\left.\begin{aligned} \tau_{ij} = 2\mu\left(\dot{\varepsilon}_{ij} - \delta_{ij}\frac{\dot{\varepsilon}_{kk}}{3}\right) = \mu\left[\left(\frac{\partial v_i}{\partial x_j} + \frac{\partial v_j}{\partial x_i}\right) - \delta_{ij}\frac{\partial v_k}{\partial x_k}\right] \\ \sigma_{ij} = \tau_{ij} - \delta_{ij}p \end{aligned}\right\} \tag{3.154}$$

Equation (3.154) of the fluid flow is similar to the constitutive model of a solid. It is worth to note that the characteristic of a fluid flow depends on viscosity (μ)

AIRO Springer Series

Volume 5

Editor-in-Chief

Daniele Vigo, Dipartimento di Ingegneria dell'Energia Elettrica e dell'Informazione "Gugliemo Marconi", Alma Mater Studiorum Università di Bologna, Bologna, Italy

Series Editors

Alessandro Agnetis, Dipartimento di Ingegneria dell'Informazione e Scienze Matematiche, Università degli Studi di Siena, Siena, Italy

Edoardo Amaldi, Dipartimento di Elettronica, Informazione e Bioingegneria (DEIB), Politecnico di Milano, Milan, Italy

Francesca Guerriero, Dipartimento di Ingegneria Meccanica, Energetica e Gestionale (DIMEG), Università della Calabria, Rende, Italy

Stefano Lucidi, Dipartimento di Ingegneria Informatica Automatica e Gestionale "Antonio Ruberti" (DIAG), Università di Roma "La Sapienza", Rome, Italy

Enza Messina, Dipartimento di Informatica Sistemistica e Comunicazione, Università degli Studi di Milano-Bicocca, Milan, Italy

Antonio Sforza, Dipartimento di Ingegneria Elettrica e Tecnologie dell'Informazione, Università degli Studi di Napoli Federico II, Naples, Italy

The AIRO Springer Series focuses on the relevance of operations research (OR) in the scientific world and in real life applications.

The series publishes peer-reviewed only works, such as contributed volumes, lectures notes, and monographs in English language resulting from workshops, conferences, courses, schools, seminars, and research activities carried out by AIRO, Associazione Italiana di Ricerca Operativa - Optimization and Decision Sciences: http://www.airo.org/index.php/it/.

The books in the series will discuss recent results and analyze new trends focusing on the following areas: Optimization and Operation Research, including Continuous, Discrete and Network Optimization, and related industrial and territorial applications. Interdisciplinary contributions, showing a fruitful collaboration of scientists with researchers from other fields to address complex applications, are welcome.

The series is aimed at providing useful reference material to students, academic and industrial researchers at an international level.

Should an author wish to submit a manuscript, please note that this can be done by directly contacting the series Editorial Board, which is in charge of the peer-review process.

More information about this series at http://www.springer.com/series/15947

Claudio Gentile • Giuseppe Stecca • Paolo Ventura
Editors

Graphs and Combinatorial Optimization: from Theory to Applications

CTW2020 Proceedings

ASSOCIAZIONE ITALIANA DI RICERCA OPERATIVA
OPTIMIZATION AND DECISION SCIENCE

Editors
Claudio Gentile
Consiglio Nazionale delle Ricerce
Istituto di Analisi dei Sistemi ed
Informatica "Antonio Ruberti"
Roma, Italy

Giuseppe Stecca
Consiglio Nazionale delle Ricerche
Istituto di Analisi dei Sistemi ed
Informatica "Antonio Ruberti"
Roma, Italy

Paolo Ventura
Consiglio Nazionale delle Ricerche
Istituto di Analisi dei Sistemi ed
Informatica "Antonio Ruberti"
Roma, Italy

ISSN 2523-7047   ISSN 2523-7055   (electronic)
AIRO Springer Series
ISBN 978-3-030-63074-4   ISBN 978-3-030-63072-0   (eBook)
https://doi.org/10.1007/978-3-030-63072-0

This Springer imprint is published by the registered company Springer Nature Switzerland AG.
The registered company address is: Gewerbestrasse 11, 6330 Cham, Switzerland

Preface

The Cologne-Twente Workshop (CTW) on Graphs and Combinatorial Optimization is a workshop series initiated by Ulrich Faigle, around the time he moved from the Twente University to the University of Cologne. After many CTW editions in Twente and Cologne, it was decided that CTWs were mature enough to move about: in 2004, the CTW was organized in Villa Vigoni (Menaggio, Como, Italy) by Francesco Maffioli (Politecnico di Milano) and Leo Liberti (CNRS-LIX). Since then, the CTW visited again Italy for three times, beyond France, Germany, the Netherlands, and Turkey. This edition is the first time that the CTW is organized with the contribution of CNR-IASI members (Claudio Gentile, Giuseppe Stecca, Paolo Ventura, Giovanni Rinaldi, and Fabio Furini) in addition to the members of University of Rome "Tor Vergata" (Andrea Pacifici), Roma Tre University (Gaia Nicosia), and CNRS & LIX Polytechnique Palaiseau (Leo Liberti).

Having been initially set up by discrete applied mathematicians, the CTW still follows the mathematical tradition. In this CTW edition (hereafter, CTW2020), for the first time we adopted two submission tracks: standard papers of at most 12 pages and traditional CTW extended abstracts of at most 4 pages.

This volume collects the standard papers that were submitted to the CTW2020. The papers underwent a standard peer review process performed by a Program Committee consisting of 30 members:[1] 17 CTW steering committee members and

---

[1] Ali Fuat Alkaya (Marmara U., Turkey), Christoph Buchheim (TU Dortmund, Germany), Francesco Carrabs (U. Salerno, Italy), Alberto Ceselli (U. Milano, Italy), Roberto Cordone (U. Milano, Italy), Ekrem Duman (Ozyegin U., Turkey), Yuri Faenza (Columbia U., USA), Bernard Gendron (IRO U. Montreal & CIRRELT, Canada), Claudio Gentile (CNR-IASI, Italy), Johann Hurink (U. Twente, The Netherlands), Ola Jabali (Politecnico di Milano, Italy), Leo Liberti (CNRS & LIX Polytechnique Palaiseau, France), Frauke Liers (FAU Erlangen-Nuremberg, Germany), Bodo Manthey (U. Twente, The Netherlands), Gaia Nicosia (U. Roma Tre, Italy), Tony Nixon (U. Lancaster, UK), Andrea Pacifici (U. Roma Tor Vergata, Italy), Ulrich Pferschy (U. Graz, Austria), Stefan Pickl (U. Bundeswehr München, Germany), Michael Poss (LIRMM U. Montpellier & CNRS, France), Bert Randerath (U. Koeln, Germany), Giovanni Righini (U. Milano, Italy), Heiko Roeglin (U. Bonn, Germany), Oliver Schaudt (RTWH Aachen U., Germany), Rainer Schrader (U. Koeln, Germany), Giuseppe Stecca (CNR-IASI, Italy), Frank Vallentin (U. Koeln, Germany),

13 guest members. PC members came from Italy, Germany, France, the USA, Canada, the Netherlands, the UK, Austria, and Turkey. We received 46 submissions of which we accepted 31 for publication in this volume with a rate of success of 67%.

The chapters of this volume present works on graph theory, discrete mathematics, combinatorial optimization, and operations research methods, with particular emphasis on coloring, graph decomposition, connectivity, distance geometry, mixed-integer programming, machine learning, heuristics, meta-heuristics, math-heuristics, and exact methods. Applications are related to logistics, production planning, energy, telecommunications, healthcare, and circular economy.

The scientific program of the CTW2020 includes the 31 standard papers in this volume, 33 extended abstracts, and two plenary invited talks. As usual for the CTW, extended abstracts were subject to a high acceptance level, allowing also papers presenting preliminary results with a particular accent to works presented by MScs, PhDs, or Postdocs. The traditional CTW extended abstracts will be published on the conference's website http://ctw2020.iasi.cnr.it, where also additional material collected during the conference will be posted.

We thank all the PC members and the subreviewers for the complex work performed to select the papers and to improve their quality considering also a possible second round of revision.

Following the CTW tradition, a special issue of Discrete Applied Mathematics (DAM) dedicated to this workshop and its main topics of interest will be edited.

Not every CTW edition features invited plenary speakers, but this one does. Two very well-known researchers accepted our invitation: Prof. Dan Bienstock (Columbia University) and Prof. Marco Sciandrone (University of Florence). Prof. Dan Bienstock works in many topics of Combinatorial Optimization, Integer and Mixed-Integer Programming, and Network Design. He is the author of many journal and conference papers and of two textbooks: "Electrical Transmission System Cascades and Vulnerability: An Operations Research Viewpoint," ISBN 978-1-611974-15-7, SIAM-MOS Series on Optimization (2015), and "Potential Function Methods for Approximately Solving Linear Programming Problems: Theory and Practice," ISBN 1-4020-7173-6, Kluwer Academic Publishers, Boston (2002). Prof. Marco Sciandrone works in Nonlinear Programming with a particular expertise in Machine Learning, Neural Networks, Multiobjective Optimization, and Nonlinear Approximation of Discrete Variables.

Finally, we thank AIRO for hosting this volume in its AIRO-Springer series. We thank both AIRO and CNR-IASI for their support to the realization of the conference.

Paolo Ventura (CNR-IASI, Italy), Maria Teresa Vespucci (U. Bergamo, Italy), and Angelika Wiegele (Alpen-Adria U. Klagenfurt, Austria).

in Eq. (3.154). If the viscosity is constant, the fluid flow is a *Newtonian fluid flow*; otherwise, the fluid flow is a non-Newtonian fluid flow whose viscosity varies with the strain rate.

3.7.5.2 Mass Conservation

The mass of a matter can be not created or vanished. Since a fluid does not hold a shape and it flows under an external force, it is convenient to evaluate the mass of the fluid in terms of the flow rate passing a given section of the container. As shown in Fig. 3.60, the mass rates of the fluid flow at two sections of an enclosed volume should be the same,

$$\dot{m}_1 = \dot{m}_2 = \rho_1 v_1 A_1 = \rho_2 v_2 A_2 \tag{3.155}$$

where $\dot{m}_1$ and $\dot{m}_2$ are the scalar mass rates; ρ_1 and ρ_2 are the densities, v_1 and v_2 are the scalar velocities, and A_1 and A_2 are the sectional areas.

In a 3-D space, the mass flow is defined as a vector shown in Fig. 3.61. An infinitesimal volume with the size of $dx \times dy \times dz$ is considered, the velocity of flow is represented by (u, v, w), the density of fluid is ρ, the rate of density change is $d\rho/dt$, and a source of mass is $\dot{S}_m$. The mass conservation is expressed as,

Fig. 3.60 Mass conservation in an enclosed volume

Fig. 3.61 Representation of the fluid flow in an infinitesimal volume

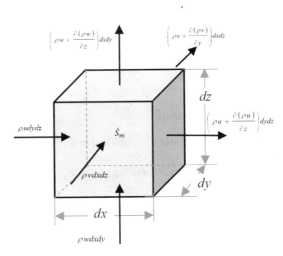

$$\dot{m}_{in} - \dot{m}_{out} = \frac{dm}{dt} \qquad (3.156)$$

where $\dot{m}_{in}$ and $\dot{m}_{out}$ are the masses flow in and out of the infinitesimal volume, and $\frac{dm}{dt}$ is the rate of mass change over time.

Expanding Eq. (3.156) with the variables in Fig. 3.61 yields,

$$\rho(udydz + vdxdz + wdxdy)$$
$$- \left(\begin{array}{l} \left(\rho u + \dfrac{(\rho u)}{x} dx \right) dydz + \left(\rho v + \dfrac{(\rho v)}{y} dy \right) dxdz + \\[2mm] \left(\rho w + \dfrac{(\rho w)}{z} dz \right) dxdy \end{array} \right)$$
$$+ \frac{d\rho}{dt} dxdydz = \dot{S}_m dxdydz \qquad (3.157)$$

Equation (3.156) can be further rewritten as,

$$\nabla(\rho v) + \frac{d\rho}{dt} = \dot{S}_m \qquad (3.158)$$

where $\nabla(\rho v) = \frac{\partial(\rho u)}{\partial x} + \frac{\partial(\rho v)}{\partial y} + \frac{\partial(\rho w)}{\partial t}$ is the divergence of fluid flow and $\dot{S}_m$ is the rate of mass generation.

In fluid mechanics, Eq. (3.158) is referred as the continuity equations described in a variety of forms.

3.7.5.3 Momentum Conservation

Energy may be transferred from one form to another. In a fluid flow, energy exists in three forms as below.

The first form is *potential energy* (PE) that is relevant to the altitude of the fluid as,

$$PE = W \cdot H \qquad (3.159)$$

where *PE* is the amount of potential energy, *W* and *H* are the weight and altitude of the fluid weight, respectively.

The second form is *kinetic energy* (*KE*) that is associated with the fluid movement as,

$$KE = W \cdot \frac{v^2}{2g} \qquad (3.160)$$

where KE is the amount of kinetic energy, v the velocity of the fluid flow.

The third form of energy is determined by the flow pressure (FP) as,

$$FP = \frac{W \cdot p}{g} \tag{3.161}$$

where p is pressure of the fluid, and g is the gravity acceleration.

Adding the energies in three forms in Eqs. (3.159–3.161) finds the total energy of the fluid as,

$$E = PE + KE + FP = W\left(H + \frac{v^2}{2g} + \frac{p}{g}\right) \tag{3.162}$$

When the fluid flow is steady, some energy is consumed to overcome the friction along the streamline. In such a case, the momentum of the flow should be conserved and this leads to the Bernoulli's equation as,

$$H_1 + \frac{v_1^2}{2g} + \frac{p_1}{g} = H_2 + \frac{v_2^2}{2g} + \frac{p_2}{g} + h_L \tag{3.163}$$

where $H1$ and $H2$ are the heads, p_1 and p_2 are the pressures, and $\frac{v_1^2}{2g}$ and $\frac{v_2^2}{2g}$ are velocity heads at two points over the streamline, respectively, and h_L is the head loss by friction, fittings, bends, and valves as (Acharya 2016),

$$\left.\begin{array}{l} h_L = k\frac{v^2}{2g} \quad (minor\ loss\ from\ valves,\ fitting\ and\ bents) \\ h_L = f\frac{L}{D}\frac{v^2}{2g} \quad major\ loss\ from\ friction\ and\ pumping \end{array}\right\} \tag{3.164}$$

where k and f is the friction factors, L is the distance of two sections, v is the average velocity, and g is the gravity acceleration.

In an open space, the condition of momentum conservation can be expressed as,

$$\left.\begin{array}{l} \frac{\partial(\rho u)}{\partial t} + \frac{\partial(\rho u^2)}{\partial x} + \frac{\partial(\rho uv)}{\partial y} + \frac{\partial(\rho uw)}{\partial z} - \frac{\partial\tau_{xx}}{\partial x} - \frac{\partial\tau_{xy}}{\partial y} - \frac{\partial\tau_{xz}}{\partial z} + \frac{\partial p}{\partial x} - \rho f_x = 0 \\ \frac{\partial(\rho v)}{\partial} + \frac{\partial(\rho v^2)}{\partial y} + \frac{\partial(\rho uv)}{\partial x} + \frac{\partial(\rho vw)}{\partial z} - \frac{\partial\tau_{yy}}{\partial y} - \frac{\partial\tau_{xy}}{\partial x} - \frac{\partial\tau_{yz}}{\partial z} + \frac{\partial p}{\partial xy} - \rho f_y = 0 \\ \frac{\partial(\rho w)}{\partial} + \frac{\partial(\rho w^2)}{\partial z} + \frac{\partial(\rho uw)}{\partial x} + \frac{\partial(\rho vw)}{\partial y} - \frac{\partial\tau_{zz}}{\partial z} - \frac{\partial\tau_{xz}}{\partial x} - \frac{\partial\tau_{yz}}{\partial y} + \frac{\partial p}{\partial z} - \rho f_z = 0 \end{array}\right\} \tag{3.165}$$

3.7.5.4 Energy Conservation

Equation (3.165) is applicable to an isothermal state where no heat transfer occurs. If a fluid flow involves in the temperature change, the thermal energy must be taken into the consideration in the energy conservation. When temperature changes, density (ρ) depends on pressure (p) and temperature (T) as,

$$\rho = \rho(p, T) \tag{3.166}$$

The density function Eq. (3.166) depends on the fluid type. For example, the density of ideal gas can be expressed explicitly as,

$$\rho = \frac{p}{R \cdot T} \tag{3.167}$$

where R is the universal gas constant.

There are other three types of energy (1) *intrinsic energy e*, (2) the kinetic energy relevant to the fluid motion, and (3) the energy relevant to pressure of fluid; therefore, the equivalent heat of the system is found as,

$$H = E + \frac{p}{\rho} = e(p, T) + \frac{v_i v_j}{2} + \frac{p}{\rho} \tag{3.168}$$

where E is the sum of intrinsic energy and kinematic energy; H called as is *the enthalpy* of the system, $e(p, T)$ is the intrinsic energy, and v_i ($i = 1, 2, 3$) corresponds to u, v, and w in Fig. 3.61.

The substantial derivative of the total energy is found as,

$$\frac{d(\rho E)}{dt} = E \left(\frac{\partial \rho}{\partial} + \frac{\partial \rho v_i}{\partial x_i} \right) + \left(\frac{\partial \rho E}{\partial t} + \frac{\partial \rho E v_i}{\partial x_i} \right) \tag{3.169}$$

Heat energy can be also transferred by convection, conduction, and radiation. For example, the heat flux q_i relates to the temperature gradient as,

$$q_i = -k \frac{\partial T}{\partial x_i} \tag{3.170}$$

where k is the conductivity coefficient, and x_i ($i = 1, 2, 3$) correspond to x, y, z axis, respectively.

In addition, energy can be dissipated due to internal stresses that can be calculated as,

$$\frac{\partial}{\partial x_i} (\sigma_{ij} v_j) = \frac{\partial}{\partial x_i} (\tau_{ij} v_j) - \frac{\partial p_i}{\partial x_j} (p v_j) \tag{3.171}$$

As the result, the energy conservation with the consideration of thermal energy becomes,

$$\frac{\partial(\rho E)}{\partial} + \frac{\partial(\rho v_i H)}{\partial x_i} - \frac{\partial}{\partial x_i}\left(k\frac{\partial T}{\partial x_i}\right) - \frac{\partial}{\partial x_i}\left(\tau_{ij}v_j\right) - \rho g_i v_i - q_H = 0 \quad (3.172)$$

Note that in many cases, not all of the equations for mass, momentum, and energy conservations are needed to find the solutions to fluid mechanics problems.

3.7.5.5 SolidWorks Flow Simulation

Fluid mechanic problems can be solved in SolidWorks Flow Simulation, which is an intuitive computational fluid dynamics (CFD) solution to fluid flows. It adopts the Cartesian-based meshes for solid–fluid and soli-solid interfaces to increase the flexibility of meshing process, and the meshes are automatically based on specified control parameters. To illustrate the application of the CAE tools in solving various fluid mechanics problems, the SolidWorks Flow Simulation is used to evaluate the fuel efficiency of car body.

Example 3.13 Analyze the fluid flow around an example racecar model shown in Fig. 3.56a and estimate the drag force when the racecar run at the speed of 90 m/s (~200 miles per hour).

Solution. The SolidWorks Flow Simulation is used to simulate the airflow around the rackcar body. *The flow Simulation* tool is included in the *add-ins* of the SolidWorks. It can be activated by selecting the tool in the list of add-ins under *Options* → *Add-ins* as shown in Fig. 3.62. The SolidWorks Flow Simulation provides the project wizard to guide users in defining a flow simulation model by following the main steps shown in Fig. 3.63. *Firstly,* the CAD model is imported, the configuration of interest is specified, and the name of flow simulation is created. *Secondly,* the preferable units are specified for all of parameters and variables used in the simulation model.

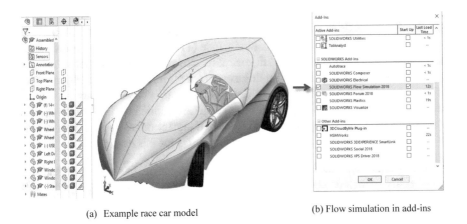

(a) Example race car model   (b) Flow simulation in add-ins

Fig. 3.62 Create a flow simulation model for an example racecar body

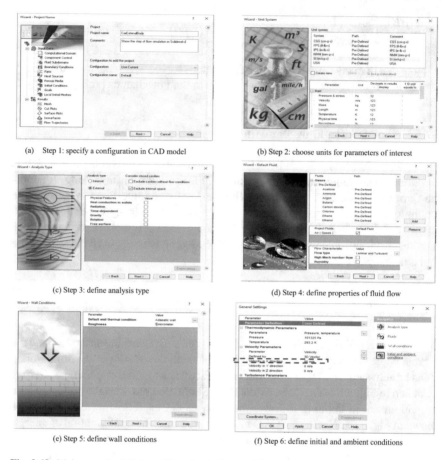

(a) Step 1: specify a configuration in CAD model

(b) Step 2: choose units for parameters of interest

(c) Step 3: define analysis type

(d) Step 4: define properties of fluid flow

(e) Step 5: define wall conditions

(f) Step 6: define initial and ambient conditions

Fig. 3.63 Main steps in defining a flow simulation model

Thirdly, the analysis type is given; the flow simulation can be for an *internal* or *external* flow; moreover, it is optional to include (1) cavities or internal surfaces, (2) heat transfer and radiation, and (3) historical data over time. An external simulation with excluded cavities is defined in this example. *Fourthly*, the types and properties of fluid are defined; the software include a design library that have many commonly used fluids and gases. *Fifthly*, the wall conditions are defined for fluid–solid contact surfaces. *Sixthly*, initial conditions of fluid are defined; the relative speed (i.e., 90 m/s) of the airflow and racecar is defined along the x-axis for this example.

A flow simulation model involves in a large number of state variables, and it is unnecessary to include all of derived variables in the solving process. Figure 3.64a shows the interface where the variables of interest can be specified as the analysis goals. In addition, the computational volume can be should be adjusted with the consideration of computation time and accuracy (see Fig. 3.64b); users can also

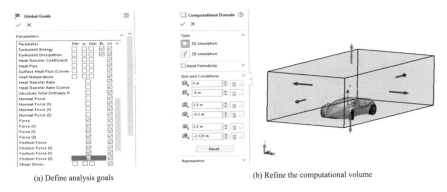

(a) Define analysis goals   (b) Refine the computational volume

Fig. 3.64 Define analysis goals and refine computational volume

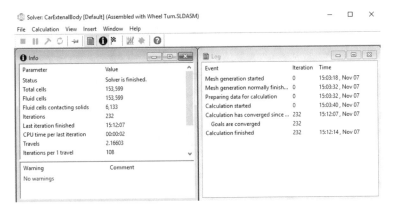

Fig. 3.65 The statistics of meshes and the solving process

specify the levels or sizes of meshes or apply control meshes in the areas of interest when they are needed.

Figure 3.65 gives the statistic data of solid and fluid meshes and the solving process. Figure 3.66 shows the example plots of velocity and pressure distribution over a selected cutting plane. The goals of the flow simulation is shown in Table 3.14. It is found that the drag force along moving direction (x-axis) is 4802 (N).

3.8 Summary

An engineering problem can be usually formulated into a design problem with the objectives, inputs, system parameters, outputs, and constraints from the perspective of system. To solve a design problem, the possible solutions in the design space are

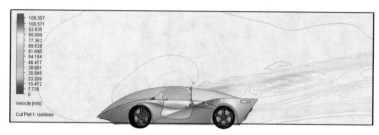

(a) Velocity distribution

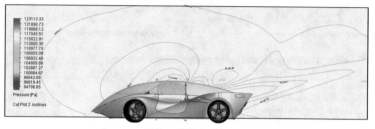

(b) Pressure distribution

Fig. 3.66 Visualization of velocity and pressure distribution over a cut plane

Table 3.14 The results of analysis goals in flow simulatioon

Goal Name	Unit	Averaged value	Minimum value	Maximum value
GG Bulk Av Velocity 1	[m/s]	87.28900777	87.28787613	87.29089342
GG Bulk Av Velocity (X) 1	[m/s]	86.92918179	86.92779739	86.93108198
GG Bulk Av Velocity (Y) 1	[m/s]	2.036041808	2.034343274	2.037097925
GG Bulk Av Velocity (Z) 1	[m/s]	0.038781473	0.036952233	0.040201268
GG Force 1	[N]	4923.384152	4919.574211	4928.137101
GG Force (X) 1	[N]	4797.504651	4794.285348	**4802.072032**
GG Force (Y) 1	[N]	−1089.135225	−1092.571206	−1085.872086
GG Force (Z) 1	[N]	−193.4805561	−196.6832158	−187.0958669

analyzed and compared, and design synthesis is performed to obtain the optimized solution. When the scope and complexity of a design problem increases, engineering design goes far beyond the capabilities of analytical or experimental approaches, the numerical methods such as CAE tools become imperative to model, simulate, and evaluate engineering solutions. In this chapter, the fundamental theory of numerical simulation is introduced, and the focus lies on the implementation of finite element analysis (FEA). The SolidWork Simulation is used as an example CAE tool to analyze engineering problems in different disciplines. From this chapter, students should master the knowledge and skills of (1) formulating an engineering problem as a design problem and (2) utilizing the CAE tools for design analysis and synthesis of any engineering systems.

Design Problems

Problem 3.1 Formulate the following real-world problems as engineering design problems that include design objectives, inputs, system parameters, outputs, and design constraints.

(a) Due to a high strength-weight ratio, composite materials have been widely adopted in aerospace products, and industrial robots are used for light machining of composite parts. However, one challenge is to collect the dwarf and dust generated from machining process in open space, which is shown in Fig. 3.67; since the dissipated dusts will pollute working environment (Bi 2010). Explore an engineering solution to collect dusts at sources.

(b) An Industrial robot with an open kinematic chain usually has its limitation of low workload for machining process. In contrast, a parallel kinematic machine with closed kinematic chains is capable of offering a high workload for light machining (see Fig. 3.68a). However, the rigidity of a parallel kinematic machine varies from one place or orientation to another (Bi 2014). Explore an engineering solution to predict the accuracy of a parallel kinematic machine in real time.

(c) Many machine elements are used to transfer motions and power in machines. The wears of machine elements determine their lifespans. The lifespan of a machine element relates to many factors such as loads, lubrications, material properties, surface properties, pressure, and temperature. Explore an engineering solution which is capable of predict fatigue life of a type actuators shown in Fig. 3.69a.

Problem 3.2 Determine the workspace of the 2-DOF robot shown in Fig. 3.70. Note that the motion ranges of joints A and B are $\theta_A = (-80°,\ 80°)$ and $\theta_B = (-60°,\ 60°)$

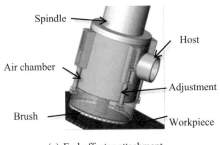

(a) End-effector attachment

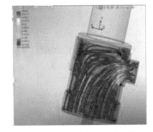

(b) Simulation of dust collection

Fig. 3.67 Illustration of design Problem 3.1(a) (Bi 2010)

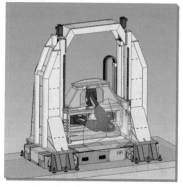

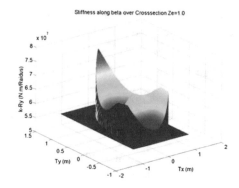

(a) Parallel kinematic machine   (b) Example of stiffness distribution

Fig. 3.68 Illustration of Design Problem 3.1(b) (Bi 2014)

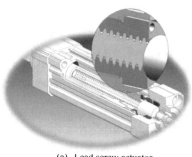

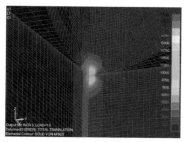

(a) Lead screw actuator   (b) Fatigue analysis

Fig. 3.69 Illustration of Design Problem 3.1(c) (Bi and Meruva (Bi and Meruva 2019)

Problem 3.3 Determine the voltage drop in each resistor of the circuit shown in Fig. 3.71.

Problem 3.4 Determine the maximum displacement in the truss structure in Fig. 3.72; note that the element model of a truss member in 2-D space is shown below,

$$\frac{EA}{L}\begin{bmatrix} l^2 & lm & -l^2 & -lm \\ lm & m^2 & -lm & -m^2 \\ -l^2 & -lm & l^2 & lm \\ -lm & -m^2 & lm & m^2 \end{bmatrix}\begin{Bmatrix} u_i \\ v_i \\ u_j \\ v_j \end{Bmatrix} = \begin{Bmatrix} f_{ix} \\ f_{iy} \\ f_{jx} \\ f_{jy} \end{Bmatrix}$$

where E, A, L are the Young module, the cross-section area, and the length of truss element, $l = \cos\theta$, $m = \sin\theta$, and θ is the direction of truss with respect to x-axis.

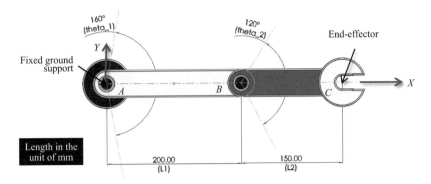

Fig. 3.70 A two dimensional robot in Design Problem 3.2

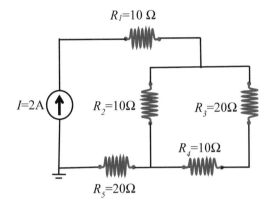

Fig. 3.71 The circuit diagram in Design Problem 3.3

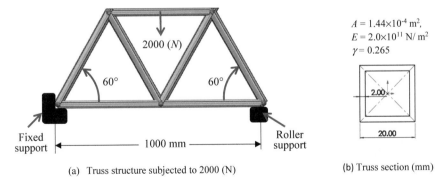

(a) Truss structure subjected to 2000 (N)

(b) Truss section (mm)

Fig. 3.72 The truss structure in Design Problem 3.4

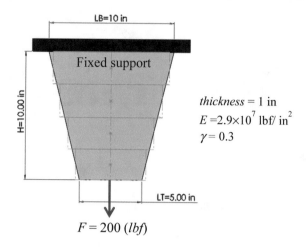

Fig. 3.73 A two-dimensional thin plate subjected to an axial load

Problem 3.5 Figure 3.73 shows a thin plate subjected to an axial load F, create a one-dimensional model with 4 axial members to find the displacement at free end. Note that the element model of an axial member is,

$$\frac{EA}{L}\begin{bmatrix} 1 & -1 \\ -1 & 1 \end{bmatrix}\begin{Bmatrix} u_i \\ u_j \end{Bmatrix} = \begin{Bmatrix} f_i \\ f_j \end{Bmatrix}$$

Problem 3.6 Find a weak solution to the following boundary value problem:

$$\left.\begin{array}{c} \frac{d^2y}{dx^2} + 2x - y = 0 \quad 0 \le x \le 1 \\ \text{subjected to}: \ y(0) = 1 \\ \frac{dy(0)}{dx} = 0 \end{array}\right\}$$

Problem 3.7 Figure 3.74 shows a torsional member that consists of three segments AB, BC, and CD with different materials. Determine the angular displacement at D

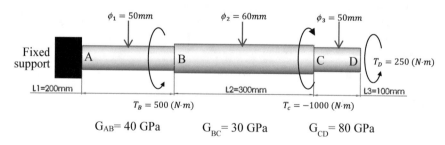

Fig. 3.74 A torsional member in Problem 3.7

relative to position A. Note that the element model of torsional member is,

$$\frac{GJ}{L}\begin{bmatrix} 1 & -1 \\ -1 & 1 \end{bmatrix}\begin{Bmatrix} \theta_i \\ \theta_j \end{Bmatrix} = \begin{Bmatrix} T_i \\ T_j \end{Bmatrix} \quad J = \frac{\pi d^4}{32}$$

Problem 3.8 Figure 3.75 shows a truss structure in three-dimensional space with the specified loads $F = (0, 50, -100)$ lbf at two nodes (G and I) and $F = (0, 0, -500)$ lbf at node H. The nodes at the ground (A, B, C, D, E, F) are all fixed. All truss members use the same material—Gray Case Iron, which has the Young's modulus of $E = 9.598 \times 10^6$ psi and the Poisson' ratio of 0.27. The yield strength is $S_y = 2.20 \times 10^4$ psi. The cross-section areas of all truss members are given as $A = 1$ in^2. Predict the maximal stress and deflection of the truss structure.

Problem 3.9 Figure 3.76 shows a thin plate with a thickness of 0.2-in. The plate includes three discontinuities of geometries, a Φ 1.5-in hole, a Φ 2.0-in hole, and a 0.5-in fillet at the shoulder. The plate uses Aluminum 1060 alloy with the elastic modulus of $E = 1.0 \times 10^7$ psi, the Poisson' ratio $\nu = 0.33$, yield strength $S_y = 3.999 \times 10^3$ psi. Determine stress concentration factors at three discontinuity sections under a bending load.

Problem 3.10 Figure 3.77 shows a truss structure in two-dimensional space. Nodes A and B are fixed. All truss members use the same materials Gray Case Iron, which has the Young's modulus $E = 9.598 \times 10^6$ psi and the Poisson' ratio of 0.27. The yield strength is $S_y = 2.20 \times 10^4$ psi. The cross-section areas of all truss members are given as $A = 1.0$ in^2. Use the Solidworks Simulation to estimate the first 4 natural frequencies.

Problem 3.11 Figure 3.78 shows a fin to transfer the heat from base by convection; the dimensions of lengths are in inch. The temperature of fluid flow is 20 °C, and heat transfer coefficient $h = 0.1$ Btu/in$^2 \cdot$s$\cdot$°C. The coefficient of conduction of fin is $k = 3.0$ Btu/in$\cdot$s$\cdot$°C, determine the temperature distribution in the fin.

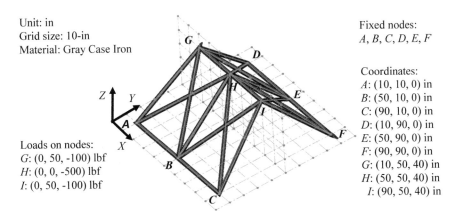

Unit: in
Grid size: 10-in
Material: Gray Case Iron

Loads on nodes:
G: (0, 50, -100) lbf
H: (0, 0, -500) lbf
I: (0, 50, -100) lbf

Fixed nodes:
A, B, C, D, E, F

Coordinates:
A: (10, 10, 0) in
B: (50, 10, 0) in
C: (90, 10, 0) in
D: (10, 90, 0) in
E: (50, 90, 0) in
F: (90, 90, 0) in
G: (10, 50, 40) in
H: (50, 50, 40) in
I: (90, 50, 40) in

Fig. 3.75 A truss structure in Problem 3.8

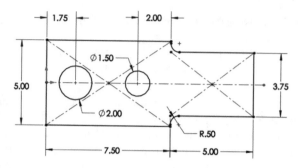

Fig. 3.76 A plane stress plate with a thickness of 0.2-in

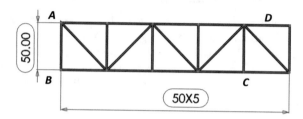

Fig. 3.77 A Truss Structure in Problem 3.10

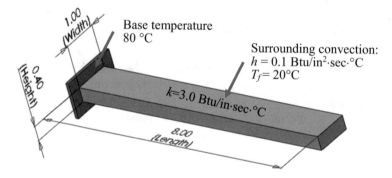

Fig. 3.78 Example of heat transfer problem (length unit: inch)

Problem 3.12 Figure 3.79 shows a wall consisting of two materials, i.e., the first layer is 2-cm with the conductivity of $k_1 = 0.2$ W/cm·s·°C and the second later is 6-cm with the conductivity of $k_2 = 0.05$ W/cm·s·°C. Other parameters related to the boundary conditions are shown in Fig. 3.79. Create a simplified model in SolidWorks Simulation to estimate the distribution of temperature in the wall.

Problem 3.13 A 2-D domain with an incompressible irrotational fluid flow is shown in Fig. 3.80. The length is 2 inches and the height is 1 inch, and there is the obstacles

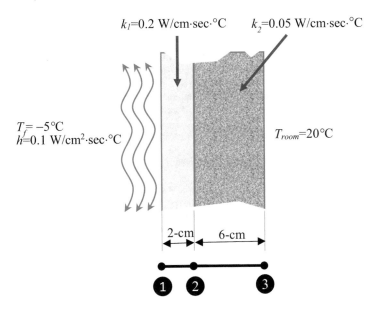

Fig. 3.79 1-D heat transfer through wall

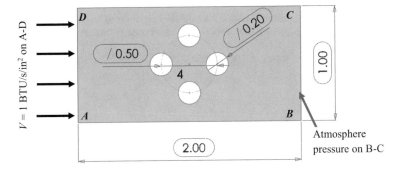

Fig. 3.80 Ideal fluid flow in a 2-D domain for problem 3.13 (length unit: inch)

in the middle with 4 circles of ϕ 0.2-in on a circle of ϕ 0.5-in. Assume that the reference atmosphere pressure at the outlet is and the fluid speed at the inlet is 1.0 BUT/s/in^2. Create an FEA model and determine the distribution of the velocity in the fluid domain.

References

Acharya S (2016) Analysis and FEM simulation of flow of fluids in pipes. Arcada University of Applied Science, Finland. Available online, https://www.theseus.fi/bitstream/handle/10024/106 991/Acharya_Saroj.pdf?sequence=1 (accessed on April 18, 2021)

Bahman AS (2018) Chapter 8 Computer-aided engineering simulations, in Wide Bandgap Power Semiconductor Packaging: Materials, Components, and Reliability. Elsevier, ISSN 0922–3444, pp 199–223

Bi ZM (2010) Design and simulation of dust extraction for composite drilling. Int J Adv Manuf Technol 54(5):629–638

Bi ZM (2014) Kinetostatic modeling of Exechon parallel kinematic machine for stiffness analysis. Int J Adv Manuf Technol 71:325–335

Bi ZM (2018) Finite element analysis applications: a systematic and practical approach, 1st edn. Academic Press. ISBN-10 018099526

Bi Z, Luo C, Miao Z, Zhang B, Zhang CWJ (2020) Automatic robotic recharging systems–development and challenges, Industrial Robot, Vol. ahead-of-print No. ahead-of-print. https://doi.org/10.1108/IR-05-2020-0109

Bi ZM, Meruva K (2019) Modeling and prediction of fatigue life of robotic components in intelligent manufacturing. J Intell Manuf 30(7):2575–2585

Chandrupatla TR, and Belegundu AD (2012) Introduction to finite elements in engineering. 4th Edn. Pearson, ISBN-10: 0132162741

Espon (2020) Synthis T6 all-in-one SCARA robots. https://files.support.epson.com/docid/cpd5/cpd 55682.pdf.

Fuelmatics (2020) Fuelmatics systems. https://ibem-management.com/onewebmedia/FUELMA TICS%20SYSTEMS%20(IBEM).pdf

Krahe C, Iberl M, Jacob A, Lanza G (2019) AI-based computer aided engineering for automated production design–a first approach with a multi-view based classification. Procedia CIRP 86:104–109

Suh NP (2005) Complexity: theory and applications, 1st edn. Oxford University Press, New York, p 2005

Chapter 4
Computer-Aided Manufacturing (CAM)

Abstract In a manufacturing system, *manufacturing processes* are the operations through which raw materials are transformed into final products. Manufacturing processes use machines, tools, and labors to implement the transformation of materials. Design of manufacturing processes is generally complex since it involves in *machines*, *tools*, *materials*, and numerous *operating parameters*. A manufacturing process is evaluated comprehensively based on a number of conflict criteria such as *cost*, *accuracy*, *productivity*, *flexibility*, and *adaptability*. Traditional intuitive or experimental designs of manufacturing processes have their limitations in (1) exploring a wide scope of alternative processes and (2) ensuring the first-time-right for virtual design to physical implementation. In this chapter, *computer-aided manufacturing* (CAM) is introduced to model and evaluate manufacturing processes in the virtual environment, the theory and enabling techniques of CAM are discussed, and the applications of CAM are explored in *designs of materials, geometric dimensioning and tolerance* (GD&T), designs and simulation of *fixtures, molds and dies, machining processes*, and *machining programming*.

Keywords Manufacturing processes · Computer-aided manufacturing (CAM) · Modeling and simulation · Geometric dimensioning and tolerance (GD&T) · DimXpert · Toansist · Computer-aided fixture design (CAFD) · Design for manufacturing (DFM) · Computer numerical controls (CNC) · Machining programming

4.1 Introduction

Manufacturing processes are essential activities in the secondary industry. Manufacturing processes transform starting materials from *the primary industry* into the finished products used as capital goods in *the secondary* or *third industry* or customers products for end-users. Artificial goods in our modern society are mostly produced from manufacturing processes.

Design of manufacturing processes follows product design, and manufacturing processes determine the way how a virtual product model is transformed into a physical product. Therefore, design of manufacturing processes covers many aspects including (1) the *selections of materials, processing types, machines,* and *tools*; (2) *geometric dimensioning and tolerancing* (GD&T); (3) *work-holding and fixturing*; (4) *operating conditions*; (5) *machining programming*; and (6) *quality controls*. Computer-aided manufacturing (CAM) has been widely used to support the decision-making activities in these aspects.

4.2   Material Characteristic, Structures, and Properties

As shown in Fig. 4.1a, the characteristic of a material consists of structures, properties, processes over the material, and the exhibited performance in applications. Figure 4.1b uses the steel material as an example to illustrate the contents of its characteristic: it has its metallic bonds in structure; it supports the processing types such as costing, forging, and particulate processes. It has the properties of high elasticity, density, yield strength, and hardness. It exhibits the performance of high strength, ductility, and thermal and electric conductivities in applications.

Four elements in a material tetrahedron are dependent on each other. For example, material structure has the great impact on material properties and applicable processes. As shown in Fig. 4.2, different materials exhibit distinct properties, and materials should be appropriately chosen for given applications. Steel materials have high strengths and have affordable costs that have widely been applied in large-scale construction (see Fig. 4.2a). Products in musculoskeletal healthcare are embedded in human bodies; this requires the materials be corrosion resistant (see Fig. 4.2b). Machine elements are mostly involved in the motions at contact surfaces; the parts with oil impregnation from powder metallurgy (PM) will reduce wear and prolong fatigue lives of products (see Fig. 4.2c). It will be helpful for engineers to understand material structures so that right materials can be chosen to meet the functional requirements for products.

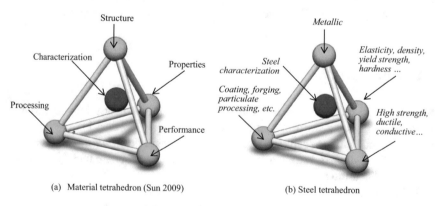

(a) Material tetrahedron (Sun 2009)   (b) Steel tetrahedron

Fig. 4.1 Material tetrahedron and example

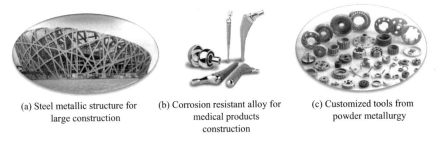

(a) Steel metallic structure for large construction

(b) Corrosion resistant alloy for medical products construction

(c) Customized tools from powder metallurgy

Fig. 4.2 Example of selecting right material and processing for products

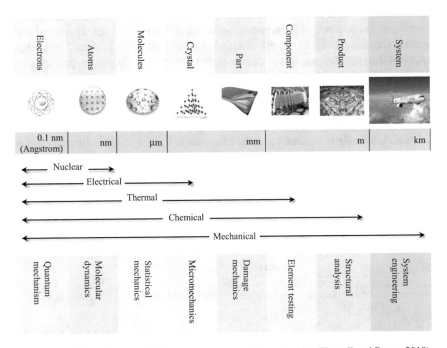

Fig. 4.3 Disciplines for material characterizations at different scales (Kreculj and Rasuo 2018)

In a manufacturing process, certain change occurs to the material, and some common changes of the materials include *reshaped geometries, improved material properties*, and *joined structures*. It is clear that the material properties (1) determine the types of manufacturing processes that can be applied to the given material and (2) affect operating conditions when certain manufacturing process is applied. Taking an example of shaping a part geometry by machining, since unwanted material must be cut away from a workpiece, the material properties affect the selections of machines, cutting tools, and operating conditions such as cutting speed, feed rate, and depth.

As shown in Fig. 4.3, material structures and properties are characterized in a varying scale from the sub-atom level to the system level. Nuclear properties are determined by atomic structures, while the mechanical properties will be determined by all of the characteristics and attributes cross over the scales. The study on material structures and properties is interdisciplinary; the materials at different scales fall in the different disciplines relevant to the material science.

At molecule level, numerous materials are formed by combing protons, neutrons, and electrons, and the material properties at that level are direct results of (*i*) constitutive structures and (*ii*) processing of property enhancement to change material structures. *Atoms* are the basic structure of matter. An atom consists of *a nucleus* with positive charges and a set of *electrons* with negative charges. *A planetary model* by Niels Bohr can be used to represent an atomic structure. As shown in Fig. 4.4a, electrons orbit around the nucleus at certain distances called as orbits or shells, and the maximum number of electrons in the *n*-th shell is $2n^2$. Note that a nucleus consists of neutrons and protons, and the positive charges of the nucleus are associated with protons. The similarities of the elements are determined by their atomic structures, Fig. 4.4b–d shows the examples of nonmetal, metal, and noble gas, respectively.

Atoms are distinguished by *atomic numbers* and *elements*. Up to now, 118 elements have been discovered. As shown in Fig. 4.5, the discovered elements have been classified into element families based on their similarities and relations. Elements are organized in columns and rows. The elements in the same column have their similarities. For example, the elements in the very right column including helium, neon, argon, krypton, xenon, and radon are *noble gases*; these elements exhibit the great chemical stability. The elements in the left and middle sections of the table are *metallic*. The elements at the right section are *nonmetal* and the elements

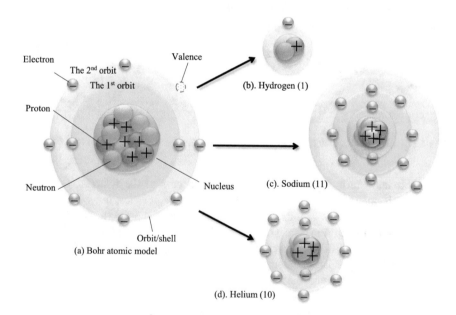

Fig. 4.4 Description of atomic structure and examples

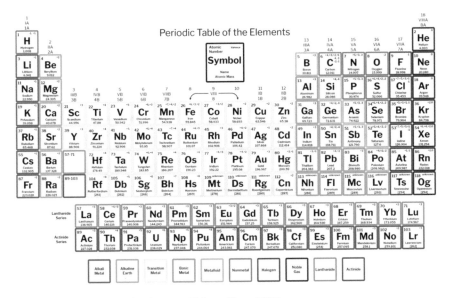

Fig. 4.5 Periodic table of the elements (Science Notes 2020)

in the diagonal transition zone of metal and nonmetal elements are *metalloids* or *semimetals* (Science Notes 2020).

Molecules are built from atoms by bonds. Therefore, the material strengths at the molecular level are determined by the types of bonding relations. *Bonds* can be classified into (1) *primary bonds* for the formulation of molecules and (2) *secondary bonds* for the connections in a set of molecules. Valence electrons in the outermost shell of atoms are involved in bonding.

The atoms in a primary bond exchange the valence electrons with a strong atom-to-atom attraction. In *an ionic bond* shown in Fig. 4.6a, the electron(s) at the outer shell of the atom are moved to the outer shell of the other atom to form the mutual attraction. Material properties with ionic bonds have low electrical conductivity and poor ductility. In a covalent bond shown in Fig. 4.6b, the electrons at the outer shells of two atoms are shared to bond two atoms. The material properties with covalent bonds have high harnesses and low electrical conductivity. In a metallic bond shown in Fig. 4.6b, the valence electrons of metallic atoms are moved freely to form an electron cloud that generates the strong attractions among the atoms. The material properties with metallic bonding have high electrical and thermal conductivities, high strength, and high melting points.

Secondary bonds generate the attraction forces of molecules. In contrast to primary bonds, there is no electron transfer or sharing; therefore, the secondary bonds are weaker than primary bonds. Secondary bonds can be classified into *dipole force*, *London force*, and *hydrogen bonding*. As shown in Fig. 4.7a, the dipole force is formed for the molecule with positive and negative poles. Therefore, the positive pole of one molecule is attracted by the negative pole of the other molecule. As shown

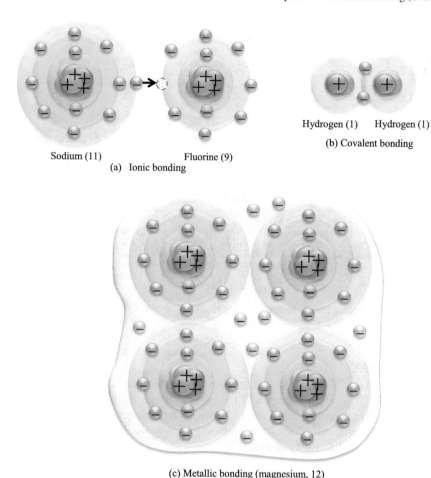

(c) Metallic bonding (magnesium, 12)

Fig. 4.6 Three primary bonding types

in Fig. 4.7b, even though molecule does not have a permanent pole, it generates an instanton pole due to the high movement of electrons. A *London force* reflects the attraction of molecules with an instanton pole. As shown in Fig. 4.7c, hydrogen bonding refers to the intermolecular force of the molecules with hydrogen atoms.

Microstructures of materials can be *crystalline* or *amorphous*. In a crystalline structure, molecules are arranged in three-dimensional lattices, and Fig. 4.8 shows a comparison of three basic lattice structures, i.e., body-centered cubic (BCC), face-centered cubic (FCCn), and hexagonal closed packed (HCP). In an amorphous structure, atoms are not arranged in a lattice structure. Note that microstructure of material can be changed by certain manufacturing processes. For example, both crystalline

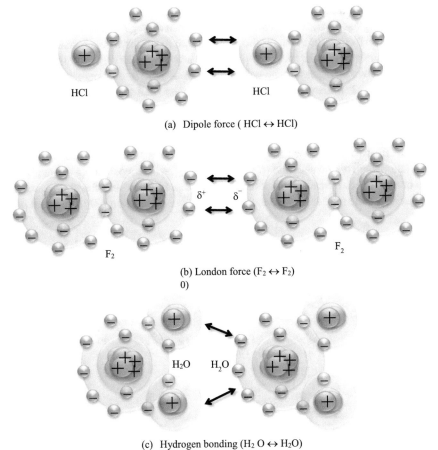

(a) Dipole force (HCl $\leftrightarrow$ HCl)

(b) London force (F$_2$ $\leftrightarrow$ F$_2$)
0)

(c) Hydrogen bonding (H$_2$O $\leftrightarrow$ H$_2$O)

Fig. 4.7 Three secondary bonding types

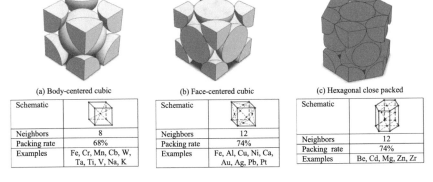

(a) Body-centered cubic	
Schematic	
Neighbors	8
Packing rate	68%
Examples	Fe, Cr, Mn, Cb, W, Ta, Ti, V, Na, K

(b) Face-centered cubic	
Schematic	
Neighbors	12
Packing rate	74%
Examples	Fe, Al, Cu, Ni, Ca, Au, Ag, Pb, Pt

(c) Hexagonal close packed	
Schematic	
Neighbors	12
Packing rate	74%
Examples	Be, Cd, Mg, Zn, Zr

Fig. 4.8 Microstructures of materials

and amorphous structures exist in glass, the higher level the crystalline is, the better strength the glass can achieve, and the higher density the glass has.

The imperfections are usually the most critical locations where a fracture of the material is mostly initialized. An imperfection involves in the discontinuity of material to pass over the stress on the neighboring areas. Micro-level defects include (1) *point defects*, (2) *line defects*, and (3) *surface deflects* that are illustrated in Fig. 4.9. *Point defects* include '*vacancy*' where an atom is missed, '*interstitial*' where a relatively small atom is filled in the gap of other atoms, and '*substitutional*' where an atom is replaced by another atom type. When the crystal structure of an alloy is considered, two other types of point defects include a '*Frenkel defect*' where the small atoms are dislocated, and '*ion-pair-vacancy*' where a pair of an ion and atom has been missed. *Line defects* include '*edge dislocation*' where the atoms along a line of lattice are missed and '*screw dislocation*' where the atoms along a spiral path

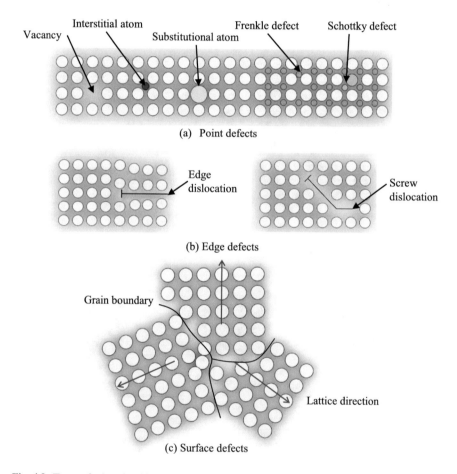

Fig. 4.9 Types of micro-level imperfections (Kailas 2020)

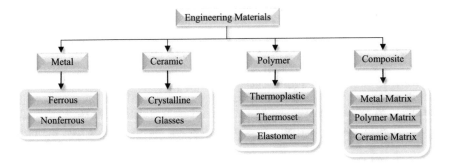

Fig. 4.10 Classification of engineering materials

in the lattice are missed. *Surface defects* are mainly for the boundaries of grains; when crystalline grains along different orientations are met, irregularities occur to the boundaries.

In applications, engineering materials can be classified in different ways. For example, Fig. 4.10 shows that materials can be classified into *metallic, ceramic,* and *polymeric* based on the types of atomic bonding forces. In addition, different materials can be structured to form *composite* materials. Moreover, the materials in each category can be further grouped based on chemical compositions, mechanical properties, and physical properties.

Metals are classified into (1) *ferrous* metals that contain iron as one of constitutive elements and *nonferrous* metals that are free of iron element; a ferrous metal is generally magnetic in nature. Many other metals fall into the group of nonferrous metals, the examples of nonferrous metals are *copper, aluminum,* and *lead.*

Ceramic is composed of inorganic and nonmetallic elements through the processes of heating and consequent solidification. Ceramic is brittle and with a high melting point, high elastic module, and high strength. Ceramic materials are used to make cutting tools such as drills, cutting chips, and grinding wheels. However, ceramic products are also vulnerable to be broken under shock loads. Depending on the level of crystalline, ceramics can be classified into crystalline ceramics and glasses.

Polymer is composed of recurring molecular structures as macromolecules. Polymers are classified as *thermoplastic, thermoset,* and *elastomer.* Due to the difference of molecular structures, polymers are highly diversified in the mechanical properties such as strength, toughness, and hardness.

A composite material is formed by combing two or more distinct materials, and each constitutive material retains its properties. Materials are combined to create a new composite material with the properties that cannot be achieved by any of constitutive components alone. Composite materials differ from conventional materials since the material properties depend on hown different materials are composed in manufacturing processes. Therefore, design and manufacture of composite materials are discussed in the following section.

4.3 Composite Materials

A composite material consists of two phase types, i.e., *reinforcing phase* and *matrix phase*. The materials for the reinforcing phase are usually light and strong; reinforcing phases are present in the form of particles, fibers, or sheets. The materials for the matrix phase are tough and ductile since the matrix material needs to ensure the deflection to pass external forces cross the matrix and protect reinforcing materials.

Figure 4.11 shows that composite materials are classified based on the types of reinforcing and matrix materials. All conventional materials including *metal*, *polymer*, and *ceramic* can be used as matrix materials. Aluminum, titanium, and magnesium are widely used as the materials of metal matrix. Polymer matrix materials can be *thermoset* and *thermoplastic*; thermoset polymers are stronger and stiffer than thermoplastic polymers; however, thermoplastic polymers have better ductility and are reusable. The materials for ceramic matrix include Si_3N_4 and Al_2O_3. A ceramic material has very high specific modulus and mechanical strength at high temperature.

In Fig. 4.11, reinforcing materials are classified in terms of the shapes and material types. The shapes of the reinforcing material particles, flakes, whiskers, and long fibers are shown in Fig. 4.12. Common materials used as the reinforcements in the composites include carbon, glass, aramid, graphite, silicon carbide, and boron.

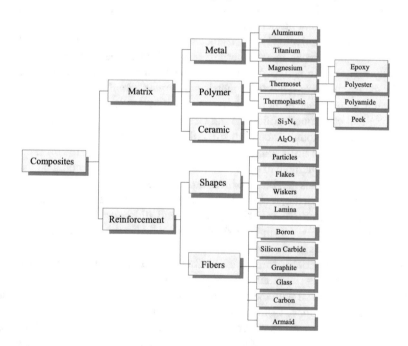

Fig. 4.11 The classification of composite materials (Bi et al. 2009)

(a) particles (b) flakes (c) whiskers (a) fibers

Fig. 4.12 Shapes of reinforcing materials

The composites with the reinforcing materials of Fig. 4.12a–c can be modeled as isotropic materials that exhibit the similar properties in any direction. For isotropic material, the generic Hooke's law is used to describe the relation of stress and strain as

$$
\begin{bmatrix} \sigma_{11} \\ \sigma_{22} \\ \sigma_{33} \\ \sigma_{23} \\ \sigma_{13} \\ \sigma_{12} \end{bmatrix} = \frac{E}{(1+v)(1-2v)} \begin{bmatrix} 1-v & v & v & 0 & 0 & 0 \\ v & 1-v & v & 0 & 0 & 0 \\ v & v & 1-v & 0 & 0 & 0 \\ 0 & 0 & 0 & 1-2v & 0 & 0 \\ 0 & 0 & 0 & 0 & 1-2v & 0 \\ 0 & 0 & 0 & 0 & 0 & 1-2v \end{bmatrix} \begin{bmatrix} \varepsilon_{11} \\ \varepsilon_{22} \\ \varepsilon_{33} \\ \varepsilon_{23} \\ \varepsilon_{13} \\ \varepsilon_{12} \end{bmatrix}
$$

$$(4.1)$$

where $[\sigma]$ and $[\varepsilon]$ are the vectors for stresses and strains, E is Young's modules, and v is Poisson's ratio.

A composite with the reinforcing materials of Fig. 4.12d is usually referred as *laminated composite*, and plies are basic building elements of a laminated composite. A ply is a thin orthotropic material which is characterized by four parameters, i.e., (1) longitudinal stiffness E_{11}, (2) transverse stiffness E_{22}, (3) shear stiffness G_{12}, and (4) Poisson's ration v_{12}. Accordingly, the constitutive model of a ply becomes

$$
\begin{bmatrix} \sigma_{11} \\ \sigma_{22} \\ \sigma_{12} \end{bmatrix} = \begin{bmatrix} Q_{11} & Q_{12} & 0 \\ Q_{21} & Q_{22} & 0 \\ 0 & 0 & Q_{66} \end{bmatrix} \begin{bmatrix} \varepsilon_{11} \\ \varepsilon_{22} \\ \varepsilon_{12} \end{bmatrix}
$$

$$(4.2)$$

where

$$
\left.\begin{aligned}
Q_{11} &= E_{11}/\left(1 - v_{12}^2\right) \\
Q_{22} &= E_{22}/\left(1 - v_{12}^2\right) \\
Q_{12} &= v_{12}E_{22} \\
Q_{21} &= Q_{12} \\
Q_{66} &= G_{12}
\end{aligned}\right\}
$$

$$(4.3)$$

As shown in Fig. 4.13a, a laminated composite is built-up from plies, the constitutive model of the laminated composite is developed based on that of plies. As shown in Fig. 4.13b, the transformation of the constitutive model of a ply is determined by the direction (θ) of the ply coordinate system in the laminate coordinate system. Let

(a) Plies in laminates   (b) Coordinate systems

Fig. 4.13 Plies in laminated composite

$m = \cos\theta$ and $n = \sin\theta$, we have

$$
\left.
\begin{aligned}
\overline{Q}_{11} &= Q_{11}m^4 + 2(Q_{12} + 2Q_{66})m^2n^2 + Q_{22}n^4 \\
\overline{Q}_{12} &= (Q_{11} + Q_{22} - 4Q_{66})m^2n^2 + Q_{12}(m^4 + n^4) \\
\overline{Q}_{22} &= Q_{11}n^4 + 2(Q_{12} + 2Q_{66})m^2n^2 + Q_{22}m^4 \\
\overline{Q}_{16} &= (Q_{11} - Q_{22} - 2Q_{66})m^3n + (Q_{12} - Q_{22} + 2Q_{66})mn^3 \\
\overline{Q}_{26} &= (Q_{11} - Q_{22} - 2Q_{66})mn^3 + (Q_{12} - Q_{22} + 2Q_{66})m^3n \\
\overline{Q}_{66} &= (Q_{11} + Q_{22} - 2Q_{12} - 4Q_{66})m^2n^2 + Q_{66}(m^4 + n^4)
\end{aligned}
\right\}
\tag{4.4}
$$

Figure 4.14 shows the free-body diagram at a point of composite material; the constitutive relation of stress and strain of laminated material can be found as

$$
\begin{bmatrix} N \\ M \end{bmatrix} \leftarrow
\begin{bmatrix} \sigma_{11} \\ \sigma_{22} \\ \sigma_{33} \\ \sigma_{23} \\ \sigma_{13} \\ \sigma_{12} \end{bmatrix}
=
\begin{bmatrix}
C_{11} & C_{12} & C_{13} & C_{14} & C_{15} & C_{16} \\
C_{21} & C_{22} & C_{23} & C_{24} & C_{25} & C_{26} \\
C_{31} & C_{32} & C_{33} & C_{34} & C_{35} & C_{36} \\
C_{41} & C_{42} & C_{43} & C_{44} & C_{45} & C_{46} \\
C_{51} & C_{52} & C_{53} & C_{54} & C_{55} & C_{56} \\
C_{61} & C_{62} & C_{63} & C_{64} & C_{65} & C_{66}
\end{bmatrix}
\begin{bmatrix} \varepsilon_{11} \\ \varepsilon_{22} \\ \varepsilon_{33} \\ \varepsilon_{23} \\ \varepsilon_{13} \\ \varepsilon_{12} \end{bmatrix}
\rightarrow
\begin{bmatrix} A & B \\ B & D \end{bmatrix}
\begin{bmatrix} \varepsilon \\ K \end{bmatrix}
\tag{4.5}
$$

where $[C]$ is stiffness matrix;

$N = \begin{bmatrix} \sigma_{11} \\ \sigma_{22} \\ \sigma_{33} \end{bmatrix}$ and $M = \begin{bmatrix} \sigma_{23} \\ \sigma_{13} \\ \sigma_{12} \end{bmatrix}$ are the vectors of membrane loads and bending loads, respectively;

$\varepsilon = \begin{bmatrix} \varepsilon_{11} \\ \varepsilon_{22} \\ \varepsilon_{33} \end{bmatrix}$ and $K = \begin{bmatrix} \varepsilon_{23} \\ \varepsilon_{13} \\ \varepsilon_{12} \end{bmatrix}$ are the vectors of strains and curvatures, respectively;

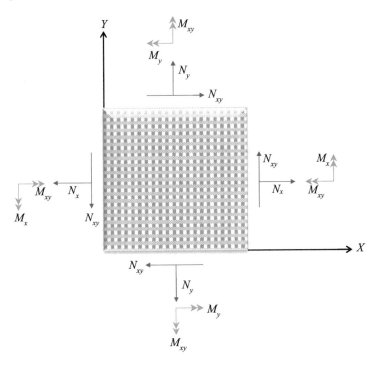

Fig. 4.14 Free-body diagram of a material point in composite

$$A = \begin{bmatrix} C_{11} & C_{12} & C_{13} \\ C_{21} & C_{22} & C_{23} \\ C_{31} & C_{32} & C_{33} \end{bmatrix}, \quad B = \begin{bmatrix} C_{14} & C_{15} & C_{16} \\ C_{24} & C_{25} & C_{26} \\ C_{34} & C_{35} & C_{36} \end{bmatrix} = \begin{bmatrix} C_{41} & C_{42} & C_{43} \\ C_{51} & C_{52} & C_{53} \\ C_{61} & C_{62} & C_{63} \end{bmatrix}, \text{ and } D = \begin{bmatrix} C_{44} & C_{45} & C_{46} \\ C_{54} & C_{55} & C_{56} \\ C_{64} & C_{65} & C_{66} \end{bmatrix}.$$

In Eq. (4.5), A, B, and D are extensional stiffness matrix, stretching-bending coupling matrix, and flexural stiffness matrix that are assembled from the constitutive models of plies (Eq. 4.4) as

$$\left. \begin{aligned} A_{ij} &= \sum_{k=1}^{n} \left(\overline{Q}_{ij}\right)_k (h_k - h_{k-1}) \\ B_{ij} &= \frac{1}{2} \sum_{k=1}^{n} \left(\overline{Q}_{ij}\right)_k \left(h_k^2 - h_{k-1}^2\right) \\ D_{ij} &= \frac{1}{3} \sum_{k=1}^{n} \left(\overline{Q}_{ij}\right)_k \left(h_k^3 - h_{k-1}^3\right) \end{aligned} \right\} \tag{4.6}$$

To use laminated materials safely, the materials should be analyzed at (1) micro-level and (2) macro-level. *Micro-level analysis* is at the scale of fiber diameter; the reinforcing and matrix materials are treated as isotropic and modeled, respectively. It is mainly used to investigate the stress distribution at the interface of fiber and matrix where a crack may likely be initiated.

In general, micro-level analysis involves in the intensive computation. Therefore, unit cells are often defined and analyzed to make the amount of computation manageable in simulation. Figure 4.15 shows some common models of unit cells in representing composites. Since both fiber and matrix materials are treated as isotropic materials, respectively, conventional failure criteria (see Fig. 4.16) can be used in micro-level analysis of composites.

Macro-level analysis takes into consideration the part geometry which is made of composite materials; the plies in laminates are treaded as homogeneous anisotropic materials. Consequentially, the strengths of the laminated composite are characterized as σ_{11}^{fT}, σ_{11}^{fC}, σ_{22}^{fT}, σ_{22}^{fC}, and τ_{12}^{f}, and Table 4.1 shows three commonly used failure criteria for composites in a macro-level analysis.

Example 4.1 In the spider lift shown in Fig. 4.17a, a few of the highlighted components are made of composite materials. These components are adopted to avoid a direct circuit flow from a power line to ground. The parts are prototyped and tested in the real-world application environment. However, a failure occurred to one prototyped component as shown in Fig. 4.17b, and the user wanted to find out the cause of failure in the product design (Bi and Mueller 2016). The CAD model of the failed part is shown in Fig. 4.7c; establish the modeling procedure of FEA for the analysis of composite parts.

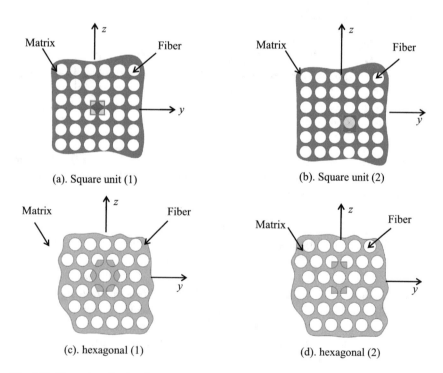

(a). Square unit (1)

(b). Square unit (2)

(c). hexagonal (1)

(d). hexagonal (2)

Fig. 4.15 Examples of unit cell models for micro-level composite analysis (Tuttle 2020)

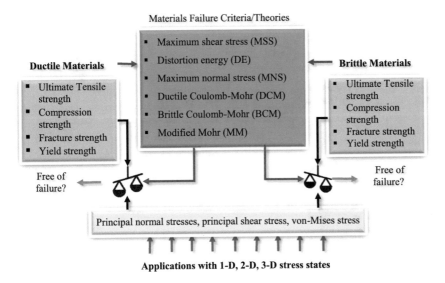

Fig. 4.16 Failure criteria for isotropic materials (Pilkey et al. 2020)

Table 4.1 Commonly used failure criteria of composite materials

Failure criterion	Formula	Strengths and stresses
Maximum stress failure criterion	$-\sigma_{11}^{fC} < \sigma_{11} < \sigma_{11}^{fT}$ $-\sigma_{22}^{fC} < \sigma_{22} < \sigma_{22}^{fT}$ $\lvert\tau_{12}\rvert < \tau_{12}^{f}$	σ_{11}^{fT} and σ_{11}^{fC} are tensile and compression strengths along longitude direction
Tsai-Hill failure criterion	$\dfrac{(\sigma_{11})^2}{\left(\sigma_{11}^{fT}\right)^2} + \dfrac{(\sigma_{22})^2}{\left(\sigma_{22}^{fT}\right)^2} + \dfrac{(\tau_{12})^2}{\left(\tau_{12}^{f}\right)^2} - \dfrac{\sigma_{11}\sigma_{22}}{\left(\sigma_{11}^{fT}\right)^2} < 1$	σ_{22}^{fT} and σ_{22}^{fC} are tensile and compression strengths along transverse direction τ_{12}^{f} is the shear strength in the laminate plane
Tsai-Wu failure criterion	$X_1\sigma_{11} + X_2\sigma_{22} + X_{11}\sigma_{11}^2 +$ $X_{22}\sigma_{22}^2 + X_{66}\tau_{12}^2 +$ $2X_{12}\sigma_{11}\sigma_{22} < 1$	σ_{11}, σ_{22}, and τ_{12} are the longitude stress, and transverse stress, and shear stress, respectively $X_1, X_2, X_{11}, X_{12}, X_{22}$, and X_{66} are experimental determined material strength parameters

Solution The Solidwork Simulation is used to analyze a composite part as follows.

Firstly, the properties of the composites are defined. Figure 4.18 shows the interface to define new material properties. '*Linear elastic orthotropic*' is selected as *the model type* to specify the attributes of composite materials.

Secondly, the laminates of the composite material are configured. Figure 4.19a shows the interface to define a surface which is made of composites. Figure 4.19b

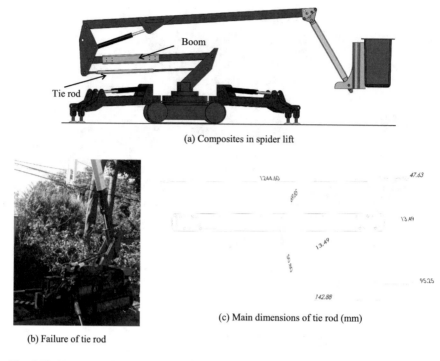

(a) Composites in spider lift

(b) Failure of tie rod

(c) Main dimensions of tie rod (mm)

Fig. 4.17 Tie rod in a spider lift (Bi and Mueller 2016)

Fig. 4.18 Interface for definition of composite materials

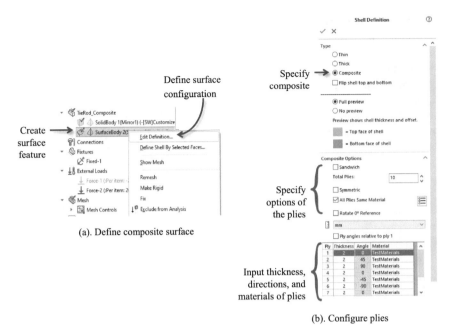

(a). Define composite surface

(b). Configure plies

Fig. 4.19 Configure the plies of composite materials on surface

Fig. 4.20 Failure criteria of composites

shows its main attributes, which include the number, the thicknesses, and the materials of plies; the attributes also include the directions of plies in each layer.

Thirdly, the failure criteria are specified for each solid body. As shown in Fig. 4.20, the Solidworks Simulation supports all three options of failure criteria in Table 4.1 to define *the factor of safety* for composite materials. In addition, the software supports the automatic selection of failure criteria based on the assigned material properties.

Fourthly, the steps in FEA modeling are followed to (1) create part geometry, (2) assign material properties, (3) create meshes, (4) define boundary conditions and loads, (5) run the model, and (6) post-process the simulation results. Figure 4.21a–c shows the outcome examples for the distributions of displacement, stress, and the factor of safety in the first ply of the part surface.

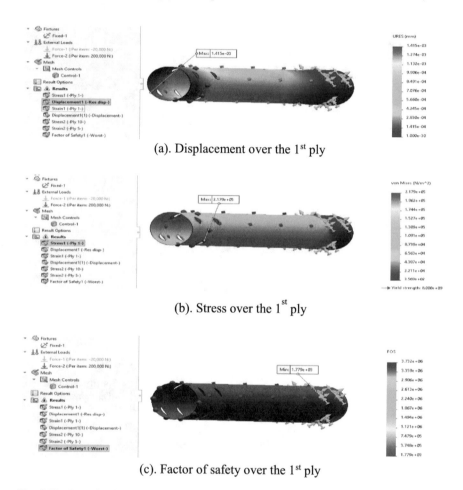

(a). Displacement over the 1st ply

(b). Stress over the 1st ply

(c). Factor of safety over the 1st ply

Fig. 4.21 Example of simulation results over plies

4.4 Geometric Dimensioning and Tolerancing (GD&T)

Geometric dimensioning and *tolerancing* (GD&T) specify the desired size and accuracy of a feature on part; GD&T serves as the specifications for dimensional and shape controls of parts and fits. GD&T is critical since that the selection and option of a manufacturing process must meet the GD&T specifications for quality control of product. Figure 4.22 shows the relations of common manufacturing processes and typical tolerance ranges, and Fig. 4.23 shows the impact of the selection of manufacturing processes on the production times (Vorburger and Raja 1990), which are directly related to manufacturing costs.

Proper application of GD&T ensures that (1) parts are interchangeable in mass production; (2) right references are used to define the attributes of part features; (3) avoid the ambiguity in understanding the functions of machined features; (4) eliminate a tolerance stack-up for better part designs; and (5) clarify the criteria

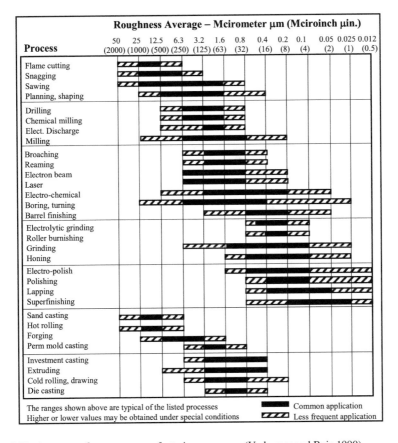

Fig. 4.22 Accuracy of common manufacturing processes (Vorburger and Raja 1990)

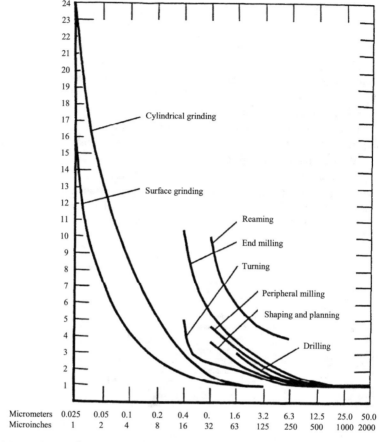

Fig. 4.23 Typical relations of surface finish to production time (Vorburger and Raja 1990)

for inspection and quality controls. GD&T communicates dimensional and tolerance requirements for the implementation of manufacturing and assembly processes (Steeves 2016). GD&T practices are governed by standards, especially the standards from ASME (Y14.5 M-1994 for 2D, Y14.41–2003 for 3D) and ISO (ISO 1101 for 2D and ISO 16,792 for 3D). While GD&T provides the necessary information for the inspections in product quality control, a tight tolerance should only be applied to some critical part features.

4.4.1 Datum Systems

Most of the GD&T specifications are defined based on the specified reference(s), which are commonly known as the datums in GD&T. A *datum* is an entity such as

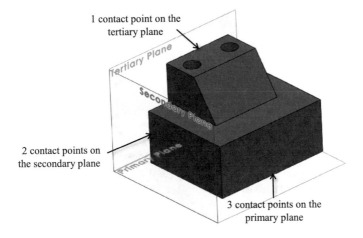

1 contact point on the
tertiary plane

2 contact points on
the secondary plane

3 contact points on the
primary plane

Fig. 4.24 The 3–2-1 rule and three orthogonal datum planes

a *point, axis, line, plane,* or *coordinate system* that is used as a reference when the part is dimensioned, measured, manufactured, or inspected. To locate and orientate a machined feature on part, a datum system with six degrees of freedom (DOF) must be applied. *The 3–2-1 rule* uses the minimum number of contact points as the references to locate and orientate the features on a part. As shown in Fig. 4.24, a datum system based on the 3-2-1 rule consists of three datum planes, i.e., *the primary plane, the secondary plane,* and *the tertiary plane.* When such a datum system is used to locate and orient a part, there will be 1, 2, and 3 points on the part that make the contacts to the tertiary plane, secondary plane, and the primary plane, respectively.

4.4.2   Geometric Tolerancing

The degrees of accuracy of geometric features such as lines, arcs, curves, planes, surfaces, and cylinders are specified by geometric *control symbols* and associated tolerances. As shown in Fig. 4.25a, the ASME standards have a total of 14 control symbols, which are used to control the geometric tolerances in *form* (F), *orientation* (O), *profile* (P), *runout* (R), and *location* (L), respectively. The main attributes of each control symbol have been described in Fig. 4.25b. A control symbol has its graphic icon, specifies the inspection method, indicates the datum when it is needed, and provides the link to the corresponding section in ASME Y14.5 M-1994 for the detailed explanation of the control symbol. Figure 4.26 gives a full table of GD&T control symbol developed by Allsup (2009) that helps users to be familiar with all control symbols and their relations. Figures 4.27, 4.28, 4.29, 4.30 and 4.31 show the examples of (1) how to use control symbols and (2) how to measure the attributes of features to validate the required degrees of accuracy.

F
FORM
O
ORIENTATION
P
PROFILE
R
RUNOUT
L
LOCATION
Indicate whether a total wide tolerance zone can be modified to a cylindrical shape
Specify the datum when it is needed
Legend
C
M
D
Symbol
Name
Z
A
Show the zone to measure the total tolerance
Specify the corresponding ASME Y14.5 M-1994 section

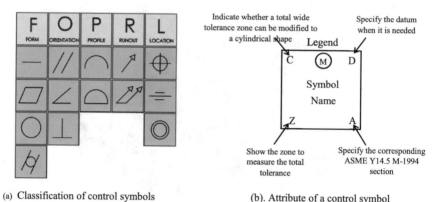

(a) Classification of control symbols (b). Attribute of a control symbol

Fig. 4.25 14 control symbols and the template for their attributes

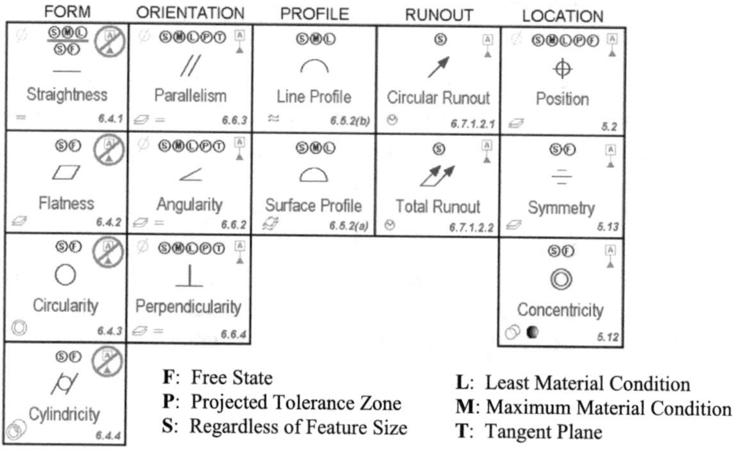

FORM
ORIENTATION
PROFILE
RUNOUT
LOCATION
Straightness
6.4.1
Parallelism
6.6.3
Line Profile
6.5.2(b)
Circular Runout
6.7.1.2.1
Position
5.2
Flatness
6.4.2
Angularity
6.6.2
Surface Profile
6.5.2(a)
Total Runout
6.7.1.2.2
Symmetry
5.13
Circularity
6.4.3
Perpendicularity
6.6.4
Concentricity
5.12
Cylindricity
6.4.4
F: Free State
P: Projected Tolerance Zone
S: Regardless of Feature Size
L: Least Material Condition
M: Maximum Material Condition
T: Tangent Plane

Fig. 4.26 Table of geometric control symbols by Allsup (2009)

4.4.3 Basic Concepts of Dimensioning

Dimensions define the nominal geometries and allowable variations for the features of parts and assemblies. A dimension should have a tolerance since every manufactured feature is subject to variation. The limits of an allowable variation must be specified to control the product quality. An engineering drawing should be dimensioned to reflect the functions of the features of parts, and the interpretation of a dimension must be unambitious.

A basic size is a nominal size which is used as a reference of ideal size from which the limits are calculated. A basic size is common to two fitting parts at their interface. A *maximum material condition* (MMC) is a dimension of a part feature

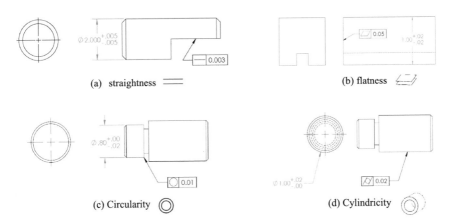

Fig. 4.27 *Form* control symbols and measurements

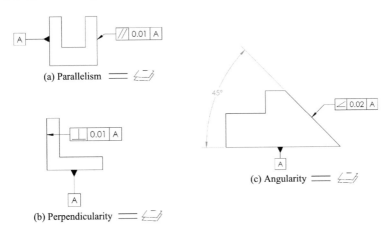

Fig. 4.28 *Orientation* control symbols and measurements

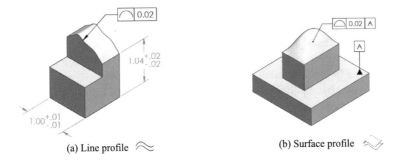

Fig. 4.29 *Profile* control symbols and measurements

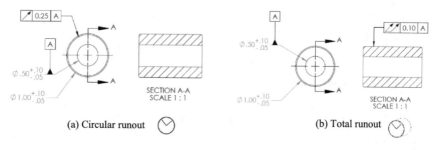

Fig. 4.30 *Runout* control symbols and measurements

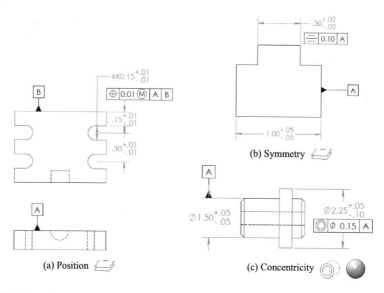

Fig. 4.31 *Location* control symbols and measurements

when it is composed of the most of material, and *a least material condition* (LMC) is a dimension of a part feature when it is composed of the least material. *Limits* are the extreme values of a dimension; *tolerance* is a difference of the lower and upper limits. A tolerance can be specified by (1) the limits from the basic size or (2) the general notes in block or other places in engineering drawing.

When a part has multiple dimensions in one direction, either chain dimensioning or baseline dimensioning can be applied. As shown in Fig. 4.32, *chain dimensioning* defines multiple dimensions in a sequence in which the next dimension is placed directly adjacent to the previous one, while *baseline dimensioning* defines multiple dimensions concurrently based on the same reference.

Tolerance stacking refers to the combination of the tolerances of multiple dimensions. As shown in Fig. 4.32a, the tolerance of *an indirect dimension* of two entities

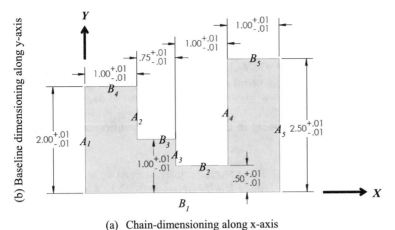

(a) Chain-dimensioning along x-axis

Fig. 4.32 The schematics of chain and baseline dimensioning

(A_1A_5) is a stacked-up from those of the tolerances of the dimensions $(A_1A_2, A_2A_3, A_3A_4, A_4A_5)$ from where the indirect dimension is formed, i.e.,

$$\left. \begin{array}{l} Norminal\ of\ A_1A_5 = A_1A_2 + A_2A_3 + A_3A_4 + A_4A_5 = 3.75 \\ Tolerance\ of\ A_1A_5 = \dfrac{Upper\ limits\ of\ A_1A_2 + A_2A_3 + A_3A_4 + A_4A_5}{Lower\ limits\ of\ A_1A_2 + A_2A_3 + A_3A_4 + A_4A_5} = \begin{array}{l} +0.04 \\ -0.04 \end{array} \end{array} \right\} \quad (4.7)$$

In the schematic of the baseline dimensioning in Fig. 4.32b, the common reference (B_1) is used to dimension the entities (B_2, B_3, B_4, B_5) along the Y-axis, and no tolerance stack-up occurs since each entity is dimensioned from the common reference.

Example 4.2 Determine the dimensioning schematic to obtain a better accuracy of the dimension A_2A_5 in Fig. 4.33.

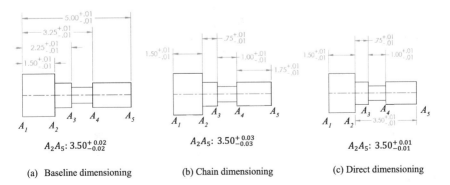

(a) Baseline dimensioning   (b) Chain dimensioning   (c) Direct dimensioning

Fig. 4.33 The schematics of chain, baseline, and direct dimensioning

Solution For the baseline dimensioning in Fig. 4.33a, the dimension of A_2A_5 is indirectly derived as

$$
\left.
\begin{array}{l}
Norminal\ of\ A_2A_5 = A_1A_5 - A_1A_2 = 3.50 \\
Tolerance\ of\ A_2A_5 = \dfrac{Upper\ limit\ of\ A_1A_5 - Lower\ limit\ of\ A_1A_2}{Lower\ limit\ of\ A_1A_5 - Upper\ limit\ of\ A_1A_2} = \dfrac{+0.02}{-0.02}
\end{array}
\right\} \quad (4.8)
$$

For the chain dimensioning in Fig. 4.33b, the dimension of A_2A_5 is indirectly derived as

$$
\left.
\begin{array}{l}
Norminal\ of\ A_2A_5 = A_2A_3 + A_3A_4 + A_4A_5 = 3.50 \\
Tolerance\ of\ A_2A_5 = \dfrac{Upper\ limits\ of\ A_2A_3 + A_3A_4 + A_4A_5}{Lower\ limits\ of\ A_2A_3 + A_3A_4 + A_4A_5} = \dfrac{+0.03}{-0.03}
\end{array}
\right\}
$$
$$(4.9)$$

For the direct dimensioning in Fig. 4.33c, the dimension of A_2A_5 is directly obtained as $A_2A_5 : 3.50^{+0.01}_{-0.01}$. By comparing the tolerances of A_2A_5 in three schematics, direct dimensioning yields the most accuracy for the dimension A_2A_5.

4.4.4 Engineering Fits

When parts are assembled into products, *engineering fits* are used to represent the relations of matting parts. The amount of an engineering fit determines whether or not two parts can move or rotate with each other independently. An engineering fit is generally referred as a pair of *shaft and hole*. As shown in Table 4.2, engineering fits can be classified into four types, i.e., *clearance fits, interference fits, transition fits*, and *line fits*.

Engineering fits are classified based on the maximum and minimum clearances. *The maximum clearance* refers to the maximum amount of space that occurs at the interface of two fitting parts, and *the minimum clearance* is the minimum amount of space that occurs at the interface of two fitting parts. The minimum clearance is also called as *allowance. The clearances of a fit* are evaluated based on the shaft and hole dimensions under MMC and LMC conditions as

Table 4.2 Types of engineering fits and conditions

Fit type	Description	Condition
Clearance fit	There is always space at the interface of two parts	*Min. clearance > 0*
Interference fit	There never be a space at the interface	*Max. clearance ≤ 0*
Transition fit	A shaft and hole may be either of interfered or shaped	*Max. clearance > 0* *Min. clearance < 0*
Line fit	There is a space or a contact at the interface	*Max. clearance > 0* *Min. clearance = 0*

$$\left.\begin{array}{l} Max.\ Clearance = LMC_{hole} - LMC_{shaft} \\ Min.\ Clearance = MMC_{hole} - MMC_{shaft} \end{array}\right\} \qquad (4.10)$$

where MMC_{hole} and LMC_{hole} are the hole sizes and MMC_{shaft} and LMC_{shaft} are the shaft sizes under MMC and LMC, respectively.

The third column in Table 4.2 gives the conditions of the minimum and maximum clearances for different fit types.

Example 4.3 Determine the fit types of shafts and holes in Table 4.3.

Solution Equation (4.10) is used to evaluate the maximum and minimum clearances, and the clearances are then used determine the fit types, and the results are shown in Table 4.4 (Fig. 4.34).

In ASME standards, *running or sliding* (RC) fits are classified into 9 levels from 1 for the tightest to 9 for loosest. The locational fits are classified into (1) *locational clearance* (LC) fits with 11 levels, (2) *locational transition* (LT) fits with 6 levels, and (3) *locational interference* (LN) fits in 5 levels. The corresponding tolerances are standardized. Table 4.5 gives the excerpted tolerances for the fits of RC5 to RC9 and a basic size range of (0.0 in, 19.69 in). In the table, the first column is the range of basic size, the other columns are for standard limits for shaft and hole fits.

Example 4.4 Use Table 4.5 to determine the tolerances of hole and shaft fit with RC8 for a basic size of 0.50.

Table 4.3 The shaft and hole dimensions (Example 4.3)

Case	Shaft	Hole
1	$1.5000^{0.0}_{-0.005}$	$1.5000^{0.0}_{-0.005}$
2	$0.850^{+0.007}_{+0.002}$	$0.850^{0.0}_{-0.005}$
3	$0.575^{+0.003}_{-0.002}$	$0.575^{0.0}_{-0.004}$
4	$1.250^{0.0}_{-0.003}$	$1.250^{+0.005}_{-0.0}$

Table 4.4 Determining fit types by maximum and minimum clearances

Case	LMC_{shaft}	MMC_{shaft}	LMC_{hole}	MMC_{hole}	Max clearance	Min clearance	Fit type
1	1.495	1.500	1.506	1.501	0.001	0.011	Clearance
2	0.852	0.857	0.850	0.845	−0.002	−0.012	Interference
3	0.573	0.578	0.575	0.571	0.002	−0.007	Transition
4	1.247	1.250	1.255	1.250	0.005	0	Line

Note that an engineering fit is for the relation of two mating parts; either a hole-basis system or a shaft-basis system can be used to specify the fit. As shown in Fig. 4.44, in *a hole-basis system*, the hole size remains constant to determine shaft sizes based on the specified fits. In *a shaft-basis system*, the shaft size remains constant to determine hole sizes based on the specified fits

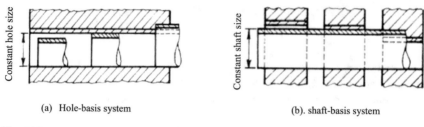

(a) Hole-basis system   (b). shaft-basis system

Fig. 4.34 Hole-basis system and shaft-basis system

Solution *Firstly*, the basic size is used to locate the row where the size falls in the given range: 0.5(0.40, 0.71). *Secondly*, the fit RC8 is used to locate the column. *Thirdly*, the tolerances at the intersections of specified column and row are interpolated as the dimensions of hole and shaft as

$$Hole: \quad 0.5000^{0.0028}_{0.0000} \quad Shaft: \quad 0.5000^{-0.0035}_{-0.0051}$$

Note the limits in Table 4.5 has a unit of thousandth of an inch.

In the ISO system, an alphanumeric code called the *international tolerance* grade number (IT#) is used to specify the tolerance ranges of fits, an uppercase represents

Table 4.5 Tolerance of RC5 to RC9 fits

Nominal Size Range Inches		Class RC5		Class RC6		Class RC7		Class RC8		Class RC9	
		Standard Limits		Standard Limits		Standard Limits		Standard Limits		Standard Limits	
Over	To	Hole	Shaft	Hole	Shaft	Hole	Shaft	Hole	Shaft	Hole	Shaft
0	−0.12	+0.6 / 0	−0.6 / −1.0	+1.0 / 0	−0.6 / −1.2	+1.0 / 0	−1.0 / −1.6	+1.6 / 0	−2.5 / −3.5	+2.5 / 0	−4.0 / −5.6
0.12	−0.24	+0.7 / 0	−0.8 / −1.3	+1.2 / 0	−0.8 / −1.5	+1.2 / 0	−1.2 / −1.9	+1.8 / 0	−2.8 / −4.0	+3.0 / 0	−4.5 / −6.0
0.24	−0.40	+0.9 / 0	−1.0 / −1.6	+1.4 / 0	−1.0 / −1.9	+1.4 / 0	−1.6 / −2.4	+2.2 / 0	−3.0 / −4.4	+3.5 / 0	−5.0 / −7.2
0.40	−0.71	+1.0 / 0	−1.2 / −1.9	+1.6 / 0	−1.2 / −2.2	+1.6 / 0	−2.0 / −3.0	+2.8 / 0	−3.5 / −5.1	+4.0 / 0	−6.0 / −8.8
0.71	−1.19	+1.2 / 0	−1.6 / −2.4	+2.0 / 0	−1.6 / −2.8	+2.0 / 0	−2.5 / −3.7	+3.5 / 0	−4.5 / −6.5	+5.0 / 0	−7.0 / −10.5
1.19	−1.97	+1.6 / 0	−2.0 / −3.0	+2.5 / 0	−2.0 / −3.6	+2.5 / 0	−3.0 / −4.6	+4.0 / 0	−5.0 / −7.5	+6.0 / 0	−8.0 / −12.0
1.97	−3.15	+1.8 / 0	−2.5 / −3.7	+3.0 / 0	−2.5 / −4.3	+3.0 / 0	−4.0 / −5.8	+4.5 / 0	−6.0 / −9.0	+7.0 / 0	−9.0 / −13.5
3.15	−4.73	+2.2 / 0	−3.0 / −4.4	+3.5 / 0	−3.0 / −5.2	+3.5 / 0	−5.0 / −7.2	+5.0 / 0	−7.0 / −10.5	+9.0 / 0	−10.0 / −15.0
4.73	−7.09	+2.5 / 0	−3.5 / −5.1	+4.0 / 0	−3.5 / −6.0	+4.0 / 0	−6.0 / −8.5	+6.0 / 0	−8.0 / −12.0	+10.0 / 0	−12.0 / −18.0
7.09	−9.85	+2.8 / 0	−4.0 / −5.8	+4.5 / 0	−4.0 / −6.8	+4.5 / 0	−7.0 / −9.8	+7.0 / 0	−10.0 / −14.5	+12.0 / 0	−15.0 / −22.0
9.85	−12.41	+3.0 / 0	−5.0 / −7.0	+5.0 / 0	−5.0 / −8.0	+5.0 / 0	−8.0 / −11.0	+8.0 / 0	−12.0 / −17.0	+12.0 / 0	−18.0 / −26.0
12.41	−15.75	+3.5 / 0	−6.0 / −8.2	+6.0 / 0	−6.0 / −9.5	+6.0 / 0	−10.0 / −13.5	+9.0 / 0	−14.0 / −20.0	+14.0 / 0	−22.0 / −31.0
15.75	−19.69	+4.0 / 0	−8.0 / −10.5	+8.0 / 0	−8.0 / −12.0	+6.0 / 0	−12.0 / −16.0	+10.0 / 0	−16.0 / −22.0	+16.0 / 0	−25.0 / −35.0

Note that unit of limit is thousandth of an inch.

Fig. 4.35 Example of fit specifications in ISO system

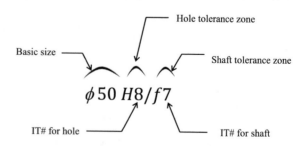

Hole tolerance zone

Shaft tolerance zone

Basic size

$\phi\,50\;H8/f7$

IT# for hole

IT# for shaft

the hole tolerance, and lowercase represents the shaft. Figure 4.35 shows an example of the fit specifications in the ISO system.

4.4.5 Computer-Aided GD&T (DimXpert)

A complete part or assembly model must include the GD&T information for all of the features and relations, while manually adding GD&T annotations in a CAD model is trivial and error-prone. *Solidworks DimXpert* is an automated GD&T tool, which is used to add both of geometric tolerances and dimensional tolerances in CAD models (see Fig. 4.36). The dimensions and tolerances by DimXpert are compatible to ASME Y14.5, Y14.41, and equivalent ISO standards. DimXpert supports engineers in using dimensions and constraints to express the design intents and define the specifications for manufacturing processes and inspections of parts.

DimXpert supports the full annotations in 3D models which are viewable as *edrewing*. It helps to avoid manual errors and maintains the consistency of GD&T settings. The *TolAnalyst* tool can be used to identify *over- or under-toleranced* parts graphically, and it can analyze tolerance stacks automatically. Especially, DimXpert is an ideal tool to create GD&T annotations for the machined parts such as those with turned, drilled, and milled features. DimXpert has been widely used to annotate sheet metals, castings, and machined elements such as gears, threads, and cams.

DimXpert can be used to automatically identify machined features. As shown in Fig. 4.27, the identified features will be listed in *the Features Tree* under the DimXpertManager. Figure 4.28 shows that identifiable machined features include *hole, cone, countersink, counterbore, plane, slot, notch, boss, cylinder, width, pocket, surface, chamfer*, and *fillet* (Figs. 4.37 and 4.38).

The *ShowToleranceStatus* tool in DimXpert is capable of detecting over-dimensioned or under-dimensioned feature. As shown in Fig. 4.39, running this tool in a CAD model colors the features in such a way that the features in a *red*

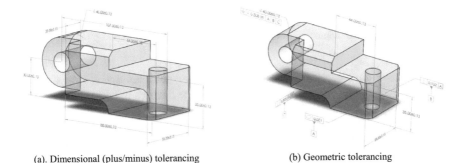

(a). Dimensional (plus/minus) tolerancing   (b) Geometric tolerancing

Fig. 4.36 DimXpert for GD&T (Steeves 2016)

DimXpertManager

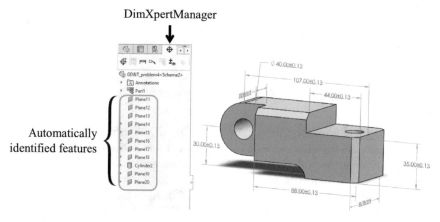

Fig. 4.37 Manufacturing features identified by DimXpert

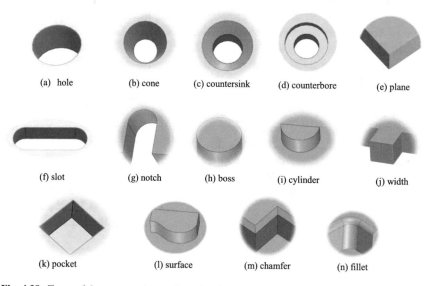

(a) hole   (b) cone   (c) countersink   (d) counterbore   (e) plane

(f) slot   (g) notch   (h) boss   (i) cylinder   (j) width

(k) pocket   (l) surface   (m) chamfer   (n) fillet

Fig. 4.38 Types of the supported manufacturing features in DimXpert

color are *over-dimensioned*, the features in a *yellow* color are under-dimensioned, the features in a *green* color are well defined, and the features in other colors are not detectable by DimXpert.

In creating GD&T annotations for CAD models, the settings of DimXpert can be customized by setting the *DocumentProperties* in Solidworks as shown in Fig. 4.40a. One has the options of (1) GD&T standards and (2) the methods to specify the tolerances. In addition, the detailed settings for the methods and styles of *Size Dimension*,

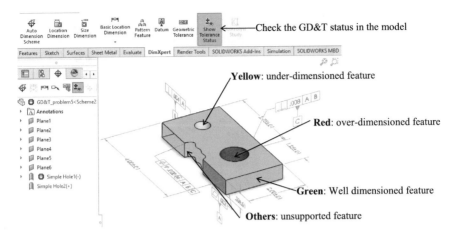

Fig. 4.39 Using the *ShowToleranceStatus* tool to detect under- and over-dimensioned features

Location Dimension, Chain Dimension, Geometric Tolerance, Chamfer Controls, and *Displace Options* can be made through the interfaces as shown in Fig. 4.40 b–g.

Example 4.5   Figure 4.41a shows the dimensions of a machined part. Use the *AutoDimensionScheme* tool tin DimXpert o annotate GD&T information automatically.

Solution *Firstly*, the part is modeled based on the dimensions specified in Fig. 4.41a. *Secondly*, the AutoDimensionScheme tool is used to specify (1) the part type; (2) tolerance type; (3) the pattern dimensioning; and most importantly, the primary, secondary, and tertiary references (see Fig. 4.41b). *Thirdly*, the AutoDimension-Scheme tool is executed to create G&D annotations automatically.

4.5   Manufacturing Processes

Manufacturing processes aim to transform raw materials into finish products. In a manufacturing system, each manufacturing process is a value-added step in the whole transformation from starting materials to final products. For *discrete manufacturing* in which discrete parts are made, a manufacturing process is to change the geometry, properties, and appearance of materials. Figure 4.42 shows the input and output of a manufacturing process. The input is a workpiece with the materials at the starting state, the manufacturing resources such as machinery, tooling, and labor are applied to transform the workpiece into new state. In addition, a manufacturing process involves some side effects such as wastes. A manufacturing process is also an economic activity since it adds the value to workpiece.

As shown in Fig. 4.43, manufacturing processes can be classified in terms of the transformation types (Groover 2012). At the highest level, manufacturing processes

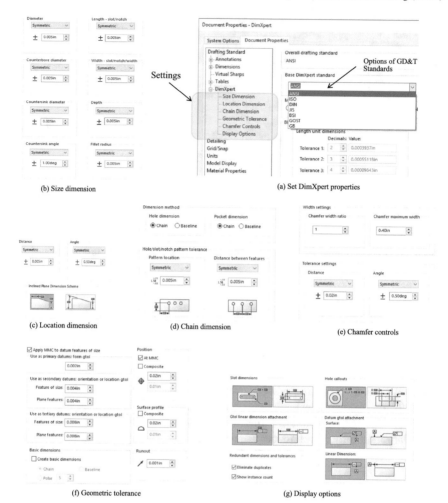

Fig. 4.40 Customize DimXpert settings

are classified into *processing operations, assembly operations,* and other *non-value-added processes.* A processing operation applies to individual parts, an assembly operation applies to two of a group of parts or components, and a non-value-added process such as inspection or transportation may apply to parts or components.

The manufacturing processes in the category of processing operations are further classified into *shaping processes, surface treatments,* and *property enhancement.* A *shaping process* is to create geometric features on parts, and geometric features can be created by *solidification, deforming processes, particulate processes, additive manufacturing,* and *material removal processes.*

Depending on the amount of geometric changes over parts, shaping processes can also be classified into primary shaping processes and secondary shaping processes.

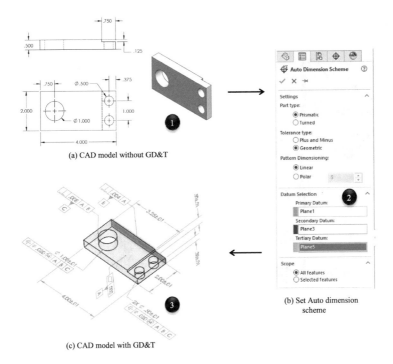

(a) CAD model without GD&T

(b) Set Auto dimension scheme

(c) CAD model with GD&T

Fig. 4.41 Example of auto dimension scheme

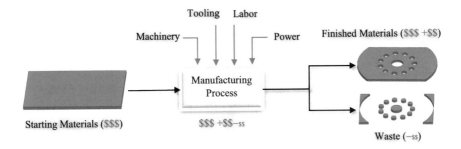

Fig. 4.42 Description of manufacturing process

A primary shaping process is to create a rough geometric shape of part; common primary shaping processes are *casting, powder metallurgy, injection modeling, metal sheet forming,* and *forging. A secondary shaping process* is to refine geometric dimensions of parts; common secondary shaping processes are *turning, drilling, milling, threading, boring, shaping, planning, sawing, broaching, hobbing,* and *grinding.* In addition, materials can be added or removed by some unconventional ways. Therefore, shaping processes include other types of machining processes such as *electrochemical machining* (ECM) and *laser beam machining* (LBM).

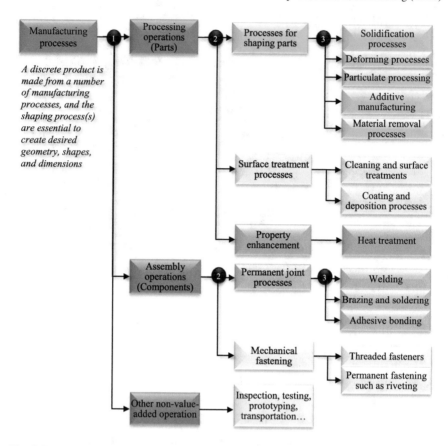

Fig. 4.43 Manufacturing processes—classification

Surface treatment aims to change the material properties over the surfaces of parts. Surface properties can be enhanced by *cleaning, coating,* and *deposition.* Some common surface treatments are *polishing, cleaning, tumbling, sanding, electroplating, deburring, buffing, painting, anodizing, coating, sand blasting, galvanizing, anodizing, honing,* and *lapping. Property enhancement* is to enhance the material properties in the bodies of parts. Material properties can be changed by the heat treatments since temperature has the great impact on material structures and characteristics. Some common heat treatments are *hardening, annealing, tempering, normalizing, grain refining,* and *shot peening.*

An assembly operation is to join two or more parts or components as new component. If an assembly is at an intermediate state of a product, it is called *a subassembly* or *component.* Parts can be joined either *permanently* or *non-permanently.* Therefore, assembly processes are classified into two types—the processes for nonpermanent and permanent joints, respectively. *Nonpermanent joints* are usually mechanical joints which are formed by some machine elements such as screws and bolts, snap fits, and shrink fits. Permanent joints can be formed by chemical bonding or mechanical means such as *welding, brazing soldering, riveting, press fitting, sintering, shrink filling, stitching,* and *stapling.*

Other manufacturing processes do not change workpieces; therefore, they do not add values to products. The examples of non-value-added processes are *inspection, testing, prototyping, transportation,* and *material handling.*

4.5.1 Shaping Processes

A shaping process is to generate desired geometric shape or features on parts. Part geometry can be generated in various ways. In a shaping process, it is desirable that the raw material has a *lower strength* and *higher formability.* Since the strength and formability relate closely to the temperature, the operating temperature is usually the most critical factor in a shaping process: the higher the operating temperature is, the lower strength and higher formability the material is, and the better the material can be shaped. Therefore, shaping processes are classified based on the temperature range when a shaping process is performed.

Figure 4.44 shows the classification of shaping processes based on the operating temperature in comparison with the melting temperature T_m of the raw material. *Cold working, warm working, hot working, sintering,* and *casting* are performed at the temperature ranges of below $0.3 \cdot T_m$, $(0.3 \cdot T_m, 0.5 \cdot T_m)$, $(0.5 \cdot T_m, 0.75 \cdot T_m)$, $(0.75 \cdot T_m, T_m)$, and above T_m, respectively. The operating temperature is used to distinguish one type of the shaping process from another. For example, when an operating temperature exceeds $0.5 \cdot T_m$, the material can be shaped by *recrystallization* or *solidification.*

Due to the differences of operating temperatures and amount of geometric changes, the raw materials have different starting conditions for various shaping

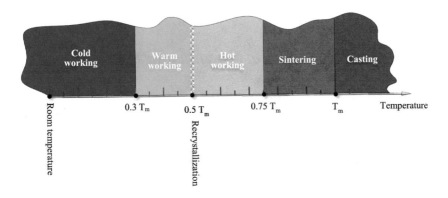

Fig. 4.44 Classification of shaping processes based on operating temperature

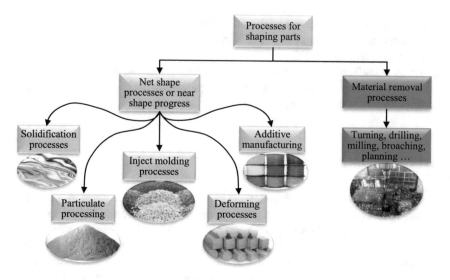

Fig. 4.45 Raw materials in different shaping processes

processes. As shown in Fig. 4.45, the starting material in *a casting process* is in the liquid state and the material does not hold the desired shape until it has been solidified in a cavity. The starting material in *a particulate process* is in the powder state, and the starting material is pressed in a mold and heated until that sintering occurs to form the crystallized geometry. The starting material in *an injection molding* is small particles and the particles are heated and pressed into a mold and then solidified as a finished part. The starting material in *a deforming process* is in a bulk or sheet form from primary shaping processes. The material is deformed mechanically with/without thermal loads. The starting material in *additive manufacturing* can be in the state of powder, strip, or fluid state. The material is heated, solidified, or sintered *layer by layer* to create part geometry.

4.5.2   Design and Planning of Manufacturing Process

Design of manufacturing processes is to select manufacturing methods and design their implementations. Manufacturing processes should be designed to make products economical and competitive. *Planning of manufacturing processes* is to transform the specifications and features of products into the instructions of machine operations. Typical tasks in planning a manufacturing process include *selection of machines and tools, programming of machines,* and *determination of process parameters.*

Design of manufacturing processes is critical and greatly affected by other design activities in the product lifecycle. Product development is usually an iterative process

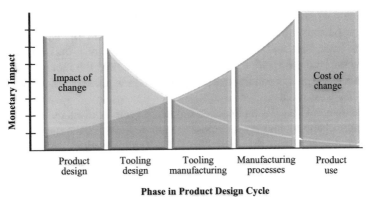

Fig. 4.46 Computer-aided design in reducing manufacturing cost

where continuous changes are made to satisfy design constraints at different design stages. As shown in Fig. 4.46, the requirements of manufacturing processes should be considered at the earliest possible stage to reduce cost in changing the solutions relevant to manufacturing. Similar to CAD tools, computer-aided techniques help to identify manufacturing defects at the stage of product design.

4.6 Simulation of Manufacturing Processes—Mold Filling Analysis

In designing a manufacturing process, it is challenging to avoid defects and optimize process parameters due to many sources of uncertainties. Computer simulation helps designers to understand the impact of process parameters, and optimize manufacturing processes in the digital environment. In this section, some basic analyses of injection modeling processes are discussed to illustrate how computer-aided manufacturing (CAM) can be deployed for design and simulation of manufacturing processes.

4.6.1 Injection Molding and Machines

Injection molding is one of the commonly used methods to make net-shape parts. In injection molding, polymer is heated to form a plastic flow and then fill in a mold cavity under pressure. The material in the cavity is solidified as molded part(s). Injection molding is widely used to make parts with complex shapes. To improve productivity, one mold may contain multiple cavities, so multiple moldings can be made in a production cycle.

Figure 4.47 shows the basic components of an injection molding machine. Two main components are an injection unit and a clamping unit. *The injection unit* has a *barrel* to keep, heat, and feed the material and a *screw* to mix, pressurize, and inject the material into the mold. *The clamping unit* is to hold the mold when the mold is filled and the part is solidified; the clamping unit is released when the part(s) are ejected from the mold.

As shown in Fig. 4.48, an injection modeling cycle involves the following steps. Firstly, the clamping unit is set to close the mold. *Secondly,* the raw material is fed, heated, and injected into the mold. *Thirdly,* the melt in cavity is solidified when the

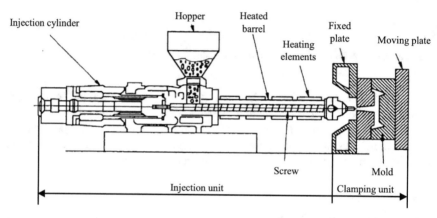

Fig. 4.47 Description of injection molding machines

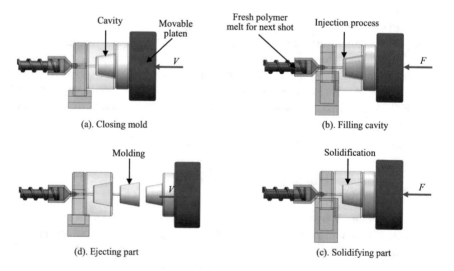

Fig. 4.48 Main steps of injection molding process

mold is filled. *Fourthly*, the clamping unit is released to open the mold, and the part is ejected from the cavity. The ejected part may need the post-processing to remove unwanted materials such as solidified materials in runners from the part. Note that a runner is a path by which the melt is filled into the enclosed cavity.

4.6.2 Moldability and Design Factors

Moldability measures how well the part geometry may conform to the mold when the molten plastic is injected to the mold. An injection molding process should be designed to avoid any defects when the material is filled into the mold cavity. Figure 4.49 shows the impact of operating temperature and pressure on the moldability. These two variables are most critical since they determine the feasible working range (area 1) of an injection modeling process. Too low or high pressure will cause the defects of *short shot* (area 2) and *flash* (area 3), respectively. Too low or high temperature will cause the defects of *melting* (area 4) and *thermal degradation* (area 5), respectively.

The cycle time increases when the mold temperature increases, and this takes a longer time to crystallize the melt in the mold. Therefore, mold temperature affects the cooling rate. A high cooling rate helps in reducing the crystallization time and reducing shrinkage of the molded parts. However, if too less time is used in cooling the part, it is prone to shrink or causes warpage after the molding process.

A high pressure helps to fill the mold quickly to avoid premature freezing of the flow of melt. Note that the clamping mechanism is used to hold the model when a high pressure is applied, and the clamping force is determined by the pressure and the effective area of cross section of the melt flow.

Fig. 4.49 Temperature and pressure for *moldability* of injection molding processes

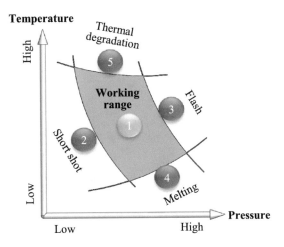

4.6.3   Design of Injection Molding Systems

Figure 4.50 shows the procedure of designing an injection molding system; it consists of five major tasks as given below.

(1).   *Firstly*, parts are modeled and analyzed to clarify the requirements of injection modeling; more specifically, the parameters such as *one-shot weight, required cores, number of cavities, injection locations,* and *clamping force* and *stroke* should be determined.

(2).   *Secondly*, the mold cavities are designed to (i) select molding types such as *hot runners, cold runners,* and *conventional sprues*; (ii) select the materials for mold *bases, inserts, plates,* and *cores*; and (iii) tailor the mold cavities to geometric shapes of parts.

(3).   *Thirdly*, the mold cavities are optimally placed, common placement options are *star, symmetrical,* and *in-line arrangement.* The mold assembly is modeled, simulated, and verified.

(4).   *Fourthly*, other components for injecting and cooling melts and ejecting parts are designed. Especially, the gating system is designed; common layouts of a gating system are *conventional, pinpoint, sub-marine, flash, tab, disk,* and *diaphragm.* The ejection component is designed; some common options are

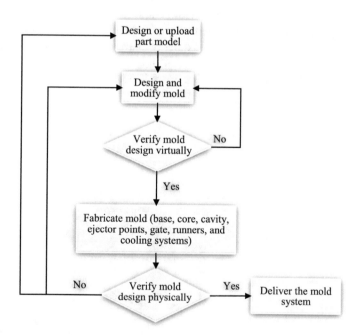

Fig. 4.50 Design of injection molding process

pins, stripper rings, stripper plates, slides, and *side cores*). The type of venting parting lines is specified.

(5).   *Fifthly,* the injection molding system is reviewed as a whole against the specified functional requirements; this ensures that all design constraints are satisfied before the injection molding system is fabricated.

4.6.4   Design Variables and Considerations

The discussion in Sect. 4.5.2 shows that two main factors to affect the moldability are operating temperature and pressure. However, injection modeling involves numerous other design variables, which might affect the success of the injection modeling process.

(1)   Shrinkage

The material in an injection modeling process involves in the change of temperature in a wide range. The thermal expansion coefficient of the material varies with the temperature change. The higher the temperature is, the higher the thermal expansion coefficient is. Accordingly, the dimensions of part in the mold cavity will be shrunk in crystallization, and such shrinkage has to be compensated in the design process. The amount of shrinkage depends on (1) the type of the material and (2) the range of temperature changes.

Table 4.6 shows the percentage of the shrinkage for four commonly used materials. To obtain an expected dimension, the corresponding dimension of a mold cavity is compensated as

$$D_c = D_p\left(1 + S + S^2\right) \tag{4.11}$$

where D_c and D_p are the corresponding dimensions on the mold and the part, respectively, and S is the percentage of the shrinkage given in Table 4.6. To yield better dimensional accuracy, the following approaches can be adopted:

(1)   Increasing the operating pressure to accelerate the injection of the melt into the mold cavity.

(2)   Increasing the operating temperature to reduce the viscosity of the polymer melt and increase the density of packing in the mold.

Table 4.6 Shrinkage of four commonly used plastics

Plastics	Typical shrinkage, mm/mm (in/in)
Nylon-6	0.020
Polyethylene	0.025
Polystyrene	0.004
polyvinyl chloride (PVC)	0.005

(3)   Increasing the compaction time and supplying the melt in the cooling process.

(2)   **Economic factor**

The cost for molds and tooling is a significant part of the cost to manufacturing processes, since the parts with different geometries or features require different molds and tooling, which are fabricated in a dimensional tolerance more strict than the parts (Bi et al. 2001). In designing an injection molding process, various process types are available, and these processes have to be analyzed and compared from the cost aspects. A high tooling cost is only allowed when the volume of certain parts is high.

Overall, injection molding is economically sound when the volume of a product is over 10,000. When the product volume is lower than 1000, *a compression molding* process should be used; when the product volume is in the range of (1000, 10,000), *a transfer molding process* should be taken into consideration.

(3)   **Design features of molded parts**

The more complex a part is, the high cost the mold and tooling are. From this perspective, adopting injection modeling is beneficial since multiple functional parts with same materials can be made as an assembly when it becomes feasible.

Design for manufacturing (DfM) can be applied in designing a molded part to minimize injection defects such as *molded-in stress, flashes, sink marks*, and *surface blemishes*. This is achieved by satisfying the requirements of an injection molding process for *flow lengths, weld line locations, injection pressures, clamping requirements, scrap rate, easiness of part assembly*, and *the needs of secondary operations* such as degating, painting, and drilling. In designing a molded part, the following features must be paid a special attention:

(1).   *Wall thickness*: A thick wall consumes more materials and tends to cause *warping* since the shrinkage along the depth is uneven; it takes a long time to harden the materials. It will be desirable to have thin walls with evenly distributed thickness.

(2).   *Reinforcing ribs:* To enhance the stiffness and strength at certain cross sections, adding reinforcing ribs is a preferable option. However, the thickness of a reinforcing rib must be less than the wall to reduce sink marks on walls.

(3).   *Corner radii and fillets*: A sharp external or internal corner causes the stress concentration in the application, and it affects the smoothness of melt flow in injection molding. Sharper corners must be avoided to eliminate surface defects on mold parts.

(4).   *Holes:* It is mostly feasible to include holes in a molded part, while the impact of having holes on the complexity of the mold and part ejection must be taken into consideration.

(5).   *Drafts:* To eject parts from the mold, the part must have the drafts on lateral surfaces. The drafts recommended for thermosetting and thermoplastic materials are ~1/2° to 1° and ~1/8° to 1/2°, respectively.

(6). *Tolerances:* Shrinkage is predictable under closely controlled conditions; however, loose dimensional tolerances are preferable since that the uncertainties of process parameters and variations of part geometries affect the shrinkage and distribution over surfaces.

4.6.5   Defects of Molded Parts

Similar to other manufacturing processes, injection molding involves many design factors. It is very challenging to develop a realistic mathematic model for process optimization. Therefore, DfM is an iterative process to make the trade-offs among some mutually conflicted design criteria, and the basic requirement is to minimize the defects of parts, which might be caused by different reasons such as the design flaws of parts, molds, and tools, not well-prepared materials, and inappropriate operating conditions. Injection modeling leads to the following common defects.

(1).   **Weldlines**

Weldline is an ill-filled line when two or more melt flows meet and remain undissolved. The weldline is a boundary of two melt flows. Figure 4.51 shows the process in which a weldline is formed by two flows. To alleviate weldline defects, the following methods can be used: (1) increasing the injection flow rate and increasing the temperature of the mold; (2) reducing the temperature of the melt and increasing the injection pressure; and (3) optimizing the gate positions and the melt flows with less or insignificant merging locations.

(2).   **Flashes**

Flashes are formed at the parting line/surface of the mold or at some locations where the ejectors are installed. Figure 4.52 shows the case where the flashes are developed at the parting plane. A flash is a phenomenon where the molten polymer smears out or sticks to the seam or gap between two surfaces. The common causes of flashes are

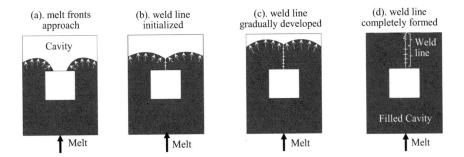

Fig. 4.51 Process of forming a weldline by two flows

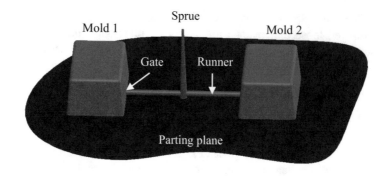

(a). without flash on parting plane

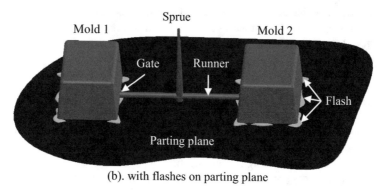

(b). with flashes on parting plane

Fig. 4.52 Flashes in injection molding

(1) poor conformance at two contact surfaces; (2) low viscosity of molten polymer; (3) high injection pressure; and (4) low clamping force.

The following methods can be used to reduce flashes (1) avoiding excessive thickness difference in part design; (2) reducing injection flow rate; (3) applying well-balanced pressure and clamping force to the mold assembly; (4) increasing clamping force; (5) improving the surface quality of the parting plane, ejector pins, and holes.

(3). **Short Shots**

Short shots are voids that are not fully filled by molten materials. Figure 4.53 shows the difference of parts with and without short shots. A short shot may occur at a location with a long flow distance to the gate. Common causes of short shots are (1) insufficient one-shot weight, (2) a low injection pressure, and (3) a low injection speed; this leads to the situation that the melt flow freezes before the mold is fully filled. Short shots can be improved by (1) increasing the injection pressure; (2) using air vents or degassing device for a high flow pressure drop; and (3) optimizing the mold design to smooth the melt flow.

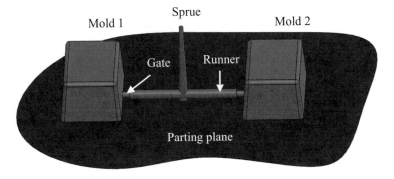

(a). without short shorts

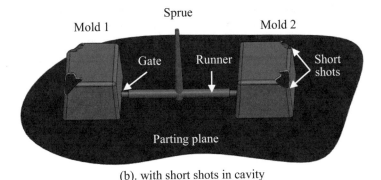

(b). with short shots in cavity

Fig. 4.53 Part with short shots

(4). **Warpages**

A warpage is formed when the part is removed from the mold. Figure 4.54 shows a part with warpages, which might be caused by (1) uneven shrinkage due to the thickness difference or temperature derivation and (2) low injection pressure and insufficient packing. To reduce warpages, the following methods can be used: (1) extending the cooling time before the part is ejected, (2) optimizing the positions of ejection pins,

Fig. 4.54 Part with warpages

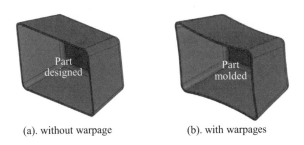

(a). without warpage   (b). with warpages

Fig. 4.55 Part with sink marks

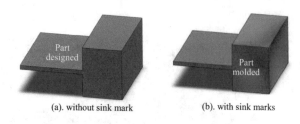

(a). without sink mark   (b). with sink marks

(3) enlarging draft angles, (4) balancing cooling lines, and (5) increasing packing pressure.

(5).   Sink marks

Sink marks may be generated when the melt in some areas (such as ribs or bosses) shrinks more than adjacent areas. Figure 4.55 shows the part with sink marks, which are caused by insufficient melts in the area before the flow is closed due to the solidification. The methods to reduce sink marks include to (1) increase packing pressure for more melt filling into the cavity, (2) keep the gates open for a longer time, and (3) reduce the temperature of the mold to strengthen the surface.

4.6.6   Mold Flow Analysis

Design of injection molding processes used to rely on designers' experiences with the aid of the trial-and-error methods. Nowadays, computer simulation tools provide the scientific methods to support the designs of injection molding processes. *Mold flow analysis* aims to simulate an injection modeling process and optimize the design of molds and process variables.

Figure 4.56 shows the inputs and outputs of *Solidworks Plastics* for mold flow analysis (Eastman 2019). The system inputs include *part model, material properties, thermal properties, melt and mold temperatures, runners, gating systems,* and *cooling systems.* The system outputs include the simulated results for *flows and fills, locations of weldlines, pressure to fill, pressure patterns, clamping force, temperature patterns, shear patterns, filling, temperature distribution, shear thinning, freezing and reheating,* and *temporary stoppages of flow.* Mold flow analysis can be used for the simulation-based optimization of injection molding processes.

As shown in Fig. 4.57, SolidWorks Plastics is integrated in the 3D solid modeling environment, and the part model can be directly utilized for mold flow analysis. A user-friendly wizard is provided to define and run a mold flow analysis step by step. Figure 4.57a shows the access to the SolidWorks Plastics, which is an *add-ins* of the SolidWorks. Figure 4.57b shows that the wizard for mold flow analysis is run to proceed automatic meshing process.

Following the wizard, analyzing the injection process for a mold part straightforward. Figure 4.58 shows six critical steps in a mold flow analysis: (1) generate

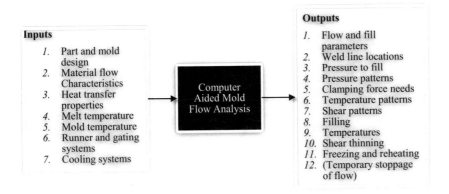

Inputs

1. Part and mold design
2. Material flow Characteristics
3. Heat transfer properties
4. Melt temperature
5. Mold temperature
6. Runner and gating systems
7. Cooling systems

Computer Aided Mold Flow Analysis

Outputs

1. Flow and fill parameters
2. Weld line locations
3. Pressure to fill
4. Pressure patterns
5. Clamping force needs
6. Temperature patterns
7. Shear patterns
8. Filling
9. Temperatures
10. Shear thinning
11. Freezing and reheating
12. (Temporary stoppage of flow)

Fig. 4.56 *Solidworks Plastics* for mold flow analysis

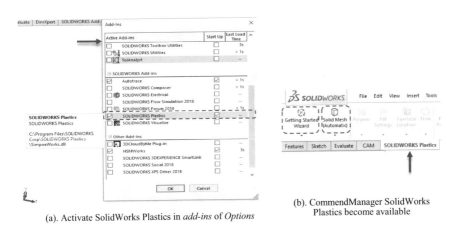

(a). Activate SolidWorks Plastics in *add-ins* of *Options*

(b). CommendManager SolidWorks Plastics become available

Fig. 4.57 Activate SolidWorks Plastics tool

mesh, (2) define material, (3) specify process parameters, (4) define the locations of injection, (5) run simulation, and (6) review the results. Figure 4.59a shows that the Solidworks Plastics has its capability of generating a mesh automatically on a surface or solid model. Figure 4.59b shows that injection locations may be automatically defined at step 4. Through the simulation, the designers can view and retrieve any type of the simulated data related to pack, flow, and warp. Figure 4.60 shows the list of attributes relevant to the mold flow. Figure 4.61 shows some examples of critical data from mold flow analysis that include the distribution of shrinkage, cooling time, pressure, mold temperature, and sink marks.

The SolidWorks Plastics also provides a set of advanced tools to control the process variables that the mold flows. Figure 4.62 shows the interface to access these

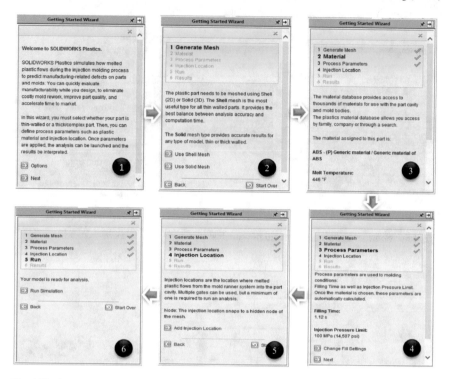

Fig. 4.58   Steps in mold analysis wizard

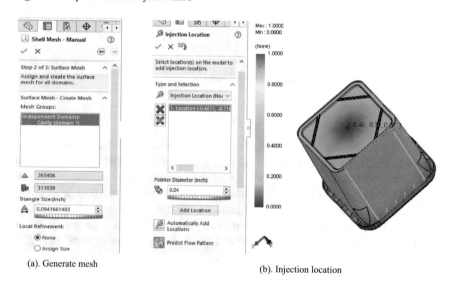

(a). Generate mesh

(b). Injection location

Fig. 4.59   Interfaces in the steps of *generate mesh* and *injection location*

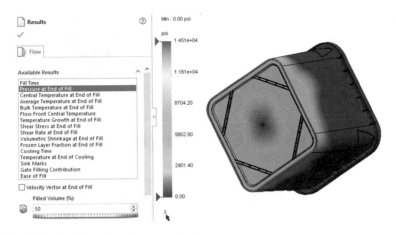

Fig. 4.60 List of attributes in mold flow analysis

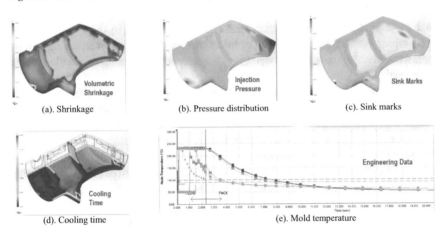

(a). Shrinkage   (b). Pressure distribution   (c). Sink marks

(d). Cooling time   (e). Mold temperature

Fig. 4.61 Critical attributes in mold flow analysis

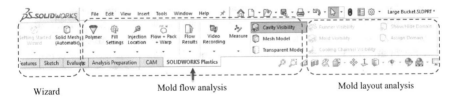

Wizard   Mold flow analysis   Mold layout analysis

Fig. 4.62 Advanced tools in SolidWorks Plastics

tools: (1) *polymer* to change material properties; (2) *filling settings* to setup injection molding parameters such as filling time, melt temperature, mold temperature, injection pressure limit, and clamp force limit; (3) *injection location* to add or change injection locations; (4) *flow + pack + warp* to specify the simulation type of interest; and (5) *flow results* to visualize the results.

4.7 Designs of Tools, Dies, and Molds (TDM)

Tools, dies, and molds (TDM) are essential to most of manufacturing processes such as casting, forging, injection modeling, and material removal processes. The geometries and motions of TDM determine part geometries. In particular, *a mold* or *die* should have the same exterior surfaces to the parts at their contact areas. From this perspective, the concepts of dies and molds are similar and they are used alternatively. However, a *die* is used to shape the materials in a deforming process such as forging and stamping operations, and a *mold* is used to shape the materials in a forming process such as casting or injection modeling. Moreover, a mold differs from a die in sense that a mold assembly is taken apart horizontally, but a die is taken apart vertically.

Due to low volumes and highly diversified TDM products, toolmakers are mostly *small- and medium-sized enterprises* (SMEs) who need highly skilled employees to design and fabricate tools. However, tooling is essential to all manufactured products, especially nearly half of tooling consumptions are in automotive and defense-related manufacturing. TDM companies offer the services to design, fabricate, and test machining tools for manufacturing enterprises. The growth of the manufacturing industry is greatly influenced by that of the TDM industry. Canie (2010) provided the data of import and export trades of TDM products in the Unites States from 1997 to 2010 in Table 4.7; it showed that the trade deficit was increased from $ 897 million to $ 4, 391 million. Currently, the companies in the United States mainly rely on foreign suppliers to make custom dies and molds. The shrinkage of TDM industry relates to the shrinkage of whole manufacturing industry. The shrinkage of the US manufacturing industry has been evidenced by many indicators such as employment rates, annual payrolls, and the number of newly established companies.

To strengthen a nation's manufacturing, expanding the TDM industry is critical, while US manufacturing is facing the challenge in attracting the talents to TDM

Table 4.7 Import and export trades of TDM from 1997 to 2010 in USA (Canie 2012)

Product	Year	Exports	Imports	Deficit
Molds	1997	$648 million	$1,291 million	$643 million
	2010	$717 million	$4,745 million	$4,028 million
Tools, dies and jigs	1997	$442 million	$696 million	$254 million
	2010	$488 million	$851 million	$363 million
Total	1997	$1,090 million	$1,987 million	$897 million
	2010	$1,205 million	$5,596 million	$4,391 million

industry. The statistics from the National Tooling & Machining Association (NTMA) showed that 95% of toolmakers had openings even in the economy with a high unemployment rate. Using new technologies was expected to improve the efficiency as much as 20% in TDM companies.

4.7.1 Design Criteria of TDM

Tooling design is a branch of manufacturing engineering that is composed of the methods and procedures to analyze, plan, design, make, and use tools, dies, and molds in manufacturing. TDM products are highly diversified since each product has to be tailored to specified functions, geometries, and manufacturing processes. While the functions of TDM products are highly diversified, the following design criteria are commonly applicable to any TDM products.

(1). **Cost**

Cost is an essential indicator to any manufactured products including TDM products. Since TDM products usually have low volumes, the unit cost is often one of the primary concerns. This is mostly critical to the companies who make household appliances, electronics, power hand tools, and housewares. Attractive price was identified as the leading competitiveness by toolmakers (United States International Trade Commission 2002).

The total cost of TDM is associated with the tasks involved in different phases of its product lifecycle. Figure 4.63 gives an estimation of cost at different phases. It is clear that adopting new technologies in designing, manufacturing, and assembling of TDM is expected to reduce cost. Therefore, computer-aided technologies have been widely used to improve productivity and competitiveness, increase machining capacity, and replace highly skilled labor.

(2). **Lead Time**

TDM products are mostly customized and manufactured in *pull production*. From the time when a user places an order to the time when the product is made and

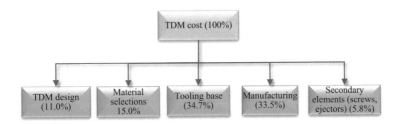

Fig. 4.63 Cost breakdown of TDM products

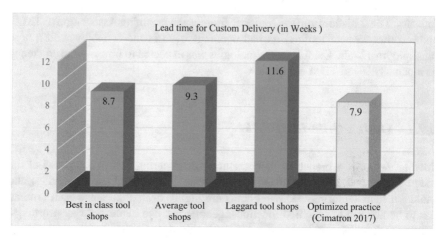

Fig. 4.64 Average lead times of TDM manufacturers

delivered to user, there is a period of the waiting time called a *lead time*. Due to the complexity and high-quality requirements of TDM products, average lead times of TDM products are long and usually in weeks. Figure 4.64 shows the performance indicators of the toolmakers based on the average lead time of their TDM products (Cimatron 2017); the toolmakers with the optimized practice can gain a time window of 3.7 weeks in capturing emerging business opportunities in comparison with less competitive toolmakers.

(3). Complexity

TDM products are highly diversified in terms of sizes, geometries, and applied manufacturing processes, but toolmakers have to make these at the same facilities to sustain manufacturing capabilities. (1) One toolmaker usually serves diversified customers in different industries; the requested TDM products are very different even if they are made from same manufacturing facilities. (2) Toolmakers are forced to encompass a wide range of tooling of a particular construction from low-end, simple fabrication, to highly intricate and technologically advanced products. The complexity of TDM products and manufacturing processes brings the challenge in (1) defining the commonalities of products for mass production and (2) reusing knowledge for new product development.

(4). Precision

The dimensional accuracy of the manufactured product relies on the precision of the toolings in the manufacturing processes. From this perspective, the precision of selected tooling should be determined by the required accuracy of products. Tooling with a low precision is suitable to some products such as plastic buckets in which the dimensional variants do not affect product uses. An average precision of tooling is applicable to the parts such as electronic components, keyboards, or clock covers. A high precision of tooling is required for parts that be assembled with other components under the tight tolerances such as cellphone housing and closures of food containers. The tooling error will be passed over to products whose accuracy is adversely affected. Typically, a tooling for high-precision products is expected to have a repeatable dimensional tolerance of ±0.00005 inch.

(5). **Quality**

Quality of a TDM product is measured by *durability, maintainability,* and *productivity.* The durability is measured by the product life span in making parts without the quality concern on parts or tools. The productivity is evaluated by the downtime and the throughput of parts in a unit time. The maintainability is assessed by the cost and easiness to prolong the life span, repair wears, prevent premature failure, and meet new requirements.

(6). **Materials**

Many factors should be taken into consideration in selecting materials for TDM products. Some common selection criteria are *strengths, toughness, hardness, hot hardness, machinability,* and *wear and corrosion resistance.* These factors might be conflicted with each other, for example, the criterion for a high strength and toughness is conflicted to the criterion for better machinability. Some trade-offs must be made among these criteria in selecting appropriate materials for TDM products. For example, if a TDM must sustain very high hot strength in its application, it likely be brittle that may be broken easily subjected to a dynamic load.

4.7.2 Computer-Aided Mold Design

The geometry of a die or mold must be designed to match the geometry of parts to be made. In this section, molds and dies are designed based on part models.

(1) **Mold Assembly**

The geometry of a part is given by the cavity in the mold assembly. Figure 4.65 shows an example of a mold assembly that consists of *a core* (male part), *a cavity* (female parts), and *a parting line or surface.* Multiple cores might be needed to define complex surfaces or some special features such as pockets, holes, and undercut features. A parting line or surface is to match or split the cores from the cavity in mold assembly or disassembly.

(2) **Computer-Aided Mold Design**

In this section, the *SolidWorks Mold Tool* is used as the computer-aided mold design tool to illustrate the procedure and steps in designing a mold assembly based on given part models. Figure 4.66 shows the interface of the Mold Tool commander

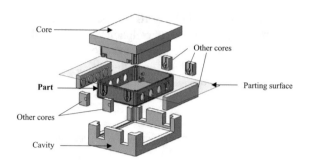

Fig. 4.65 Main components of a mold assembly

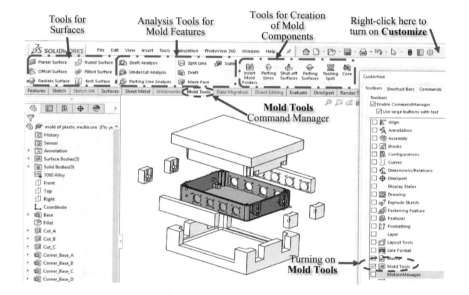

Fig. 4.66 *Mold Tool* command manager

manager. If the mold tool is unavailable in the default layout, it can be activated by right-clicking any item in the commend manager group and then checking the mold tool in the popped list of the command groups. The user may also add the mold tool by customizing the layout as shown in the right side of Fig. 4.66.

The mold tool commands are organized into three groups: (1) the tools used to modify, knit, delete, and create surfaces, and fix surfaces by filling holes; (2) the tools used to analyze the features of mold such as *drafts*, *undercuts*, and *parting lines*; and (3) the tools used to create the mold components such as cores and cavities. The procedure of computer-aided mold design is discussed in the next section.

Fig. 4.67 Procedure of computer-aided mold design

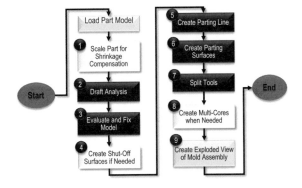

(3). **Design procedure**

Figure 4.67 shows the procedure of designing a mold based on the given part model:

(1) the part is modeled and scaled to compensate the shrinkage in the manufacturing process;
(2) the draft angles of lateral surfaces are analyzed to verify if the part can be taken out from the mold assembly;
(3) the faces with inappropriate drafts are fixed to ensure right drafts are placed;
(4) if a part has open space(s), add the boundary surfaces to make the core part and the cavity part separable by using the *Shut-Off Surfaces* tool;
(5) define a parting line to split the core and cavity by using the *Parting Line* tool;
(6) use the parting line to generate parting surfaces, so that the geometries and shapes of the core and cavity are defined, respectively;
(7) knit the parting surfaces with the other boundary surfaces as the solid for the core and cavity;
(8) split the solid into the core and cavity parts; and
(9) check if the model has an undercut(s), and define additional core(s) for the undercuts when it is needed.

Finally, new configurations can be defined for the exploded views of mold assembly.

4.7.3 Shrinkage Compensation

To achieve a better dimensional accuracy of molded part, the dimensions of the mold should be adjusted to compensate the shrinkage due to temperature changes. The shrinkage is represented by *a scale factor*, which corresponds to the ratio of a dimension between a virtual and physical part. A scale factor depends on material type and the shape of part. Figure 4.68 a, b shows that *the scale tool* can be accessed

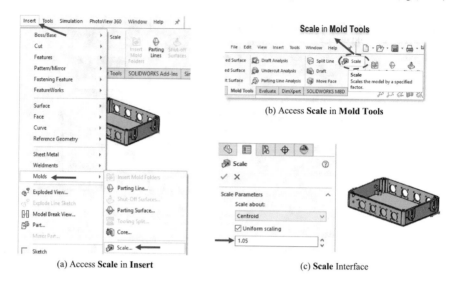

(a) Access **Scale** in **Insert**

(b) Access **Scale** in **Mold Tools**

(c) **Scale** Interface

Fig. 4.68 Scale a part model to compensate shrinkage

in multiple ways such as (1) click *insert > mold > scale* and (2) click *scale* under the command manager of the mold tool. Figure 4.68c shows that a part can be scaled about *centroid, origin,* or a *custom coordinate system* of parts.

4.7.4 Draft Analysis

Lateral surfaces of a molded part should be tapped to ensure that the part can be taken out from the mold correctly. The *Draft Analysis* in the commend manger can be used to evaluate draft angles of lateral surfaces. Figure 4.69 shows the application of the draft analysis tool. The direction of pulling can be defined by specifying *a planar face, linear edge,* or *axis,* and the user can make the adjustment by using a dynamic triad shown in the figure.

The draft angle of the tangential plane of each surface with respect to the pulling direction must be positive and larger than a minimum value. If such a condition is not satisfied for a surface, a draft feature should be added to the surface by using the *DraftXpert* tool shown in Fig. 4.70. The DraftXpert tool requires a number of inputs including (1) the minimum draft angle, (2) the direction of pulling, and (3) the group of selected surfaces to be drafted. It should be noted that some features such as undercuts do not require to have positive draft angles since they will be generated by other cores.

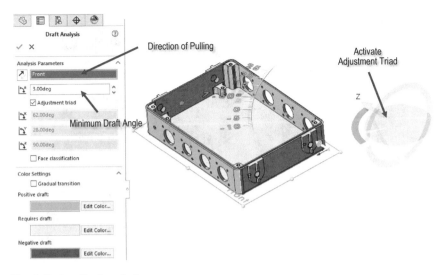

Fig. 4.69 Run *Draft* analysis

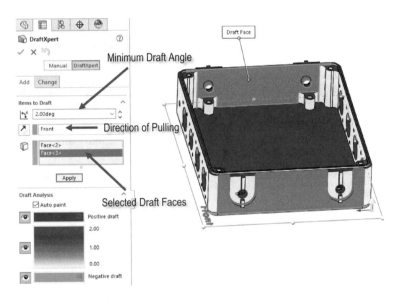

Fig. 4.70 Add positive drafts on surfaces

4.7.5   Parting Lines and Shut-Off Surfaces

A mold assembly must be separable into multiple parts. A group of the parting lines are used to define a parting surface to separate the main core from the cavity. Figure 4.71 shows the interface to define a group of the parting lines. The main inputs

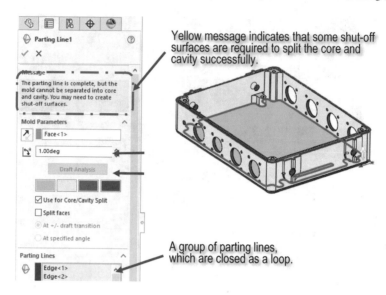

Fig. 4.71 Create parting lines

for the parting lines are (1) the direction of pulling, (2) the minimum draft angle, and (3) a list of the selected edges that form a closed loop.

The parting lines are inapplicable to internal holes or openings on the part models. *The Shut-off Surface tool* is used to create shut-off surfaces for these internal holes or openings. Figure 4.72 shows the application of the Shut-Off Surface tool where all of internal holes and openings should be selected to create virtual shut-off surfaces on these features. This ensures that the mold is separable into core(s) and cavity (Fig. 4.73).

4.7.6   Parting Surfaces

In using the mold tool, the solids of the core and the cavity are formed by knitting the boundary surfaces including the parting surfaces where the core and cavity are separated. Figure 4.74 shows that a parting surface is defined based on (1) the relation of the surface normal to the pulling direction, (2) the parting line(s), and (3) the margin distance of from the parting edges. Note that the margin distance should be as large as possible, so that the parting surface is large enough to form an enclosed volume for the core and cavity in knitting. Otherwise, the knitting process in the next step will generate an error message about knitting failure.

(a)

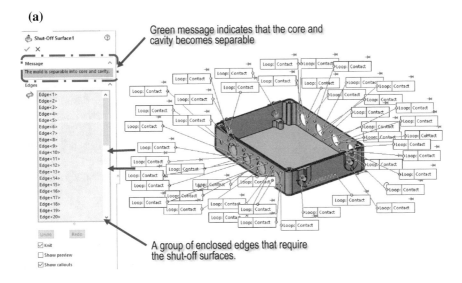

Fig. 4.72   Generate shut-off surfaces to make core and cavity separable

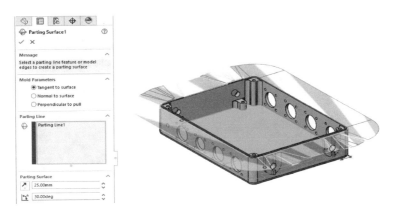

Fig. 4.73   Create a parting surface

4.7.7   Mold Components

The *Tooling Split* tool is used to create separated core and cavity. In using the tooling split tool, a ketch is defined for the outside boundary of the tooling. Note that the sketch must be within the parting surface completely to ensure the success of the knitting operations for enclosed solids. Figure 4.74 shows the interface of using the tooling split tool where the main inputs are (1) the extrude depths of the core and cavity and (2) the parting surface. The software gives the default names for the solids of core and cavity, and these solids can be renamed once they are created.

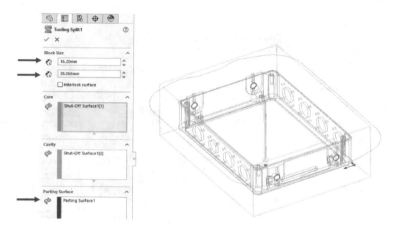

Fig. 4.74 Create *core* and *cavity* by *Tooling Split*

4.7.8   Mold Assembly

Figure 4.75a shows that many intermediate features are generated in designing a mold assembly. These features may or may not be useful in the following design stages; these features include *reference planes*, *sketches*, *surfaces*, and *solids*. It is helpful to hide the intermediate features after the mold tool is created, and it can be set by changing the hide/show settings of features, in particular, such settings can be saved as the configurations for the mold assembly.

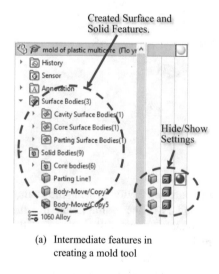

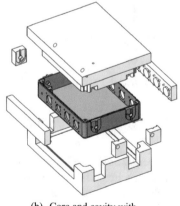

(a)   Intermediate features in   (b)   Core and cavity with
      creating a mold tool   additional features

Fig. 4.75 Review mold assembly

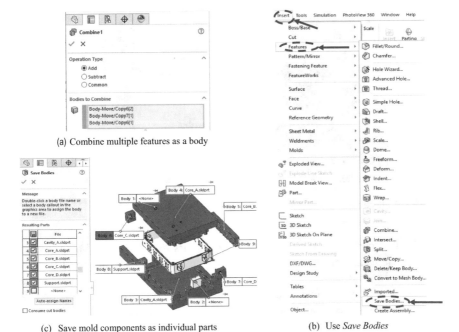

(a) Combine multiple features as a body

(c) Save mold components as individual parts

(b) Use *Save Bodies*

Fig. 4.76 Combine and save individual models of parts

Additional features can be added to the core and cavity parts as a fully functioned mold assembly as shown in Fig. 4.75b. Since a mold assembly involves a number of components, it is very helpful to save these components as individual part models. Figure 4.76 shows the *features* tool by which the solids can be edited, subtracted, combined as new graphic entities, and they can be saved as individual solids by using the Save Body tool.

4.8 Computer-Aided Fixture Design

In discrete manufacturing, parts are shaped in a series of manufacturing processes in which the parts must be fixed firmly to receive the operations (Bi and Zhang 2001). In a manufacturing process such as a material removal process, mechanical forces must be large enough to remove unwanted materials from the workpieces, and such forces also cause the deflections of machine tools and fixtures. Therefore, GD&T of parts are greatly affected by these machine tools and fixtures. Figure 4.77 illustrates the discrepancies of nominal and actual dimensions on the machine tool and the workpiece subjected to the machining force. Such discrepancies will eventually cause geometric and dimensional errors of the part. Note that a fixture is a part of the machine tool that is tailored to the workpiece.

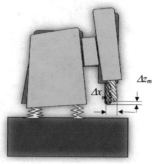
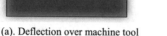

(a). Deflection over machine tool   (b). Deflection over workpiece

Fig. 4.77 Deflections of machine tool and workpiece in manufacturing process

GD&T of parts are affected by numerous factors such as material properties of machine elements and workpieces, processing parameters, and most importantly, the fixture to position and fix the workpiece. Fixture design has the direct impact on product quality and manufacturing cost. The statistic showed that the cost on fixtures typically took 10%–20% of total manufacturing cost, and around 40% of the rejected products were caused by poor fixturing solutions. As a critical component of *computer-aided manufacturing* (CAM), *computer-aided fixture design* (CAFD) has attracted a great deal of attention, and many CAFD tools and flexible fixture systems have been developed and reported recently (Wang et al. 2010).

4.8.1   Functional Requirements (FRs)

A fixture aims to secure the workpiece firmly at the specified position and orientation in a manufacturing process. A fixturing solution satisfies the following functional requirements (FRs):

(1) *positioning* and *orienting* the workpiece relative to the cutting tool;
(2) *providing* the support to reduce the deflection of the workpiece subjected to external force; and
(3) *clamping* the workpiece to resist the external loads without any rigid motion.

Figure 4.78 shows a free body in a *three-dimensional* (3D) space; it has six *degrees of freedom* (DoF), i.e., three DoF for translations (T_x, T_y, T_z) and three other DoF for the rotations (R_x, R_y, R_z) along X-, Y-, and Z-axes, respectively. A fixture usually fixes all of the rigid motions of the part. As shown in Fig. 4.78, if no constraint applies to an object in a 3D space, the object may have a total of 12 possible directions of motion. To fix a workpiece in the machining process, any rigid motion along these 12 motion directions must be eliminated. To achieve this goal, a fixturing system usually consists of three components, i.e., *base supports*, *locators*, and *clampers* as shown in Fig. 4.79.

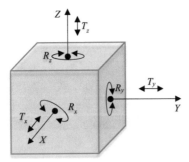

Fig. 4.78 Degree of freedoms (DoF) of a solid body

Fig. 4.79 Three types of components in a fixturing system

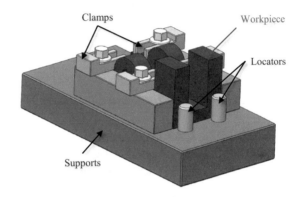

In designing these components, the constraints at the following aspects are taken into consideration, so that the fixturing system can meet the aforementioned FRs (Basha and Salunke 2013).

(1) **Geometric control**. All fixture components must be placed on datum surfaces with adequate supports. No interference is allowed between a fixture component and others such as a cutting tool. In addition, fixture components should be optimized at the aspects of loading and unloading cycle times, the number of fixture components, accessibility, and the feasibility of the supports of multiple operations.

(2) **Dimensional control**. Any solid will be deformed subjected to external loads, the deformation of a fixturing system adversely affects the accuracy of machining operations. While the deformation is unavoidable due to clamping forces or operating forces, it should be minimized to meet the tolerance requirement.

Table 4.8 Other quantifiable evaluation criteria in fixture design

Stiffness:	the measure of a fixturing system to remain original shapes subjected to given external loads.
Accuracy:	the closeness of an ideal and actual positons of workpiece to be fixed.
Repeatability:	the closeness of a normal and actual positions of workpiece in repeated operations.
Flexibility:	the measure of reconfigurability for different parts and different operations.
Setup Time:	the total amount of the time required to set up the fixing system.
Time and Cost:	the total time and cost to design and implement a fixturing system.

(3) **Mechanical control**. A workpiece is fixed by the locators of the fixturing system. GD&T of a finished part depends not only on the deflection of the fixturing system, but also on the positioning accuracy of the locators; these locators must be placed optimally for the better dimensional accuracy of parts.

To this end, the quantified criteria in Table 4.8 should be evaluated in designing a fixturing system.

4.8.2   Axioms for Geometric Control

A free body in Fig. 4.78 has 12 possible directions of motions. To immobilize a solid body, *the 3–2-1 principle* can be applied to confine these motions in which three planes are used to fix the object. A *primary plane* in Fig. 4.79 confines the rotations about x-axis or y-axis (R_x and R_y) and the translation along z-axis (T_z). A *secondary* plane in Fig. 4.80a confines the rotation about z-axis (R_z) and the translation along y-axis (T_y), and a *tertiary plane* in Fig. 4.80b confines the translation along x-axis (T_x). In addition, the primary and secondary planes are selected to counteract external forces including clamping and operating forces (Fig. 4.81).

The following axioms can be defined for geometric controls based on the 3–2-1 principle:

(1)   Six locators are essential and sufficient to fix a prismatic rigid body; more or less locators may cause the uncertainty in positioning.
(2)   One locator eliminates one DoF motion of the body.
(3)   Two directions of each DOF must be restrained to immobilize the body.
(4)   The numbers of the locators associated to the primary, the secondary, and the tertiary planes are 3, 2, and 1, respectively.
(5)   Locators should be placed as widely as possible to stabilize the object, and locators should be placed on the most accurate features to achieve the best possible accuracy.

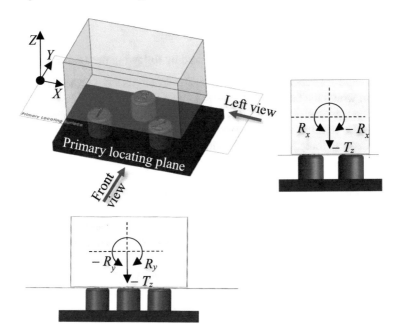

Fig. 4.80 The primary plane in fixturing system

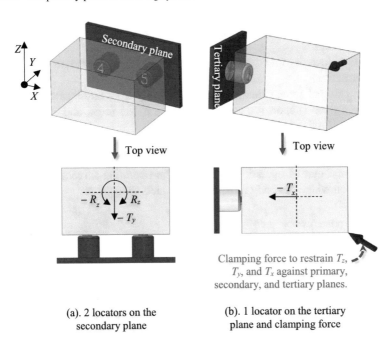

(a). 2 locators on the secondary plane

(b). 1 locator on the tertiary plane and clamping force

Clamping force to restrain T_z, T_y, and T_x against primary, secondary, and tertiary planes.

Fig. 4.81 Secondary and tertiary planes in fixturing system

4.8.3   Axioms for Dimensional Control

Better dimensional tolerance can be achieved by following the axioms given below.

(1)   To avoid tolerance stacks, a locator should be placed on a reference surface from which the body is dimensioned along this DoF.
(2)   When a geometric tolerance is defined between two planes, the primary plane must be selected as the reference planar surface; the primary plane makes its contacts to three locators shown in Fig. 4.82.
(3)   To locate the centerline of cylindrical surface, the locators must straddle the centerline.
(4)   Locators should be placed on machined surfaces to achieve better dimensional accuracy when it is possible.
(5)   When an axiom for geometric control in Sect. 8.5.2 has conflicted an axiom for dimensional control, the preference is given to the axiom for dimensional control.

4.8.4   Axioms for Mechanical Control

To minimize the deformation in the operation when a fixturing system is used, the following axioms apply:

(1)   Place a locator directly opposite to the operating force to minimize its deformation.
(2)   Place a locator directly opposite to the clamping force to minimize its deformation.
(3)   If an external force cannot be reacted directly from any locator, consider to place a fixed support (no locator) opposite to the force, and avoid an initial contact of a fixed support with the body.
(4)   Apply a clamping force toward locators.
(5)   The moment by a clamping force about any of possible axes must sufficiently overcome that of operating force, and the operating force should be in the direction where the body remains the contacts to locators.

Fig. 4.82   The primary plane for GD&T

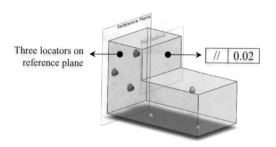

Three locators on reference plane

4.8.5 Form Closure and Force Closure

The conventional approaches to fixture design and planning rely heavily on the thumb rules, axioms, and prior experiences. A valid fixture design often needs the trial-and-error iterations. It is desirable to use more scientific methods in fixture analysis and design. *The kinematic and dynamic modeling* of a fixturing system can be an alternative approach.

In kinematic and dynamic modeling, a fixturing system is modeled as a set of *fixed rigid contacts* together with an object whose motion is restrained by rigid contacts. Figure 4.83a shows that an object is fixed by a number of contacts in a fixturing system, and Fig. 4.83b shows the free-body diagram (FBD) of the object which is subjected to external forces. All of the applied forces must meet the conditions for *form closure* and *force closure* in operations.

A fixture design should be evaluated by the criteria for *form closure* and *force closure*. For a form closure, the contacts to the object must resist all external forces to keep the object still in operations. These contacts eliminate all DoF of the object purely based on the geometric placement at the contacts. For a force closure, the object maintains the expected contacts to fixture elements subjected to external forces and moments. In practice, most of the fixtuing systems for machining processes must be force closed, and they are achieved mainly by utilizing the frictional forces to immobilize an object.

Kinematic and dynamic modeling aims to evaluate if a fixturing system quarantines the conditions for form closure and force closure so that (1) there is no possible movement under external loads; (2) the fixturing system is fully accessible and detachable when a part is loaded and unloaded; and (3) the axioms for geometric, dimensional, and mechanical controls are appropriately followed to position and secure the part with better dimensional accuracy.

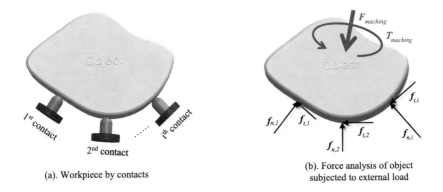

(a). Workpiece by contacts

(b). Force analysis of object subjected to external load

Fig. 4.83 The fixed object and its free-body diagram (FBD)

4.8.6   Fixture Design in Manufacturing Processes

Many manufacturing processes involve fixture designs. Figure 4.84 shows the relations of fixture design to other tasks in *manufacturing process planning*. The fixture
design is affected by machine tools, process parameters, tooling paths, and machining
programs which are tailored to the performed manufacturing processes. The tasks
of fixture design are logically followed one after another, while an iterative process
is usually needed since the information at earlier steps is insufficient to verify if
the design constraints in the following steps are satisfied. Figure 4.84 shows that a
fixture design involves four stages with the specified main tasks. It can be seen that
a fixture design requires the information workpiece and manufacturing processes,
and Fig. 4.85 shows the essential information for a fixture design includes *products*,
machine tools, *manufacturing processes*, and *verification* and *quality control*.

The fixturing solution has to be verified thoroughly since the fixture makes a
direct contact to various hardwares such as machine tools, cutters, and other auxiliary
systems (e.g., lubrication or coolant systems). Verification should be an integrated
part of the fixture design. Many components bring the constraints to a fixturing
system (see Fig. 4.85); in addition, not all information is available when the fixture

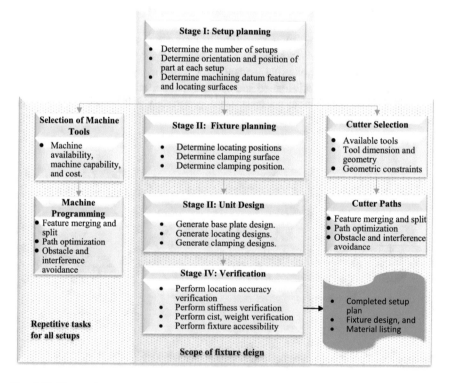

Fig. 4.84 Fixture design in manufacturing process planning

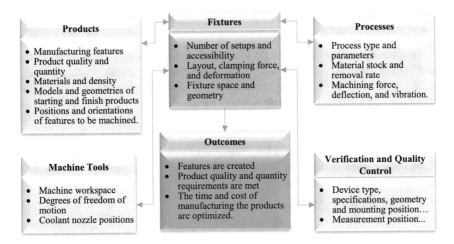

Fig. 4.85 Key design factors and criteria of fixture design

concept is developed. Therefore, the verification is needed since not all of fixturing requirements can be defined in depth from the beginning.

Figure 4.86 shows a framework for the verification of fixture design (Wang 2010). The critical tasks in the verification are *analysis of geometric constraints, tolerance analysis, stability analysis, stiffness analysis,* and *accessibility analysis.* The knowledge-based engineering (KBE) approach could be applied to accelerate the design process and eliminate design defects.

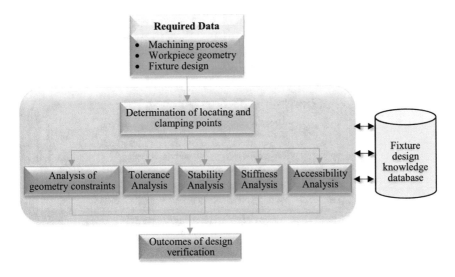

Fig. 4.86 Verification of fixture design

4.8.7 Computer-Aided Fixture Design (CAFD)

Conventionally, fixture design might take days even weeks, and the performance of the fixturing system relied heavily on the designers' experiences. Computer-aided techniques help to accelerate the process of the fixture design, and CAFD should be an essential component in the cycle time of product development. Figure 4.87 shows an integrated system for computer-aided process planning (CAPP) in which CAFD is able to obtain all of the relevant information for fixture design. Especially, part models can be directly uploaded for processing planning, and the results of CAFD can be used to plan cutting paths and program machine tools. In this section, some main aspects of CAFD are discussed below.

(1). Fixture Design Library

Fixtures are customized to the parts to be manufactured; therefore, fixtures are highly diversified for different applications. To develop customized fixturing solutions in a cost-effective way, modular structure should be adopted in which the fixturing system consists of various fixture elements such as supports, locators, and clamps, and different configurations can be assembled by selecting different elements and assembling them in different ways. By developing a fixture design library, commonly used fixture elements can be standardized, virtual models can be parametrized, and the fixtures solutions to new tasks can be obtained by reusing existing fixture elements.

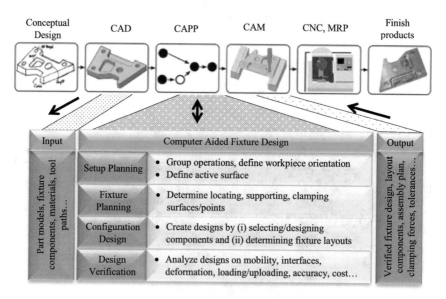

Fig. 4.87 CAFD in computer-aided process planning (CAPP)

(2). **Interference Detection**

No interference is allowed in assembling a fixture, or loading and unloading a part. To ensure no interference occurs to the fixture, *virtual analysis* could be applied to visualize assembling processes of a fixturing system, and simulate the loading and unloading process of parts. With the virtual models of parts, fixture elements, and machine tool, solid modeling tools are able to run *interference checks* to detect possible interferences before the physical fixturing system is implemented.

(3). **Accessibility Analysis**

Accessibility analysis investigates possible collisions of fixture elements to surrounding objects in (1) loading and uploading parts and (2) operating a manu-facturing process such as moving a cutting tool along the given path. To evaluate accessibility of an object, the motions of machine, tools, and parts in the reference coordinate system are considered with respect to the fixture elements. *The Solid-works Motion* can be utilized to detect possible collisions in the motions of multiple objects.

(4). **Analysis of Deformation and Accuracy**

The primary objective of a fixturing system is to hold the part when it receives the operation; therefore, the quality of part is affected by geometric and dimensional deviations of the fixturing system subjected to the operating forces. The layout of a fixturing system should be optimized to minimize the impact on the quality of parts. To estimate the dimensional accuracy, the computer-aided engineering (CAE) tool can be used to simulate the deformations of fixture elements and workpiece subjected to operating forces.

Example 4.6 Figure 4.88 shows a fixture setup for the milling operation on the part for a slot feature. The fixturing system is implemented by using the 3–2 principle. The primary plane makes the contacts to the bottom surface of the part. The movements along x-axis and y-axis are constrained by one and two locators, respectively. The position of *locator A* can be placed at $L_1 = 25$, 75, and 125 mm; the position of *locator B* can be placed at $L_2 = 175$, 225, and 275 mm; and the position of *locator C* can be placed at $H = 25, 75, 125$, and 175 mm. The clamping forces along x-axis and y-axis are -2 kN and -0.5 kN, respectively, and the cutting forces are $(F_{t\_x}, F_{t\_y}) = (-4kN, -2kN)$. Determine the positions of the locators to optimize the dimensional accuracy of the part.

Solution Based on the provided information, the workpiece and fixture elements are modeled, and an FEA model for static analysis is further defined in the Solidworks Simulation. The model for static analysis is shown in Fig. 4.89a, and an example of the workpiece deformation from the simulation is illustrated in Fig. 4.89b. To optimize the layout of three locators, a design study in Fig. 4.89c is defined for three

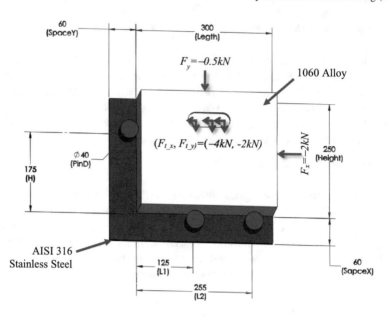

Fig. 4.88 Example of deformation analysis

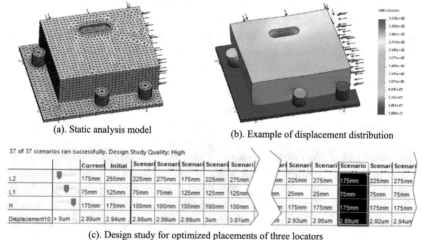

(a). Static analysis model   (b). Example of displacement distribution

(c). Design study for optimized placements of three locators

Fig. 4.89 The solution to fixture design problem in Example 4.6

variables (i.e., L_1, L_2, and H) to evaluate the displacement of workpiece. The design study finds that the optimized positions of three locators are $L_1 = 75$ mm, $L_2 = 175$ mm, and $H = 175$ mm, and the minimized displacement is 2.89 microns.

4.9 Computer-Aided Machining Programming

To automate manufacturing processes, human operators are replaced by machines, and the operation-related decisions are made by computer programs. *Computer-aided machining programming* is to generate the programs to run machine tools. In this section, machining programming is introduced, and computer numerical control (CNC) is discussed as the programming tool for automated machining processes. In this section, the programming of machining processes is focused, and the Solid-Works *High-Speed Machining* (HSM) is used to generate, simulate, and verify control programs.

Machining processes offer tight geometric and dimensional tolerances and high surface finishes. Therefore, machining processes are often used as the secondary processes to improve product quality or add machined features that are difficult to be made by other processes such as a side hole on a molded part. However, a machining process usually involves in a high cost due to a number of reasons: (1) reshaping a part by machining differs from net-shape operations such as casting and injection molding in sense that a machining process takes a long time to remove unwanted materials gradually to shape the desired features on parts; (2) a high-performance machine tool requires a high initial capital investment that increases the unit cost of products to be made; (3) cutting tools are made from the materials with high strength, durability, and toughness; however, they are consumable in machining processes; and (4) the volumes of products that are made by machining processes, such as molds and dies, are mostly very limited and impractical to be made in the mass production paradigm.

To program a machine tool, it is helpful to understand controllable variables of the machine tool in machining processes. A machining operation refers to a material removal process in which a cutting tool cuts away unwanted materials from the

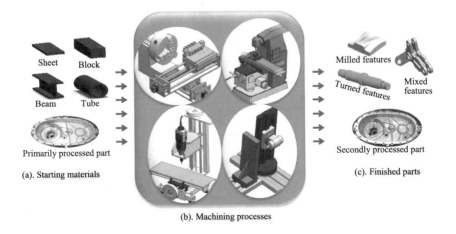

(a). Starting materials

(b). Machining processes

(c). Finished parts

Fig. 4.90 Starting materials and finished products of machining processes

part for a desired feature or dimension on parts. As shown in Fig. 4.90, the starting material of a machining process can be a primarily processed part or raw material with a simple shape such as block, bar, sheet, roll, beam, or tube. After machining processes, desired features such as turning, milling, drilling, or grinding features are made on finished parts.

Based on overall geometries, machined parts can be classified into *prismatic* parts and *cylindrical* parts, and some common machined features such as *holes, slots, pockets, flat surfaces*, and *complex surface contours*. To automate a machining process, the motions of all moving machine elements have to be controlled by programs. Figure 4.91 shows an example of machine tools in which the main moving components are worktables, spindles, axes, cutting tools, and other auxiliary equipment attached on motion axes. From the perspective of productivity, a machining process also controls all process variables relevant to *a material removal rate* (R_{MR}). R_{MR} is defined as the volume of the removed materials in the unit time; R_{MR} is evaluated based on four process variables. These process variables are illustrated in Fig. 4.92 and also have been explained.

(1) *Cutting speed* (v) is the relative speed at which a cutting tool passes over the part, and the cutting speed is measured at the contact of the cutting tool and the part. R_{MR} is affected greatly by the cutting speed v. The higher a cutting speed v is, the larger R_{MR} is, and the better the productivity is. Note that the life span of a cutting tool also depends on R_{MR}. Therefore, tool manufacturers

Fig. 4.91 Main variables in a machining process

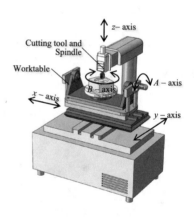

Fig. 4.92 Main variables in a machining process

usually make the recommendation of appropriate cutting speeds based on the properties of tooling and part materials.

(2) *Spindle speed* (N) is an angular velocity of the rotational axis, which is usually expected in the number of revolutions per minute (RPM). When the main motion of a machining process is from a spindle, the spindle speed N is correlated to the cutting speed v as

$$N = \frac{v}{\pi \cdot D} \tag{4.12}$$

where v is the cutting speed and D is the diameter of the part at which the contact is made between the cutting tool and the part.

(3) *The depth of cut* (d) refers to the depth that the cutting tool penetrates into the part in the material removal process.

(4) *The feed rate* (f) is the velocity to feed the cutting tool along the cutting path. It is measured by a moving distance of the cutting tool in each revolution of the spindle. The unit of a feed rate f can be either of *inches per revolution* or *millimeters per revolution*.

R_{MR} can be computed from the cutting speed v, the depth of cut d, and the feed rate f as

$$R_{MR} = v \cdot f \cdot d \tag{4.13}$$

4.9.1 Procedure of Machining Programming

Machining programming begins with the collection of the information about machined features on parts, machine tools, cutting tools, and geometric and dimensional tolerances. Figure 4.93 shows the procedure of generating a control program for a given machining process.

The procedure in Fig. 4.93 consists of eight steps, and the main tasks involved in these steps are described in Table 4.9.

4.9.2 Standards of Motion Axes

A program consists of legal statements that are written by following a set of the predefined rules called *syntax* in combining symbols and words. To program a machine tool, the nomenclatures for the motions and coordinate systems are standardized by the Electronic Industries Association (EIA) as EIA267-C standards.

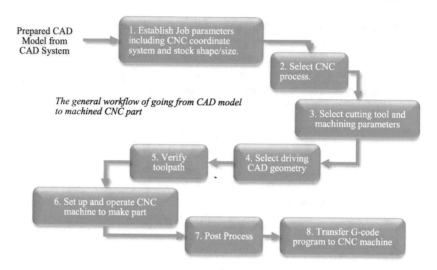

Fig. 4.93 The procedure of programming a machining process

Table 4.9 Main steps and tasks in machining programming

Step	Tasks
1)	define main parameters of a control program such as coordinate systems, sizes and shapes of starting materials and so on,
2)	specify the type of the machining operation such as *turning*, *drilling*, *milling*, and *grinding*,
3)	select a cutting tool, and determine the process parameters such as cutting speeds, feed rates, and depths of cut,
4)	analyze part features, and plan tool paths based on the a set of specified working points;
5)	verify the feasibility of toolpath(s) to ensure no interference occurs and the toolpaths pass through all working points,
6)	convert toolpaths into control programs that can be downloaded and executed on machines,
7)	post-process the control programs by considering the difference of coordinate systems for programming, machines, tools, parts, and other equipment, and
8)	download the verified program as the G-codes to control the machining system.

In the EIA267-C standards, 14 axes can be used to define the positions and movements of parts, cutting tools, or other objects involved in machining processes. Note that nine axes should be enough to describe the motions of conventional machine tools, while some advanced machine tools are equipped with more auxiliary motion axes. Without loss of generality, the following discussion considers a machine tool that has less than nine motion axes, i.e., three *primary linear axes* (X, Y, and Z); three *primary rotary axes* (A, B, and C); and three *secondary linear axes* (U, V, and W).

As shown in Fig. 4.94, a motion is defined in a given coordinate system, and the positive directions of motion axes in a coordinate system follow *the right-hand rule* as.

(1)   Three fingers (i.e., the thumb, index, and middle fingers) are placed mutually perpendicular with each other, and the origin $O(0, 0, 0)$ is set at their intersection. Each finger points to one direction of the translation, i.e., the thump

Fig. 4.94 The right-hand
rule in a coordinate system
(CS)

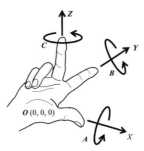

finger, the index finger, and middle finger point to the positive **X-**, **Y-**, and
Z-axes, respectively. The coordinate system is defined as $\{O - XYZ\}$.

(2) The motion axis of the main spindle is defined as **Z**-axis, and the positive
direction is toward the spindle.

(3) The longest travel slide is designated as **X**-axis, and it is perpendicular to the
Z-axis.

(4) The positive rotation around an axis is the clockwise direction of the axis.

Using the right-hand rule, the motion axes of conventional machine tools are
described in Table 4.10.

4.9.3 Default Coordinate Systems and Planes

A toolpath consists of line, arc, or curve segments in a plane or 3D space. A 2D
toolpath can be usually defined in one of three default planes (**XY, XZ**, and **YZ**)
in the Cartesian coordinate system. The default Cartesian planes are defined based
on the given coordinate axes **X, Y**, and **Z**. Figure 4.95 shows the examples of three
default planes in a mill machine. In the default **XZ**-plane, a central reference point
O is used to measure the coordinates along **X**- and **Z**-axes, and the coordinates of **O**
is set as **O** (0, 0).

As shown in Table 4.10, a turning machine has only two motion axes, i.e., a
primary axis (**Z**-axis) and a secondary axis (**X**-axis). The dimension along **Y**-axis is
given by the depth of cut. As a result, Fig. 4.96 shows a default two-axis CS. Using
default CSs helps to transfer CNC programs or measurements among different lathes.
It should be noted that the cutter on a CNC lathe is different from the cutting tools
on conventional lathes. The cutter is usually placed on the top or side of the machine
tool; alternatively, it can be mounted at the front side horizontally on a conventional
lathe.

A CNC lathe makes rotational parts; therefore, the machined features are dimensioned by *radii* or *diameters*. *Diametrical programming* uses the *x*-coordinate as a
diametrical dimension at the given Z-coordinate. For example, if a feature has a 5-in
outside diameter and *absolute* coordinates are used in programming, the x-coordinate

Table 4.10 Motion axes of conventional machine tools

Type	Description	Illustration
Lathe	On most CNC lathes, the **Z Axis** is parallel to the spindle, and the longer motion axis is defined as **X Axis**.	
Mill	In the shown mill, the spindle travels along the **Z Axis**. The other axis with a longer travel stroke is defined as the *X* **axis**, and the other axis with a shorter travel stroke is defined as the **Y Axis**.	
Knee Mill	On a common vertical knee CNC mill the spindle is stationary, and the direction of spindle axis is defined as the *Z* **axis**. The other two axes are defined as **X** and **Y** axes, respectively, based on their travel distance.	
Contour mill	On this five-axis horizontal contour milling machine, note the orientation of the **X** and **Y** axes in relation to the **Z Axis**. The rotary axes for both the **X** and **Y** axes are designated by the **A** and **B** rotary tables.	

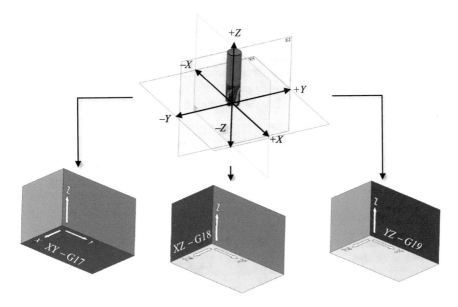

Fig. 4.95 Default coordinate planes in a mill machine

Fig. 4.96 Default coordinate system on a CNC lathe

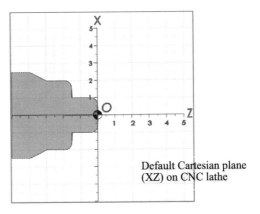

Default Cartesian plane (XZ) on CNC lathe

of the motion commend becomes X5.0. *Radial programming* uses the *x*-coordinate as a radical dimension at the given Z-coordinate. For example, if a feature has a 5-in outside diameter and *absolute* coordinates are used in programming, the x-coordinate of the motion commend becomes X2.5.

4.9.4   Machine, Part, and Tool References

Working points in a machining process are described with respect to certain references in a coordinate system, and three commonly used references are (1) *machine reference zero* (*M*), (2) *part reference zero* (*PRZ*) (*W*), and (3) *tool reference zero* (*R*). These references are used to define the coordinates of working positions as follows (Figs. 4.97 and 4.98):

(1)   The coordinates of a working point are measured from the machine reference zero *M*.

(2)   The points on workpiece are measured from the part reference zero *W*.

Fig. 4.97 Tool reference
zero (*R*), part reference zero
(*W*), and machine reference
zero (*M*)

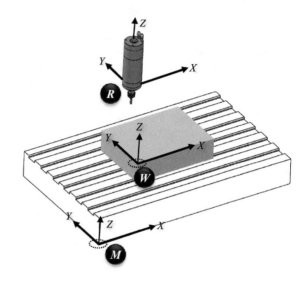

Fig. 4.98 Absolute
coordinates from the origin
of coordinate system

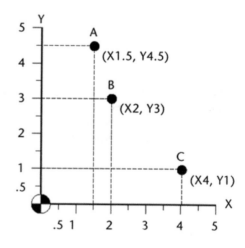

(3)   The PRZ (*W*) should be aligned to the lower left-hand corner on the top surface
      of the part stock.

4.9.5 Absolute and Incremental Coordinates

The coordinates of the working points can be either of (1) *absolute* measured from the
origin or (2) incremental measured from a pre-existing location. Figure 4.99 shows
the scenario in which *absolute coordinates* are specified to define working points.
The absolute coordinates of points *A*, *B*, and *C* are measured from the origin *O* of
the coordinate system {*O-XY*}.

Example 4.7 Determine the absolute coordinates of working points shown in
Fig. 4.99a.

Solution The absolute coordinates are measured from origin *O*; therefore, the coordinates of the working points (1, 2, ..., 7) in Fig. 4.99a are determined as (1.0, 3.0),
$(-1.0, 2.0), (-3.5, -0.5), (-3.0, -1.0), (-4.0, -2.0), (1.0, -2.5)$, and $(3.0, -2.0)$
in Fig. 4.99b, respectively.
      Absolute coordinates also apply to the default two-axis coordinate system for a
CNC lathe. For a lathe, absolute coordinates take the reference from origin $(X0, Z0)$
as shown in Fig. 4.100a. To find *Z*- and *X*-coordinates of a working point, project the
working point on *Z*-axis and *X*-axis, respectively. The absolute coordinates of the
working points P_1 to P_6 are determined in Fig. 4.100b. Note that the *X*-coordinates
can be either radical (X_R) or diametric (X_D).
      Incremental coordinates are measured from existing points. A new working point
can be defined by specifying a relative distance of an existing point. Figure 4.101
shows an example of defining incremental coordinates for working points in a
toolpath.

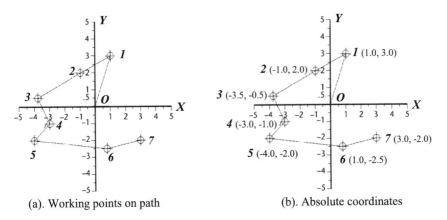

    (a). Working points on path   (b). Absolute coordinates

Fig. 4.99 Example of absolute coordinates

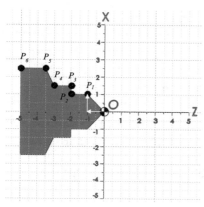

Position	Z	X_R	X_D
O	0.0	0.0	0.0
P_1	-1.0	1.0	2.0
P_2	-2.0	1.0	2.0
P_3	-2.0	1.5	3.0
P_4	-3.0	1.5	3.0
P_5	-3.5	2.5	5.0
P_6	-5.0	2.5	5.0

X_R and X_D are for radical and diametrical programming, respectively

(a). Working points on turned part   (b). Absolute coordinates on XZ plane

Fig. 4.100   Example of absolute coordinates in XZ-plane

Fig. 4.101   Working points with incremental coordinates in sequence

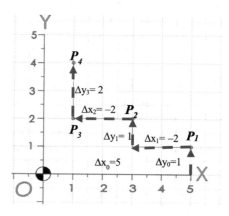

Example 4.8 Find the incremental coordinates of the working points shown in Fig. 4.101a.

Solution The incremental coordinates of a new working point are measured from an existing point one by one in sequence. Taking an example of *A*, its incremental coordinates $(-2.0, -1.0)$ are measured from origin *O*; similarly, other working points *B*, *C*, *D*, *E*, and *F* are measured from *A*, *B*, *C*, *D*, and *E* as shown in Fig. 4.102b, respectively.

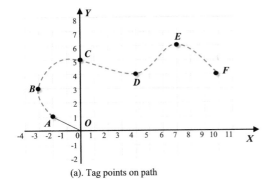

(a). Tag points on path

Position	ΔX	ΔY
O	0.0	0.0
A	-2.0	1.0
B	-1.0	2.0
C	3.0	2.0
D	4.0	-1.0
E	3.0	2.0
F	3.0	-2.0

(b). Incremental coordinates

Fig. 4.102   Incremental coordinates for Example 4.8

4.9.6   *Types of Motion Paths*

A geometric feature of part is determined by the profile and motion path of the cutting tool. The motion paths can be classified into four types at shown in Fig. 4.103. (1) A *point-to-point path* consists of a set of discrete points where the machining operations are performed; no machining operation occurs when the tool moves from one point

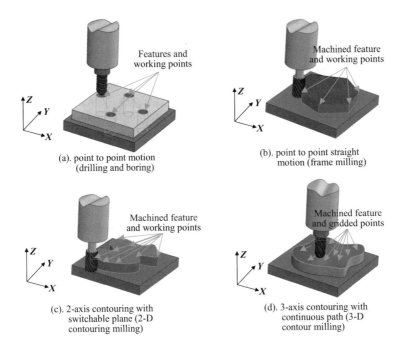

(a). point to point motion (drilling and boring)

(b). point to point straight motion (frame milling)

(c). 2-axis contouring with switchable plane (2-D contouring milling)

(d). 3-axis contouring with continuous path (3-D contour milling)

Fig. 4.103   Four types of tool paths

to the next point. A point-to-point path is used to specify the motion for a drilling or boring operation. (2) A *point-to-point straight path* consists of a set of line segments, each line segment connects two working points directly, and the machining operations are performed continuously from one end to the other end of the line segment. A point-to-point straight path is applicable to frame milling. (3) A *two-dimensional continuous path* is a smooth curve for the connections of working points on a 2D plane; it is mainly used in 2D milling. (4) A *three-dimensional continuous path* is formed by following the nodes on the grid by zigging-zagging. It is used in 3D contouring.

4.9.7 *Programming of Machining Processes*

Automated machine tools are controlled by computer programs, and three ways to program machine tools are *manual programming* (NC), *conversational programming* at shop floor, and *offline programming*.

Manual programming is to program a machine tool for simple tasks; the program is tied to certain machine tool with the optimized performance of execution. However, manual programming becomes very tedious and error-prone when the machining operation is complex. Therefore, manual programs are mainly used when (1) a machine tool is sophisticated to make parts with high volume and (2) the efficiency of machining operations is prioritized.

Conversational programming uses graphic and menu-driven interfaces to create programs for machining operations. Conversational programming allows users to check inputs, review toolpaths, and simulate the machining operations. It is very popular in small- or medium-sized enterprises (SMEs); since machine operators in SMEs often take the full responsibility of setting up machine tools, fixtures, and tooling, and creating, verifying, and running the programs. In comparison with manual programming, conversational programming is more user-friendly and it can reduce the programming time dramatically. Conversational programming provides a convenient way to write programs for parts from the same type of machines; sometimes, it is the only way of programming on a legacy machine tool.

Offline programming has gradually gained its popularity, and it supports programming at a higher level than manual and conversational programing. Offline programming has its advantages over programming techniques in (1) defining toolpaths automatically based on given working points, (2) generating the programs which can be customized to different types of machine tools, and (3) reusing functional modules by establishing and sustaining the library for reusable modules and routines.

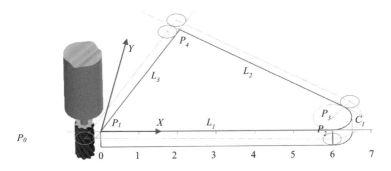

Fig. 4.104 NC programming by APT

4.9.8 Automatically Programmed Tools (APT)

One of the most popular machine programming tools is *Automatically Programmed Tool* (APT). APT was developed in 1965 by Douglas T. Ross to write the instructions for NC machine tools. APT can be used to program the machine tool for complex products. APT is still used in practice and it accounts for 5–10% of all machining processes in the defense and aerospace industries.

An NC program from APT defines a series of *lines*, *arcs*, and *points* as the entities to define the features of part to be machined. These entities are then used to generate a *cutter location* (CL) *file*. The statements in a NC program from APT can be classified into four types: (1) *geometric statements* to describe part geometry, (2) *motion statements* to describe toolpaths, (3) *post-processor statements* to specify feeds and speeds, and (4) *auxiliary statements* to define tools and tolerances.

Example 4.9 Use APT to write a NC program to mill the part shown in Fig. 4.104.

Solution By using APT, the NC program is written and shown in Table 4.11. The first part (line 1–3) is to label the program, specify machining type, and select the tool. The second part (line 4–13) is to define all geometric entities. The third part (line 14–16) is to set up operating conditions for cutting speed, feed rate, and coolant supply. The fourth part (line 17–22) is for all motion statements. The last part (line 24–26) is to reset the tool position and terminate the machining operation.

4.9.9 Computer-Aided Machining Programming

APT is suitable for manual programming, but it is tedious and error-prone in defining part geometry and working points. The technique of CAD-based programming was introduced in 1980s, CAD-based programming uses CAD models to retrieve the machined features and generate NC programs semi-automatically. Note that a programmer should have the basic understanding of the machining processes to

Table 4.11   NC program to mill the part in Fig. 4.103

	NC Code	Explanation
Line 1:	PARTNO / EXAMPLE9-3	; label the program as "EXAMLE9-4"
Line 2:	MACHIN / MILL, 1	; select the target machine and controller type
Line 3:	CUTTER/ 0.5000	; specifies the cutter diameter
Line 4:	P0=POINT/-1.0, -1.0, 0.0	
Line 5:	P1=POINT/ 0.0, 0.0, 0.0	
Line 6:	P2=POINT/ 6.0, 0.0, 0.0	
Line 7:	P3=POINT/ 6.0, 1.0, 0.0	
Line 8:	P4=POINT/ 2.0, 4.0, 0.0	; geometry statement to specify the pertinent surface of part
Line 9:	L1=LINE/P1, P2	
Line 10:	C1=CIRCLE/CENTER, P3, RADIUS, 1.0	
Line 11:	L2=LINE/P4, LEFT, TANTO, C1	
Line 12:	L3=LINE/P1, P4	
Line 13:	PL1=PLANE/P1, P2, P3	
Line 14:	SPINDLE/573	; set the spindle speed to 573 RPM
Line 15:	FEDRAT/5.39	; set the feed rate to 5.39 inch per minute
Line 16:	COOLNT/ON	; turn the coolant on
Line 17:	FROM/P0	; gives the starting position for the tool
Line 18:	GO/PAST, L3, TO, PL1, TO, L1	; initialize motion, drive, part, and check surfaces
Line 19:	GOUP/L3, PAST, L2	
Line 20:	GORGT/L2, TANTO, C1	
Line 21:	GOFWD/C1, ON, P2	; contour the part in clockwise direction
Line 22:	GOFWD/L1, PAST, L3	
Line 23:	RAPID	; move rapidly once cutting is down
Line 24:	GOTO/P0	; return the tool to home position
Line 25:	COOLNT/OFF	; turn off the coolant
Line 26:	FINI	; terminate the program

specify feed rates, speeds, and depths in a NC program. A CAD-based programming system is used to generate, edit, and simulate toolpaths, and predict the processing time and cost of machining operations. In addition, the program system usually includes a design library for standardized cutting tools and part materials. According to EIA standards, a NC program is written based on the following assumptions:

(1) The cutting tool moves relatively to the workpiece without the consideration of how such a relative motion happens. Taking an example of the motions on a shaper, the cutting tool is actually fixed, and the workpiece moves toward the cutting tool to have a relative motion with each other.

(2) The parts are positioned and oriented in a Cartesian coordinate system.

(3) The reference origin (0.0, 0.0, 0.0) is floating, so the program is flexible to align with any coordinate system of interest when the setup of a machine tool or part is changed.

(1).   Motion of Cutting Tools

In a CNC program, the cutting tool is moved in one of the following ways:

(1) A *rapid movement* to a designated location without machining operation.

(2) A *straight translation* actuated by one or multiple axes.

(3) A *circular motion* in a planer surface.

(4) A *mixed planar and linear motion* that is commonly referred as *a 2½-D motion*. A planar motion is implemented by two simultaneous motion axes in one plane.

This third motion direction is the feeding direction, which is perpendicular to the plane formed by two motion axes.

(5) A *complex motion* that is combined by the aforementioned motions. For example, the arc motion in an arbitrary plane can be modeled as a set of straight lines and implemented by three simultaneous motion axes.

(6) A *true 3D motion* along any path and orientation in the workspace of the machine tool. Due to new advancement in sensing and information technologies, most recent CNCs are able to generate true 3D motion with over 5 DoF.

(2). **Block Structure**

Each statement in a CNC program is called *a block*, and the program is a collection of blocks. A block includes one or a few of *instructions* or *words* that are separated by *space* or *tab*. A word is a character for a single function. For example, '*X*' represents a displacement along *X*-axis and '*F*' represents for the feed rate.

A block to a program is a sentence to a language. Similar to the sentences that are separated by periods in English, blocks are separated by the *End-Of-Block* (EOB) in a program. An EOB character terminates a block to show the beginning of next block. In running a program, EOB indicates to execute the commands specified in the block. An EOB character can be inserted by pressing the *Enter* or *Return Key* on keyboard. Sentences are assembled into an essay in a language, and blocks are assembled into a program in programming.

The structure of a block is shown in Fig. 4.105. '/' specifies if the following block is skipped in running the program. '*N*' is the identification of the block; the identification is only needed when it is referenced by other blocks. '*G*' is the preparatory functions for next operations. '*X*', '*Y*', '*Z*' are for the displacements in the primary *X*-, *Y*-, *Z*-axes. '*U*', '*V*', '*W*' are for the displacements in the secondary motion axes. '*A*', '*B*', '*C*' are for the angular displacements in *X*-, *Y*-, *Z*-axes. '*I*', '*J*', '*K*' are for

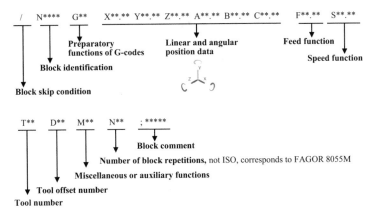

Fig. 4.105 The structure of blocks in a CNC program

the distances to the arc center or thread leads which are parallel to X-, Y-, Z-axes, respectively. 'F' is for the feed rate, 'S' is for the cutting speed, 'T' is for the tool number, 'D' is for the offset of tool, and 'M' is for the miscellaneous functions.

(3). G-Code

The programs for CNC machine tools are also called as *G-code* which has been standardized by several organizations. The US standard was RS274 *version D* developed by the Electronic Industries Alliance (EIA) in 1979. The global standards were DIN 66,025 developed in Germany, PN-73 M-55256 and PN-93/M-55251 developed in Poland, and ISO 6983 which was widely used by other countries (Wikipedia 2019). However, all standards adopt similar words and the block structure in defining toolpaths and specifying process parameters. Table 4.12 shows five types of words in G-code.

G-words in Table 4.12 are used to control the tool motions, the commonly used G-words are listed in Table 4.13.

In a CNC program, each G-word has its specific format, and the formats of some common G-words are described as follows:

Table 4.12 Five types of words in G-code

Words	Function	Block Example
N	gives a block number	N010 X70.0 Y85.5 F175 S500 (EOB)
G	control the motion of a cutting tool in a preparatory command	where
S, F, T, D, ...	specify cutting speed, feed rate, tool, offset.....	N-word = a sequence number (010) X-word = x coordinate position (70.0 mm)
X, Y, Z, U, V, W, A, B, C...	specify displacements of cutting tools	Y-word = y coordinate position (85.5 mm) F word = feed rate of 175 mm/min
M	specifies miscellaneous parameters for machine controls	S-word = spindle speed of 500 rev/min (EOB) = end of block

Table 4.13 Commonly used G-words

G-Word	Description	G-Word	Description
G00	Rapid point to point movement	G20	Input data in inches
G01	Linear motion between two points	G21	Input data in millimeters
G02	Clockwise circular motion	G28	Go to reference point
G03	Counterclockwise circular motion	G90	Absolute coordinates
G04	Tool dwell	G91	Incremental coordinates
G10	Tool offset	G94	Feed/minute in milling and drilling
G17	Selection of XY plane	G95	Feed/revolution in milling and drilling
G18	Selection of XZ plane	G98	Feed/minute in turning
G19	Selection of YZ plane	G99	Feed/revolution in turning

(1) *Rapid point-to-point motion* (G00) takes the shortest path to move the tool to the new position at the specified coordinates at a rapid feed rate:

$$\text{Format:} \quad \text{G00 X*** Y*** Z***}$$

where

$X, Y,$ and Z are the words for the displacements along X-, Y-, and Z-axes, respectively, and *** is the value of a displacement.

(2) *Linear motion* (G01) moves the tool to a new position with the specified coordinates at the given feed rate; Fig. 4.106 shows an example of a straight line generated by a G01 block:

$$\text{Format:} \quad \text{G01 X*** Y*** Z*** F***}$$

where

$X, Y,$ and Z are the words for the displacements along X-, Y-, and Z-axes, respectively; F is the word for the feed rate; and *** is the value of a displacement or a feed rate.

(3) *Closewise arc motion* (G02) moves the tool along an arc in the clockwise direction until it arrives to a new position at the given feed rate; Fig. 4.107 shows a closewise arc path in XY-, XZ-, and YZ-planes, respectively,

$$\text{Format: G17 G02 X*** Y*** I*** J*** F***;} \quad \text{Arc on XY} - \text{plane}$$
$$\text{G18 G02 X*** Z*** I*** K*** F***;} \quad \text{Arc on XZ} - \text{plane}$$
$$\text{G19 G02 Y*** Z*** J*** K*** F***;} \quad \text{Arc on YZ} - \text{plane}$$

Fig. 4.106 A straight-line path defined by G01

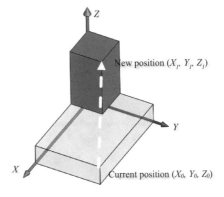

New position (X_1, Y_1, Z_1)

Current position (X_0, Y_0, Z_0)

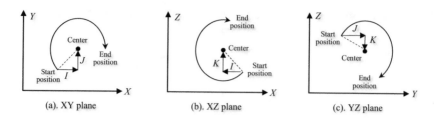

Fig. 4.107 Clockwise arc paths by G02 in *XY-*, *XZ-*, and *YZ*-planes

where

G17, *G18*, or *G19*	represents the plane where an arc lies in;
X, Y, and Z	are the words for the coordinates of the designated position;
I, J, and K	are the words of the displacements from the starting position to the arc center along X-, Y-, Z-axes, respectively;
F	is the word for the feed rate; and.
***	is the value of a coordinate, displacement, or feed rate.

'*M*' stands for 'machine' and *M* words are used to activate or deactivate miscellaneous machine functions. Table 4.14 shows the functions of some commonly used *M* words.

Example 4.10 Write the G-code for the motion control of the milling operation on the part shown in Fig. 4.108a.

Solution The milling operation is to create the profile of workpiece. The milling tool starts its motion from the home position S, and then it passes a series of working points in the order of $S \rightarrow A \rightarrow B \rightarrow C \rightarrow D \rightarrow E \rightarrow F \rightarrow G \rightarrow H \rightarrow I \rightarrow A \rightarrow S$. The toolpath includes two arc motions. The first arc is $B \rightarrow C \rightarrow D$ in the counterclockwise direction, and the second arc is $G \rightarrow H$ is in the clockwise direction. The G-code for the motion along the toolpath is written and shown in Fig. 4.108b.

Table 4.14 Commonly used M words

M-Word	Description	M-Word	Description
M02	End of program and machine stop	M09	Turn off cutting fluid
M03	Clockwise spindle start (CSS)	M10	Automatic clamping of fixture
M04	Counterclockwise spindle start (CCSS)	M11	Automatic unclamping of fixture
M05	Spindle stop	M13	CSS and turn on cutting fluid
M06	Tool change	M14	CCSS and turn on cutting fluid
M07	Turn cutting fluid on; flood mode	M17	Turn off spindle and cutting fluid
M08	Turn cutting fluid on, mist mode	M19	Turn off spindle at oriented position

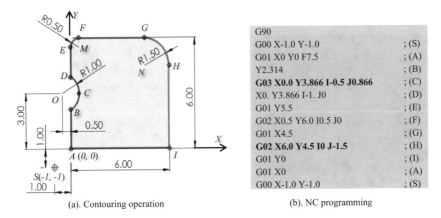

G90	
G00 X-1.0 Y-1.0	; (S)
G01 X0 Y0 F7.5	; (A)
Y2.314	; (B)
G03 X0.0 Y3.866 I-0.5 J0.866	; (C)
X0. Y3.866 I-1. J0	; (D)
G01 Y5.5	; (E)
G02 X0.5 Y6.0 I0.5 J0	; (F)
G01 X4.5	; (G)
G02 X6.0 Y4.5 I0 J-1.5	; (H)
G01 Y0	; (I)
G01 X0	; (A)
G00 X-1.0 Y-1.0	; (S)

(a). Contouring operation   (b). NC programming

Fig. 4.108 G-code example with arc motions

Computer-aided manufacturing (CAM) tools support the offline programming, so that CNC programs can be generated automatically based on part models. Here, *the SolidWorks HSMWorks* is used as an example of the CAM tools. HSMWorks supports all conventional strategies of machining operations such as contour, parallel, pocket, radial, scallop, spiral, and pencil. It is capable of generating optimal toolpath with less machining time and better surface finish, and it provides the simulation environment to verify CNC programs. Figure 4.109 shows the main features of HSMWorks to program CNCs for turning and milling operations.

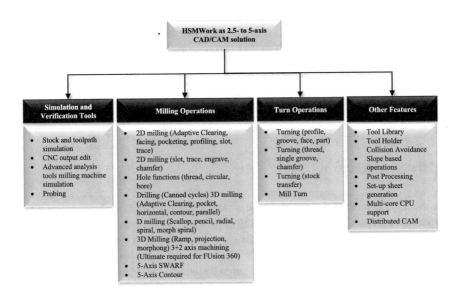

Fig. 4.109 The main features of HSMWorks for turning and milling operations

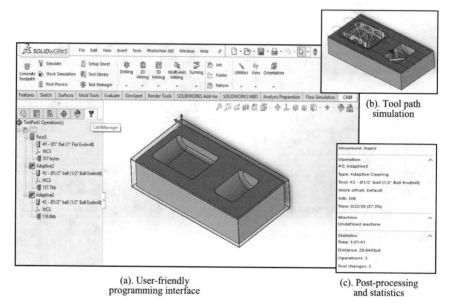

(a). User-friendly
programming interface

(b). Tool path
simulation

(c). Post-processing
and statistics

Fig. 4.110 Programming interface of HSMWorks

Figure 4.110 shows the interface of HSMWorks, which is integrated as an adds-in CAM tool in SolidWorks. A part model can directly be imported, the machined features can be automatically identified, but the operation for each machined feature is programmed, respectively. The toolpath for certain feature can be created automatically, and it can be visualized and simulated at any time as shown in Fig. 4.110b. The HSMWorks includes design libraries for programmers to select process types, machines, and cutters as shown in Fig. 4.110a. When the CNC is specified, the complete CNC program can be generated by post-processing, and the program can be downloaded to the machine to control machining operations. In addition, the performance of a CNC program and corresponding machining operation can be evaluated automatically as shown in Fig. 4.110c.

4.10 Summary

Design of manufacturing processes takes into consideration many factors including materials, machines, tools, and numerous operating parameters; in addition, manufacturing processes are evaluated against some conflict criteria such as cost, accuracy, productivity, flexibility, and adaptability. Conventional ways for design of manufacturing processes have their limits in dealing with the complexities of modern products and manufacturing processes. It becomes critical for engineers at SMEs to formulate manufacturing process design problem as a multidisciplinary engineering problem, and seek the engineering solution via modern computer-aided techniques.

Many computer-aided manufacturing (CAM) tools are available to deal with various engineering problems in design of manufacturing processes. This chapter covers the theories and tools for designs of composite materials, molds and dies, fixtures, and machining programs. Engineers should be trained to use these CAM tools in designing manufacturing processes for complex products and systems.

Design Problems

Problem 4.1 Create a model of the object with composite materials and determine the stress distribution subjected to the loads shown in Fig. 4.112. The object has a rectangle shape with the width and the height of 1 inch. The load along y-axis on the top edge is 2000 lb/in, and the loads along x-axis and y-axis on the right side are 1000 lb/in. The displacements of X, Y, Z, R_y are restrained on the left side (Fig. 4.111).

Table 4.15 shows the layup of the composites. It is composed of graphite/epoxy tape, the angles shown are relative to the global axis, i.e., the 0 degree ply 1 has its fibers running along the Y-direction. 90 degree ply 4 has its fibers running along the X-direction. The composite plies are graphite/epoxy tape with a thickness of 0.0054 in. The properties of graphite are modulus of elasticity $E = 6.96 \times 10^5$ psi, Poisson's ratio $v = 0.28$, density $\rho = 0.0809$ lb/in^3, $S_{ut} = 1.46 \times 10^4$ psi, and yield strength is $S_y = 1.75 \times 10^4$ psi.

Fig. 4.111 Composite part in Problem 4.1 (MSCsoftware 2020)

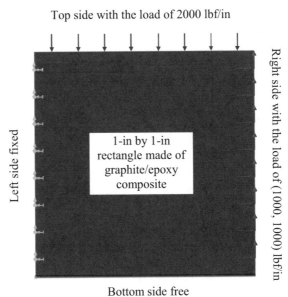

Top side with the load of 2000 lbf/in

Left side fixed

1-in by 1-in rectangle made of graphite/epoxy composite

Right side with the load of (1000, 1000) lbf/in

Bottom side free

Fig. 4.112 Part model for Problem 4.2

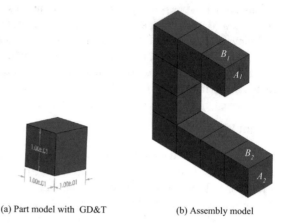

(a) Part model with GD&T   (b) Assembly model

Table 4.15 Layup of example composites in Problem 4.1

Ply no.	Materials	Thickness (inch)	Direction (degree)
1	Graphite/epoxy	0.0054	0
2		0.0054	45
3		0.0054	−45
4		0.0054	90
5		0.0054	90
6		0.0054	−45
7		0.0054	45
8		0.0054	0

Problem 4.2 Consider a cube with the specified dimension in Fig. 4.112a. For the assembly in Fig. 4.112b, what will be the dimension and tolerance (1) between the plane of A_1 and A_2 and (2) between the plane of B_1 and B_2?

Problem 4.3 Figure 4.113 shows the dimensions of a machined part. Use the *AutoDimensionScheme* tool in the Solidworks DimXpert to annotate GD&T information automatically.

Problem 4.4 Figure 4.114 shows the dimensions of a machined part. Use the *AutoDimensionScheme* tool in DimXpert to create both of plus/minus dimensioning and geometric dimensioning automatically.

Problem 4.5 For Example 4.6, let $L_1 = 100$ mm, $L_2 = 250$ mm, and $H = 150$ mm, and assume that clamping forces can be adjusted as $F_x[-2.0\,kN, 0\,kN]$; $F_y[-4.0\,kN, 0\,kN]$. Determine the optimized clamping forces to minimize the maximum displacement of workpiece in operation.

Problem 4.6 For Example 4.6, let $L_1 = 70$ mm, $L_2 = 275$ mm, $H = 175$ mm, $F_y = -2.5$ kN, $F_y = -0.25$ kN, and assume that the cutting forces are changed as

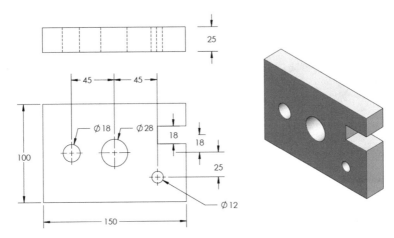

Fig. 4.113 Part model for Problem 4.3

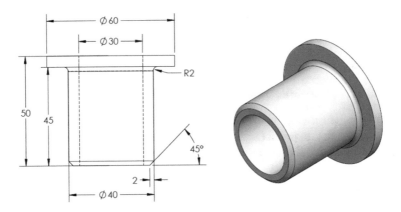

Fig. 4.114 Part model for Problem 4.4

$F_{t\_x}[-2.0\,kN, 0\,kN]$; $F_{t\_y}[-1.5\,kN, 0\,kN]$. Determine the maximum displacement of workpiece in the operation.

Problem 4.7 Assume that the materials of part and cutter are low-carbon steel and high-speed steel (HSS), respectively. The thickness of parts is 0.500 in and the tolerance is 0.020 inch for all dimensions, unit. Write the programs for the parts shown in Fig. 4.115. For each part, write two programs for (1) cutting holes and (2) contouring milling.

Design Projects

Project 4.1. (1) Create a plastic part you can reach such as the one shown in Fig. 4.116; (2) use SW Plastics package, select injection locations and filling settings; (3) run

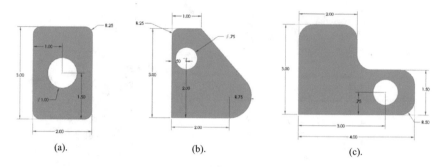

(a).   (b).   (c).

Fig. 4.115 Write CNC programs for parts in Problem 4.2

(a). bucket   (b). Hairdry part   (c). Stool

Fig. 4.116 Part example for mold filling analysis in Project 4.1

injection simulation to predict injection molding time; (4) predict defects including air traps, welding lines, and wraps; and (5) document your process and result.

Project 4.2. Create a mold assembly for injection molding of a plastic part such as a part in Fig. 4.116.

Project 4.3. Find some machined parts such as the examples shown in Fig. 4.117. Use the SolidWorks integrated system to design its machining processes and create machining programs for the part by following the steps given below:

(1)   Create CAD model by reverse engineering.
(2)   Design fixtures for all of machined features.
(3)   Create an assembly model for fixturing systems.
(4)   Create CNC program(s) for all of machined features.
(5)   Simulate the CNC program to obtain basic statistics of machining (number of setups, machining tools, the number of tools, cost....).
(6)   Post-process the CNC programs to convert them into G-code for a generic three-axis Hass machine tool.
(7)   Analyze the part to predict deflection (tolerance).
(8)   (Optional for bonus) make the part if a generic three-axis Hass machine tool is available.

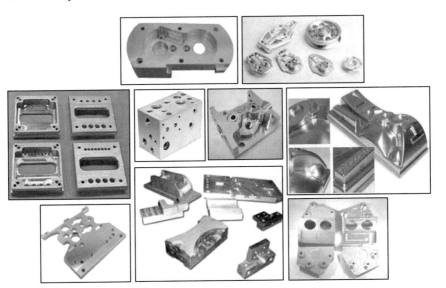

Fig. 4.117 Examples of machined parts for design Project 4.3

References

Allsup T (2009) How to speel GD&T a new way to learn GD&T. https://www.anidatech.com/How ToSpellGDT.pdf

Basha VR, Salunke JJ (2013) An advanced exploration of fixture design. Int J Eng Res Appl 6(5), part 3, pp 30–33

Bi ZM, Hinds B, Jin Y, Gibson R, McToal P (2009) Drilling processes of composites-the state of the art. In: Drilling of composite materials, Nova Science Publisher, ISBN: 978–1–60741–163–5, pp137–171

Bi ZM, Mueller D (2016) finite element analysis for diagnosis of fatigue failure of composite materials in product development. Int J Adv Manuf Technol 87(5):2245–2257

Bi ZM, Zhang WJ (2001) Flexible fixture design and automation: review, issues and future directions. Int J Product Res 39(13):2867–2894

Cimatron E Inc (2017) CAD/CAM Solution for Mold Making From Quoting to Deliver. https://www.cimatron.com/SIP_STORAGE/files/0/1760.pdf

Eastman (2019) Medical devices processing guide. https://www.eastman.com/Literature_Center/S/SPMBS3689.pdf

Kailas SV (2020) Chapter 3 imperfections in solids. https://nptel.ac.in/content/storage2/courses/112108150/pdf/Lecture_Notes/MLN_03.pdf

Kreculj D, Rasuo B (2018) Impact damage modelling in laminated composite aircraft structures, sustainable composites for aerospace applications, Woodhead Publishing Series in Composites Science and Engineering, pp 125–153

MSCsoftware (2020) Making a composite model. https://www.mscsoftware.com/exercise-modules/making-composite-model

Pilkey WD, Pilkey DF, Bi, ZM (2020) Petersons stress concentration factors, the 4th version, ISBN-13: 978–1119532514, ISBN-10: 1119532515, Wiley

Science Notes (2020) Periodic table of the elements. https://sciencenotes.org/printable-periodic-table/

Steeves M (2016) Before there was MBD, there was MBD and more! https://techday2016dotcom. files.wordpress.com/2016/04/16-q2-mi-before-there-was-mbd-there-was-mbd-and-more.pdf

Sun CC (2009) Materials science tetrahedron—useful tool for pharmaceutical research and development. J Pharm Sci 98(5):167–1687

Tuttle M (2020) Predicting failure of multiangle composite laminate: micromehanics failure analysis vs macromechanics failure analysis. https://courses.washington.edu/mengr450/LamFailures.pdf

United States International Trade Commission (2002) Tools, dies, and industrial molds: competitive conditions in the United States and selected foreign markets, Investigation No. 332–435. https:// www.usitc.gov/publications/332/pub3556.pdf

Vorburger TV, Raji J (1990) Surface finish metrology tutorial. https://www.nist.gov/system/files/ documents/calibrations/89-4088.pdf

Wang H, Rong Y, Li H, Shaun P (2010) Computer aided fixture design: recent research and trends. Comput Aided Des 42:1085–1094

Chapter 5
Computer Integrated Manufacturing (CIM)

Abstract A manufacturing system is an organization to make products; a manufacturing system consists of various functional units for designing, manufacturing, assembling, transporting, and marketing and sales. This chapter discusses the design and operation of manufacturing systems; it covers some enabling technologies including *cellular manufacturing, discrete event dynamic simulation, lifecycle assessment*, and *cost analysis*; and it also discusses how enabling technologies can be integrated for effective coordination and interaction of hardware and software systems from different vendors. Moreover, the evaluation of manufacturing systems, products, and processes is discussed from the perspective of sustainability.

Keywords Group technology (GT) · Cellular manufacturing (CM) · Discrete event dynamic system (DEDS) · Petric nets · Lifecycle assessment (LCA) · Computer integrated manufacturing (CIM) · Cost analysis · Design for sustainability

5.1 Introduction

Figure 5.1 shows a description of a manufacturing system. A manufacturing system is to transform raw materials into finished products through a series of manufacturing processes. A manufacturing system requires various resources to perform manufacturing processes. These resources can be classified into (1) *fixed assets* such as plants, machines, tools, fixtures, software tools, and human resources and (2) *flow assets* as the system inputs such as raw materials, power supplies, and capital investments at one end, and finished products, profiles, and the supports in the lifecycles of products at the other end. The manufacturing system implements its business goals by delivering value-added products to customers. In addition, manufacturing system operations are more and more influenced by the dynamics, the uncertainties, and the disturbance of the business environment, the boundary of the manufacturing system and the business environment becomes vague.

The complexity of a manufacturing system depends on the products to be manufactured in the system. Due to the growing complexity of modern products, manufacturing systems become more and more complex due to the increases of system components and variants, the interactions of system components, and their

© The Author(s), under exclusive license to Springer Nature Switzerland AG 2021
Z. Bi, *Practical Guide to Digital Manufacturing*,
https://doi.org/10.1007/978-3-030-70304-2_5

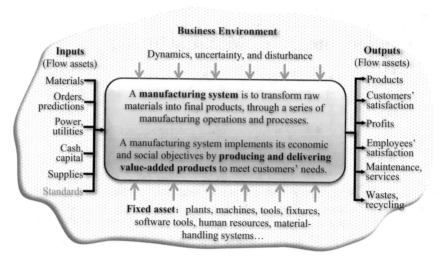

Fig. 5.1 Description of a manufacturing system

dynamic changes over time. Manufacturing systems are generally complex engineered systems. To design and operate a manufacturing system, engineers should understand basic resource types and their interactions in manufacturing. As an introduction of manufacturing systems, some basic concepts relevant to manufacturing systems are discussed in this section.

5.1.1 Continuous and Discrete Manufacturing Systems

A continuous manufacturing system transforms raw materials to finished products continuously through a series of processes. In a continuous system, raw materials are usually uncountable, which are in a state of powder, gas, or fluid. Manufacturing processes are continuous transformations such as chemical reactions or the changes of properties subjected to mechanical, heat, or other types of loads. The materials are processed continuously in motion as shown in Fig. 5.2. Since manufacturing processes are followed one after another, no inventory is needed in continuous manufacturing. Continuous manufacturing systems are widely used in chemical factories, food processing plants, and oil refineries. They are also used to produce raw materials such as metal sheets, rolls, wires, powders, plastic resin, and cement for discrete manufacturing. The advantages and disadvantages of continuous manufacturing are summarized in Table 5.1 (Knowledgiate 2017). In this chapter, continuous manufacturing systems are not discussed since their layouts are the sequential connections of processing machineries, which are relatively stable and simple in contrast to discrete manufacturing systems.

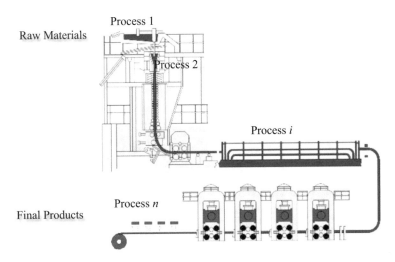

Fig. 5.2 Example of a continuous manufacturing system (HIBA 2019)

Table 5.1 Continuous manufacturing: advantages and disadvantages

Advantages	Disadvantages
• The product quality is consistent since the product goes through the same sequence of processes and machineries. • The production can be easily automated to reduce direct labor, and the system control can be simplified due to sequential processes. • No inventory is needed for sequence balancing. • The need of material handling is minimized due to the set pattern of continuous manufacturing. • Due to high-volume production, the overhead cost per unit can be reduced since the fixed costs of specialized equipment are shared by a large volume of output. Accordingly, there is quick return on investment (RoI).	• Continuous manufacturing is rigid in sense that when one machine is malfunctioned, the whole process is affected. • It requires to avoid piling up of work or any blockage on the line. • Unless the malfunctioned machine can be fixed immediately, it will force the preceding as well as the subsequent stages to be stopped completely.
Typical Applications: chemical factories, food processing plants, oil refineries, and the suppliers of conventional industrial materials.	

A *discrete manufacturing system* is characterized by making distinct products such as automobiles, computers, cellphones, toys, aircraft, and furniture items. In a discrete manufacturing system, manufacturing processes perform manufacturing operations on discrete parts. Each manufacturing process can be begun or terminated, respectively, and it is unnecessary for different machineries to have the same cycle time of manufacturing processes. A discrete manufacturing system may involve assembling processes, in which final products are assembled from parts and components. A discrete manufacturing system often needs a variety of raw materials to make different parts and components. For example, building a computer requires mainboards, CPUs, storages, keywords, monitors, and many other accessories, and these components are made from different materials or from different suppliers.

Figure 5.3 shows an example of discrete manufacturing systems which was built to manufacture aero-engines (Fang et al. 2020). The differences of continuous and discrete manufacturing were discussed by many researchers (Pritchett et al. 2000; Al-Habahbah 2015; Andrew 2019), and Table 5.2 gives a summary of the differences in *throughouts, quality measures, control variables, units,* and *characteristics.*

Fig. 5.3   A discrete manufacturing system for aero-engines (Fang et al. 2020)

Table 5.2   Continuous versus discrete manufacturing

Aspects	Continuous Manufacturing	Discrete Manufacturing
Throughout	Measured by attributes such as weight and volume	Measured by types, models, and numbers of products
Quality indicator	Consistency, concentration, free of contaminants, and conformance to specifications.	Dimensions, tolerances, surface finishes, free of defects, reliability, and lifespans.
Typical control variables	Relevant to recipes and formulas such as ingredients, temperature, volume flow rate, time, and pressure.	Relevant to shaping, assembling, or property enhancement such as position, velocity, path, acceleration, force, temperature, heat, and power.
Units of measurement	Lot, grades, potency, and shelf-life	Pieces, counts, and numbers.
Characteristic	Making '*stuff*' such as milk and milk powder	Making '*things*' such as cars, furniture, and computers.
Examples	Oil refining Milk production	Car production Computer assembly

5.1.2 Variety, Quantity, and Quality

Discrete products in a manufacturing system are by *variety* (V), *quantity* (Q), and *quality*. The product quantity Q affects the way how the resources, peoples, facilities, and procedures are organized in a manufacturing system. Accordingly, the productions in manufacturing systems are classified into 1) *low productions* ($Q = 1–100$ units), *medium productions* ($Q = 100–10,000$ units), and *high productions* ($Q > 10,000$ units).

The product variety V refers to the number of product variants, which have differences in materials, features, quality, or manufacturing processes; but they are made in the same manufacturing system. The number product variety V depends on how a difference is defined for two products. For example, El-Sherbeeny (2016) suggested classifying the products into *soft product varieties* and *hard product varieties*. The difference of products in the first catalogue can be handled by planning, controlling, and scheduling; however, different hardware equipment is needed to deal with the difference of products in the second catalogue. An example of soft product varieties is a car product family in the same brand; the products have a common platform and they can be made in the same production line. An example of hard product varieties is the difference between passenger cars and pickup trucks, different production lines are used to build cars and trucks, respectively. Finally, *product quality* describes how well a product is made.

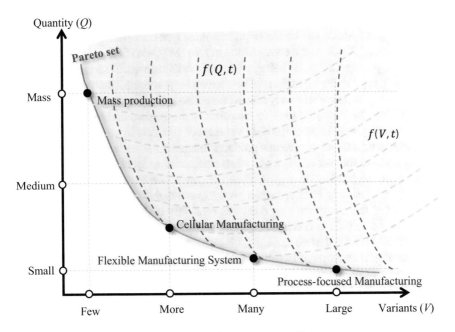

Fig. 5.4 Product variety and quantity in selecting a system paradigm

Figure 5.4 shows that product variety (V) and quantity (Q) are two criterial factors in designing a manufacturing system. Assume the total capacity of a manufacturing system is given, it is impractical to maximize both product variety and quantity; therefore, some trade-offs must be made in selecting an appropriate paradigm for a manufacturing system. The products with a low variety and a high quantity are made in *mass production*; the products with a high variety and a low quantity are made in *process-focused manufacturing*; the products with a medium variety and medium quantities are made in *cellular manufacturing, flexible manufacturing*, or *mass customization*. Design of a manufacturing system is a complex engineering problem. To optimize a system for a given design criterion, a design variable should be changed in a favorable direction (e.g., f (Q, t) and f (V, t)). However, when a number of conflicting criteria are involved, some methods, such as *the Pareto set* in Fig. 5.4, should be deployed to make the trade-offs for overall system performances.

5.1.3 Decoupled Points in Production

A manufacturing system makes final products to meet customer's needs, and manufacturing businesses are driven by customers' orders. To reduce lead times, parts, components, or even products can be made before customers' orders arrive. For complex products, manufacturers usually customize products at the assembly stage using pre-made parts and components. Therefore, designing a manufacturing system requires to decouple a production line into two stages that are for the required businesses and activities before and after customers' orders, respectively. Accordingly, manufacturing systems can be classified into three types, i.e., *make-to-stock* (MTS), *make-to-order* (MTO), and a combination of two (MTO-MTS).

In MTS, products are made before customers' orders arrive. MTS is a *push* system paradigm. It is effective when the products have a large volume, small variety, and the manufacturing costs are relatively low in comparison with the costs of raw materials. MTS requires the inventory to store materials, semi-finished and finished products. It is challenging to determine the right volumes of products in inventory; on the one hand, a reduced number of premade products reduces the inventory cost; on the other hand, it exposes the risk of stock out when customers' orders increase.

In MTO, the production begins when customers' orders arrive. MTO is a *pull* system paradigm. It is effective when products are highly diversified and with low volumes, and the manufacturing costs are relatively high in comparison with the costs of raw materials. It is critical to match manufacturing capabilities and product demands in MTO. On the one hand, reducing production capabilities helps to increase the utilization rates of manufacturing resources thus will reduce product unit costs. On the other hand, limited product capabilities expose the risks of losing customers' orders due to the need for long lead times.

In MTO-MTS, the 'push' and the 'pull' system paradigms are combined. The production lines in the manufacturing system are decoupled. The 'push' strategy is adopted for the premade parts or components, and the 'pull' strategy is adopted to

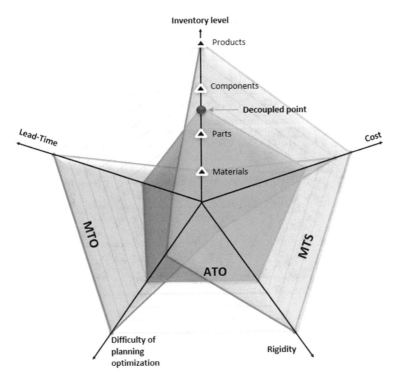

Fig. 5.5 Comparison of MTO, MTS, and ATO

customize the products to meet customers' needs. Products can be customized by the technique so-called *assemble to order* (ATO). In ATO, different parts or components are selected and assembled for different products. MTO-MTS is effective to balance the strategic goals of reducing inventory costs and shortening product lead times (Kaminsky and Kaya 2009).

Figure 5.5 shows a comparison of MTO, MTS, and ATO in terms of *inventory level, lead time*, and the difficulty of *planning and scheduling, rigidity*, and *cost*. The overall performance of a manufacturing system should be optimized by minimizing the aforementioned five indicators simultaneously. On the one hand, MTS minimizes product lead times but demands the highest inventory level of products, and MTS has poor flexibility since the system is planned and controlled based on the long-term prediction. On the other hand, MTO minimizes the inventory level of products, and it increases the system flexibility since system planning and controlling is based on customers' orders. However, products have the maximum product lead times. In contrast, ATO splits the supply chain in system planning and scheduling: before the decoupling point, parts and components are made before customers' needs arrive; after the decoupling point, the manufacturing and assembling processes are planned and scheduled after customers' orders arrive. The overall performance of the manufacturing system can be improved since all five indicators of system performance are

Table 5.3 Comparison of MTS, MTO, and ATO at other aspects

Aspect	MTO	MTS	MTO-MTS (ATO)
Data for production	Volumes, varieties, and specifications of products by customers	Volumes, varieties, and specifications of products by predication	Customer' orders and configuration management
Assumption of production planning	Engineering capacities	Projected inventory levels	Volumes and varieties, lead-times of products
System control	Adjust engineering capacities to meet customer' needs	Assure the levels of customer services	Make products in specified lead times
Sales and operations	Predicted demands and performed the designs of products and manufacturing processes	Predicted demands for designed products and manufacturing processes.	Predicated demands for all alternatives in configuration management
Assumption of master production scheduling	Actual demands	Projected demands	The combination of prediction and actual demands
Lead-time of products	Begin with design stage and specify delivery time.	Available until next inventory replenishment	A short lead time for the assemblies of existing parts and components into products.

optimized simultaneously. Table 5.3 shows the difference of MTS, MTO, and ATO at other aspects (Vollmann et al. 2004; Cruz-Mejia and Vilalta-Perdomo 2018).

5.2 Manufacturing System Architecture

A manufacturing system involves various manufacturing resources that are closely interacted with each other. To run a manufacturing system in a controlled manner, the complexity of system must be *manageable* for the smoothness of business operations. *Enterprise architecture* (EA) aims to manage system complexity by defining the structure and operation of a manufacturing system. An enterprise architecture can be defined from different aspects such as *functions, processes, businesses, information*, and *technology changes*. The functional requirements (FRs) of a manufacturing system architecture include to (Gao 2001).

(1) define the mission, strategies, methods, and functions, and utilize them to plan and operate the system;

(2) regulate the communications among functional units with standardized vocabularies.

(3) be open to adopt emerging technologies for technological advancement and upgrading;

(4) ensure the consistency, integrity, availability, promptness, and security of data sharing and information integration;

(5) seek the solution to enhance system flexibility, adaptability, and efficiency at an affordable cost;

(6)   support resource sharing for a high utilization rate of manufacturing resources in system;

(7)   expend the lifespan of EA by practicing *continuous improvement* (CI).

Numerous EAs have been proposed for different applications, and the most influential EAs are (1) *Open System Architecture for Computer Integrated Manufacturing* (CIMOSA) from the European CIM Architecture Consortium, (2) *GRAI Integrated Methodology* (GRAI-GIM) from the GRAI Laboratory in France, (3) *Purdue Enterprise Reference Architecture* from the Industry-Purdue University Consortium, and (4) *Enterprise Architecture* from the National Institute of Standards and Technology (EA-NIST) (Williams 1994).

As an example, Fig. 5.6 shows the EA-NIST which was used to represent the organization of a manufacturing system (Wikipedia 2020a). In the EA-NIST, system elements are divided into *layers* and *domains*. The architecture consists of the *business, information, applications, data,* and *technology layers*. The system has its boundaries to tell if relevant businesses are internal and external, but the influence of the business environment on the manufacturing system is reflected in *external discretionary and non-discretionary standards*.

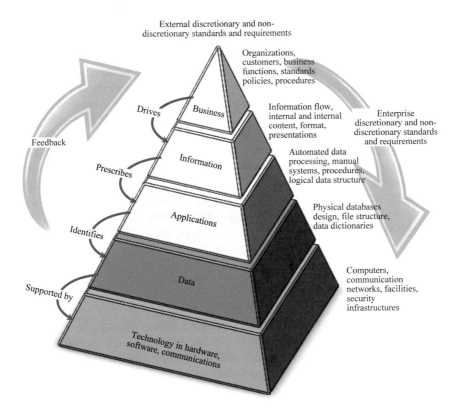

Fig. 5.6 Enterprise architecture by NIST (EA-NIST)

The system complexity can be measured by *entropy*. Entropy represents the amount of information for the operations and interactions of system elements. Therefore, system entropy is evaluated by *the number* of system elements and the *types* and *changes* of system interactions over time. For system interactions in Fig. 5.6, the elements at *the information-layer* are driven by the elements at *the business-layer*, and the elements at *the application-layer* are prescribed by the elements at the information-layer. The elements at *the application-layer* are used to define the elements at *the data-layer*. Finally, the elements at *the data-layer* are supported by the hardware, software, and communication at *the technology-layer*. The executive information flow is from the top layer to the bottom layer, and the feedback information flow is from the bottom layer to the top layer in the opposite direction.

In implementing system architecture, system elements are usually modularized. In other words, system elements are separable, and the elements at certain layers and domains are sophisticated in managing, contextualizing, and producing data for respective tasks independently. EA can benefit from modularization for its continuous improvement (CI): EA should not be affected by the changes occurring to system elements at module levels.

In a manufacturing system, the differences in manufacturing operations can be examined from different aspects. Figure 5.7 shows the classification of manufacturing businesses from the perspectives of *structural layer, information integration,* and *life-cycle* (SAC 2018). System elements can be distinguished from one to another based on the layers they are in EA; the layers corresponding to EA-NIST in Fig. 5.6 are the layers of *equipment, controls, workshops, enterprises,* and *corporations. Information integration* distinguishes system elements based on their roles in dealing with data; typical activities relevant to data are *collection, processing, utilization, communication,* and *integration,* and the capabilities of system elements can be enhanced by the information integration for self-sensing, self-learning, self-decision, self-execution, and self-adaptation; therefore, system elements are classified based on the scope of data processing into *device, interconnection and interworking, information fusion, incorporation, enterprise,* and *enterprise alliances.* System elements play their roles at different stages of product lifecycles. Therefore, they can be classified based on the stages where they execute as *design, manufacture* and *assemblies, logistics, sales* and *service,* and recycling.

5.3 Production Facilities

Production facilities make physical contacts with materials, parts, components, and products. As shown in Fig. 5.8, typical production facilities are *plants, machine tools, utility equipment, fixtures, molds, dies* and *tools, gauge inspecting tools,* and *material handling equipment,* and *system layouts* where all manufacturing resources are arranged as a system.

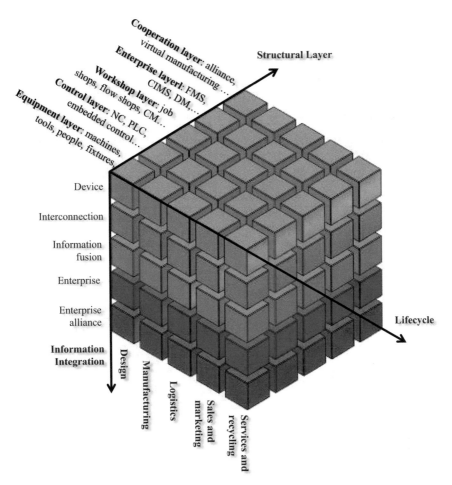

Fig. 5.7 Manufacturing system architecture

5.3.1 *Machine Tools*

Figure 5.9 shows that two important specifications of machine tools are (1) applicable manufacturing processes and (2) the strategies to pay back the initial costs. Machine tools are the core facilities to perform manufacturing processes, and machine tools can be classified based on the types of applicable manufacturing processes such as *power metallurgies*, *metal sheet forming*, *injection molding*, *machining*, and *non-conventional machining*.

A machine tool usually involves significant investment cost, which should be recovered from the sales of the products that have a manufacturing process performed on the machine tool. The more products a machine tool can make, the lower *the product unit cost* is, the better *the return of investment* (ROI) of the machine tool is, and the higher value of the machine tool contributes to products.

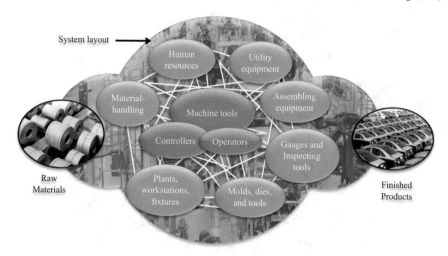

Fig. 5.8 Typical production facilities in manufacturing

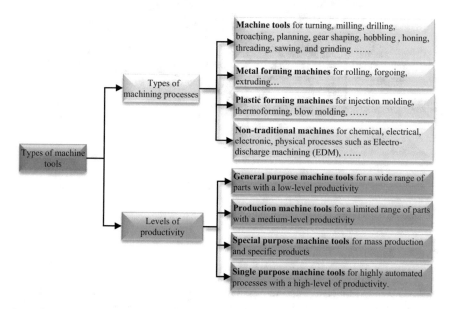

Fig. 5.9 Typical types of machine tools

Assume that a machine tool has *an initial cost (IC)*; it is designed to serve for M product varieties and with the volumes of N_i (i = 1, 2... M) for each product variety. In addition, the added value by the machine tool is v_i (i = 1, 2... M). The *total number of products (NP)* made by the machine tool is a sum of the volumes of all product varieties as

Table 5.4 Classification of machine tools based on ROC models

Types	Variety (M)	Volume (N_i)	Added value (v_i)	Initial cost (IC)
General purpose machine tools with a low-level productivity	High	low	Low	Low
Flexible machine tools with a limited productivity	Medium	Medium	Medium–High	High
Special purpose machine tools with a high-level productivity	Low	High	Low–High	High
Single purpose machine tools with a high-level productivity	One	High	Low–High	Low-Medium

$$NP = \sum_{i=1}^{M} N_i \qquad (5.1)$$

Accordingly, *the total added value* (*VP*) of the machine tool to these products are

$$VP = \sum_{i=1}^{M} N_i \cdot v_i \qquad (5.2)$$

Assume that each part takes an average machining time on the machine tool, *the unit cost* (*UC*) *to a product* relevant to the machine tool can be estimated as

$$UC = \frac{IC}{NP} \qquad (5.3)$$

The ROI of the machine tool is the difference of the total added value (*VP*) and the initial cost of

$$ROI = \sum_{i=1}^{M} N_i \cdot v_i - IC \qquad (5.4)$$

The design variables in Eq. (5.4) can be used to describe the characteristics of different machine tools shown in Table 5.4. A machine tool can be one of the following types, i.e., a *general-purpose, flexible, special purpose,* or *single-purpose* machine tool.

5.3.2 Material Handling Tools

A product has multiple features which are made on different machine tools, and the product is transported over the production line to receive the services at machine tools. *Material handling* (MH) tools are used to move materials, parts, or tools in a manufacturing system; *a material handling system* consists of all of manufacturing resources which are used to keep and transport materials, parts, components, and

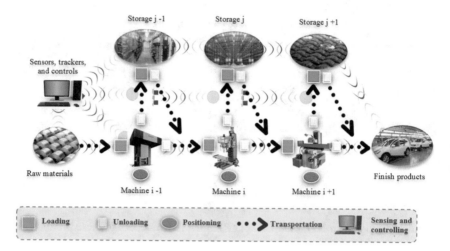

Fig. 5.10 Main components of a material handling system

products. Figure 5.10 shows the schematic of a material handling system which consists of the following five types of material handling tools.

(1) *Transportation devices* are used to transport objects in a manufacturing system. Parts should be transported to workstations to receive the services over machine tools. If a machine tool is not immediately available, the part should be temporarily stored until the machine tool becomes available for the next manufacturing process. Common transportation devices are *conveyors, cranes,* rails, *trucks,* and *forks.* It is also very common that human operators handle materials manually such as loading/unloading parts at workstations.

(2) *Positioning devices* are used to position objects for manufacturing operations such as machining, inspecting, assembling, or storing. Differing from a transportation device to tackle the movement at multiple locations, a positioning device locates the part at a single workstation.

(3) *Loading/unloading devices* are used to load or unload parts to machine tools before or after manufacturing processes. Some machine tools are equipped with sophisticated loading/unloading devices; while operators or industrial robots are used to load/unload parts for other machine tools.

(4) *Storages* are used to keep or buffer parts temporarily in a manufacturing system. Two common types of storage are *carousels* and *buffers.* In many cases, only storage floor and space are needed to store parts without an additional storage device. Storages can be integrated into a transportation system as an *automated storage and retrieval system* (AS/RS).

(5) *Identification and control systems* are used to track objects and machine tools and monitor and control material handling systems. For some simple systems, parts are located and tracked, and MH devices are operated manually.

5.3.3 Fixtures, Molds, Dies, and Tools

Fixtures, molds, dies, and tools are the type of production facilities that make direct contact with parts in manufacturing operations.

Fixtures are used to locate and hold parts in performing manufacturing processes. A fixturing system should support a smooth and fast transition for a batch of products, simplify the system setup, and sustain the consistency of product quality (Wikipedia 2020b). Fixturing system designs have been discussed in Sect. 4.8 extensively; and Fig. 5.11 shows five types of fixture elements: (1) *a tool body* is used as a frame to mount all fixture elements together as a system. A tool body should be designed to minimize the deformations of parts and fixturing system which are subjected to external loads; (2) *a support element* supports an object by direct contacts. A support element can be *adjustable* or *fixed* depending on if the contacts should be adjusted or fixed; (3) *a locator* aims to position an object when a manufacturing process is applying on object; locators and supports work together to confine the motions in all directions; (4) *a clamping system* is to secure an object in the place in a manufacturing process. A fixturing system may need other accessories such as *lifting devices, stoppers,* and *ejectors*; these accessories are used in the fixture setup.

Other than fixtures, *molds, dies,* or *tools* make direct contacts with parts as well. The geometry and the relative motion of a mold, die, or tool determines the part geometry. For example, the geometry of casting is determined by the formed cavity in a mold, the geometry of metal formed part is determined by a pair of punch and dire, and the geometry of a machined part is determined by the profile and toolpath of the cutter.

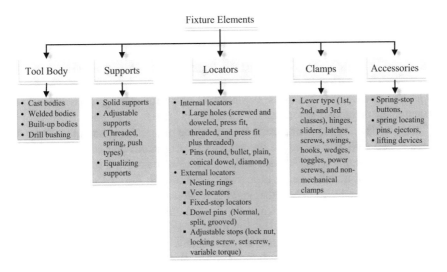

Fig. 5.11 Typical elements used in a fixturing system

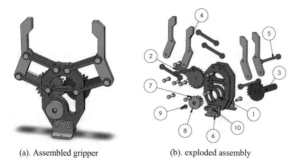

ITEM NO.	PART NUMBER	QTY.
1	Body	2
2	Gear 1	1
3	Gear 2	1
4	Gripper	4
5	Linkage	6
6	Base Plate	2
7	Pin	8
8	Motor Gear	1
9	ISO 7380 - M3 x 12 - 12N	3
10	ISO - 4035 - M3 - N	2

(a). Assembled gripper   (b). exploded assembly   (c). Bills of materials (BoM)

Fig. 5.12 Example of assembled products

5.3.4 Facilities for Other Manufacturing Operations

Depending on the complexity of products, a manufacturing system includes other value-added or non-value-added manufacturing operations processes such as assembling, inspecting, prototyping, and packaging. These manufacturing operations can also be mechanized or automated in the solutions of intelligent manufacturing (IM). Products are usually assembled from parts and components. Figure 5.12 shows an example of an assembled gripper which is assembled from 30 parts with a total of 10 different parts. Assembling facilities are used to fit and join parts and components together as a new large component or finish product.

5.3.5 Manufacturing System Layouts

A manufacturing *plant* or *factory* is an industrial site that is built to run manufacturing businesses. A plant or factory consists of *buildings, machinery, material-handling tools, capitals,* and *other resources* that are needed to make products. *The layout* of a plant or factory refers to the organization of manufacturing resources and the arrangement of production methods (Kiran 2019).

A layout is designed to maximize the utilization of machinery and minimize non-value-added activities to reduce the overall cost of system operations. The layout design of a manufacturing system must take into consideration of the complexity of products and manufacturing processes. As shown in Fig. 5.13, the layout of a manufacturing system can be either *rigid* or *flexible.* The system elements in a rigid layout are not reconfigurable; while the system elements in a flexible layout are *configurable* in production. Rigid layouts include *job shops, flow shops, project shops,* and *continuous processes.* Flexible layouts include *flexible manufacturing systems* (FMSs), *cellular manufacturing systems* (CM), *distributed manufacturing systems* (DMS), and *virtual manufacturing systems* (VMS).

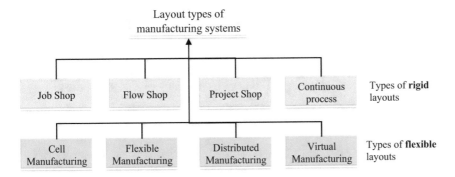

Fig. 5.13 Layout types of manufacturing systems

5.4 Cellular Manufacturing

Cellular manufacturing (CM) has a flexible system layout, and it is a hybrid solution of job shops and flow lines. Therefore, CM has the advantages of (1) a job shop for the flexibility of making a wide scope of products and (2) a *flow line* for the high productivity of making products in a production flow. *A cellular manufacturing system* (CMS) consists of a number of workcells which are logically linked based on the sequences of manufacturing processes. Figure 5.14 shows an example of CMS. In each workcell, machine tools are arranged in a production flow; however, the

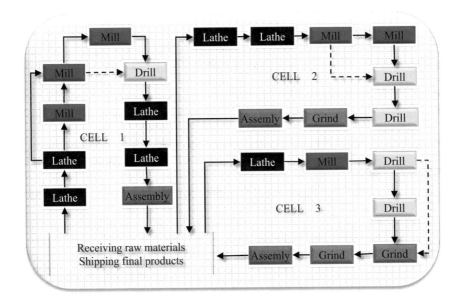

Fig. 5.14 Cellular manufacturing layout

Table 5.5 Characteristics of cellular manufacturing (Weber 2004)

Advantages	Disadvantages
• Cells are designed to make products for a family; it reduces setup time since no change is needed for machines and tools within cells to process similar parts. • With reduced setup times, the amount of work in progress can be reduced. • Each part is processed in a single cell which reduces part travelling distance & time. No effort is wasted to store, protect and control materials. • Single machine can be used to manufacture one or more products in each cell for high machine utilization. • Lead time can be used due to the reduction of setup time, work in process, and the increase of machine utilization. • Manufacturing processes become simple, and incentive and simplified processes boost worker morale.	• It reduces the manufacturing flexibility, • It is challenging to balance cells. • High mixes of low volume production can make cells impractical • Job rotation is common in CM but it can cause the problems due to changes • Hard for operators to adapt cells, there will be some resistance from operators since no rest period when parts are made • Easily underestimate training needs
Examples: U-shaped, inverted u-shaped, or straight-line cells for making discrete parts	

system elements in a workcell can be reconfigured, changed, or retooled for making different products as long as products are in the same family of products. Manufacturing operations such as loading/unloading, tooling changing, and transportations, can be fully automated or manually assisted. Table 5.5 explains the advantages and disadvantages of CMSs.

5.4.1 Design of Cellular Manufacturing System

System layouts in Fig. 5.13 were classified into two groups. A rigid layout such as a job shop, flow shop, project shop, or a continuous process does not involve in reconfiguration, while the system elements in a flexible layout such as a CMS, FMS, DM, and VM should be reconfigured over time. The reconfiguration of an FMS, DM, or VM is mainly at the software side, and it is implemented by control software or workflow composition (Viriyasitavat et al. 2019a, b). From this perspective, CMS is unique in sense that both hardware and software reconfigurations are needed to make different products in one or a few product families. In this section, the formation of CMS is discussed.

A cellular manufacturing system is designed based on two formations, i.e., *part-family formation* and *machine-cell formation*. In a part-family formation, the parts are grouped based on the similarities of part geometries and processing requirements. In a machine-cell formation, different machine tools are grouped to manufacture one or a few of part families. Note that both the part-family formation and machine-cell formation are non-deterministic polynomial (NP)-complete problems. Efficient clustering algorithms are expected to group parts and machines effectively. In the following section, *Group Technologies* (GT) is discussed to group parts based on the similarities of geometries, shapes, and processing requirements.

5.4.2 Group Technology (GT)

A system layout aims to optimize the overall performance of the manufacturing system. CMS is expected to make products in mass customization; therefore, the most critical tasks in designing a CMS are to identify the similarities of parts and machines, group parts as part families, and group machines into workcells to improve the utilizations of manufacturing resources. Here, the similarities of products are discussed, and *group technology* (GT) is introduced to identify the similarities of products.

GT is used to analyze products and manufacturing processes and identify their similarities to define product families. The outcome of GT is a set of workcells which are dedicated to make product families. GT is an ideal choice to make the products with medium variants and volumes that were conventionally made in batches. In contrast to traditional batch productions, GT is able to reduce downtimes and changeovers of machines. Moreover, if the workcell from GT becomes a long-term solution to a product family, such a solution can be referred to as a *flexible manufacturing system*.

Table 5.6 shows the similarities of products should be explored at all aspects that affect the selections and operations and organizations of machine tools. Especially the similarities of design and manufacturing attributes should be taken into consideration.

Depending on the complexity of product variants, GT can be implemented manually or automatically. Three practical techniques to define product families are *visual inspection*, *classification of products*, and *production flow analysis*.

5.4.2.1 Visual Inspection

Visual inspection is implemented manually. The materials, features, geometries of products are visually inspected to identify the similarities and group the products with the maximum similarities into product families. A *product family* is a set the products that exhibit similarities in materials, features, geometries, sizes, and manufacturing processes.

Table 5.6 Design and manufacturing attributes for similarities

Design Attributes	Manufacturing Attributes
▪ Major dimensions	▪ Major process
▪ Length/diameter ratio	▪ Operation sequence
▪ Basic external shape	▪ Batch size
▪ Basic internal shape	▪ Annual production
▪ Material type	▪ Machine tools
▪ Part function	▪ Cutting tools
▪ Tolerances	▪ Material type
▪ Surface finish	

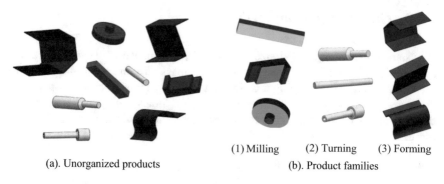

(1) Milling   (2) Turning   (3) Forming

(a). Unorganized products   (b). Product families

Fig. 5.15 Example of GT based on types of machining processes

Note that the similarities of product variants must be evaluated comprehensively. A similarity at certain aspects does not quarantine meaningful product families. For example, it does not make sense to group products with the same geometry but from different materials as a product family since the machines to process different materials are different. In a product family, the similarities of products must be significant enough to identify a set of machines that are applicable to all products. In addition, product volumes also make the difference in GT; for example, it is unlikely to find an economic solution of the machines which are able to make the products with (1) a volume of 1,000,000 units annually for a tolerance of ± 0.010 inch and (2) a volume of 100 units annually for a tolerance of ± 0.001 inch in one product family.

To practice visual inspection, engineers should analyze the features of products to be machined, and understand how these features can be made on what types of machines; then one or a few of the main criteria are used to identify the similarities of products. As shown in Fig. 5.15, the types of machined features are prioritized to group the unorganized products in Fig. 5.15a into three product families from milling, turning, and forming in Fig. 5.15b.

The machined features are usually the important measures in defining product families; however, the similarities are also critical if the same set of machines can be used to make products even though their geometries are fairly different. Figure 5.16a, b show two examples of product families that exhibit the similarities in geometries and machining processes, respectively.

5.4.2.2 Product Classification

In product classification, the coding scheme is defined to assign the codes to products, and the products are analyzed to determine the similarities based on codes. Codes can be assigned to products manually or automatically; however, a clustering process to identify product families based on the codes should be performed by computer programs. One critical task in coding products is to determine what and how the

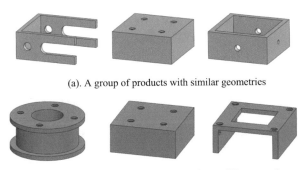

(a). A group of products with similar geometries

(b). A group of dissimilar products from milling operations

Fig. 5.16 Product families with the similarities in geometries and machining processes

features on products are coded since no universal rule exists to a wide scope of products. The common practice is to classify products based on their geometries into *rotational* and *non-rotational* products. Moreover, industrial products are surveyed to rank the parameters and features of products based on their influences on manufacturing processes. Figure 5.17 shows an example of the coding scheme for sheet metal products where major shapes, materials, and material specifications are highly ranked (Zeng 2009).

Generally, a coding scheme is required to (1) be flexible to represent and classify existing and potential future products, (2) be specific enough to identify required types of machines, and (3) be able to distinguish products by critical manufacturing attributes including materials, tolerances, and processing types.

When a coding scheme is only used in individual companies, the manufacturing attributes related to other important criteria should be taken into consideration; for examples, (1) types and capacities of manufacturing processes, (2) types and number

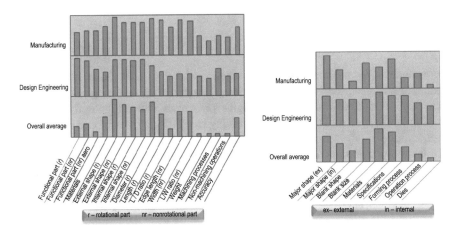

Fig. 5.17 A coding scheme for sheet metal products

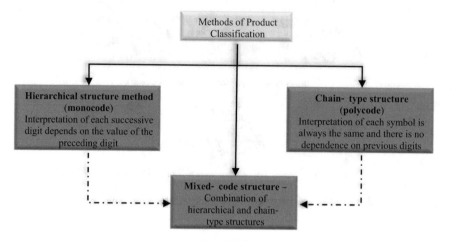

Fig. 5.18 Types of code schemes (Askin and Standridge 1993)

of tool changes, (3) types and number of machine setups, and (4) balance of utilization rates of machines in a production line. In manufacturing companies, GT does not usually lead to a permanent solution; since not all new products can be classified appropriately by an existing coding scheme. Therefore, the coding system should be continuously improved to eliminate old products and admit new products in workcells.

A cumbersome code may demand more resources for data collection and computation; therefore, a coding scheme should be concise as long as the main design and manufacturing features of products are represented. Figure 5.18a code scheme can be *hierarchical, chain-type,* or *hybrid*. A hierarchical scheme uses *monocode*, a chain-type scheme uses *polycode*, and a hybrid scheme uses both monocode and polycode.

5.4.2.3 Monocodes

Monocode is for a hierarchical code scheme, and Fig. 5.19 shows a monocode example that uses four digits to represent the main attributes of products uses a tree-like hierarchical structure. Each digit represents a main attribute; for example, the third digit describes the type of the driving mechanism; it can be one of mechanical, hydraulic, or electrical systems. The fourth digit represents the main shape in the first branch and the functions of the driving mechanism in the third branch.

In a hierarchical code scheme, the next digit amplifies the information by previous digits. Therefore, a monocode is capable of distinguishing a total number of ($n_1 \times n_2 \times \ldots \times n_n$) product variants; note that n_i is the number of possible choices at the i-th digit ($i = 1, 2, \ldots n$) and n is the number of digits in a monocode. Monocode

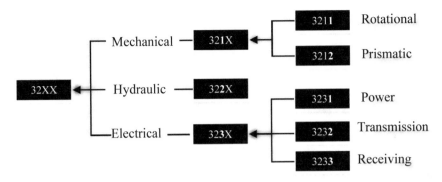

Fig. 5.19 Monocode for hieratical code structure

is effective to differentiate products in terms of geometries, materials, and sizes. However, monocode exhibits its limitations in representing the information relevant to manufacturing processes.

Figure 5.20 shows a code scheme where monocode is used to classify sheet metal products. The first three digits are for the information of raw materials; the next four digits are for the attributes of machined features, and the last digit is for the special requirement.

Example 5.1 A code scheme is defined as shown in Fig. 5.20a, determine the monocode for the product variant shown in Fig. 5.23b.

Solution. The code scheme in Fig. 5.21a has a tree-like structure with four levels. The digit at the first level is for an overall geometric shape, i.e., '0' for *cylindrical* and '1' for *block*. The digits at the second and third levels are for the dimensional ratios of main dimensions. The digit at the fourth level is for the tolerance requirements.

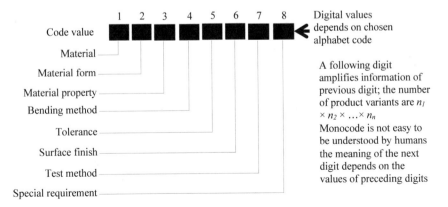

Fig. 5.20 Monocode scheme for sheet metal products

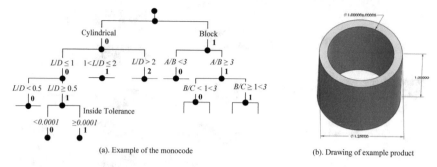

(a). Example of the monocode   (b). Drawing of example product

Fig. 5.21 Example of using monocode to distinguish products

By analyzing the attributes of the product in Fig. 5.21b, the digits at four levels of the code scheme can be determined as,

(1)   The first digit is '0' for the cylindrical body,
(2)   The second digit is '0' for the ratio of $L/D = 1/1.25$, which is 0.8 less than 1.0,
(3)   The third digit is '1' for the ration of $L/D = 0.8$, which is 0.8 larger than 0.5, and
(4)   The fourth digit is '0' is for the tolerance of 0.00005 less than 0.0001.

Therefore, the monocode for the product in Fig. 5.21b becomes '0010'.

5.4.2.4   Polycodes

A polycode is for chain-type code scheme, and a value at the specific digit of the code has the same meaning to all product variants no matter what values at other digits are. A polycode is easy to use but not very efficient since the polycode must have the digits to represent all possible attributes, while a product only has a part of these attributes; some digits in a polycode become unnecessary to those products. Table 5.7 shows an example of polycode scheme. A value at the given digit has the same meaning as any product variant; therefore, the meanings of a polycode are easily understood. However, when a product family involves in a large number of attributes, the polycode becomes very lengthy and excessive.

Table 5.7   Example of polycode scheme

Digit	Attribute	Digital value			
		1	2	3	4
1	External shape	Cylindrical without deviations	Cylindrical with deviations	boxlike	...
2	International shape	None	Centre hole	Brind center hole	...
3	Number of holes	0	1~2	3~5	...
4	Type of holes	Axial	Cross	Axial cross	...
5	Gear teeth	Worm	Internal spur	External spur	...
⋮	⋮	⋮	⋮	⋮	⋮

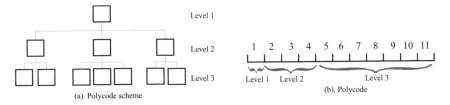

Fig. 5.22 A polycode scheme and corresponding digital representation

Figure 5.22 shows an example of writing a polycode scheme with three levels into a one-dimensional (1D) polycode. Since the digit at a certain place has been clearly explained in the code scheme, a polycode is easily understood the corresponding code scheme such as Table 5.7 is available.

Example 5.2 Table 5.8 shows the code scheme of Nissan passenger car family, write out the polycode for the car shown in Fig. 5.23 (Kirby 2019).

Solution. The code scheme in Table 5.8 includes six digits, which are used to represent type, package, color, interior finish, radio, and tire size. Based on the description of the car model in Fig. 5.23, the digits assigned to this car model are '1 1 3 4 1 1'.

Different code schemes may be used for the same product families. However, the capacities to represent product variants may be quite different. To compare the differences of monocode and polycode, it is assumed that both code schemes include

(a). Car model

	Attribute	
	Digit	Description
1	Type of car	1 for "Maxima"
2	Car package	1 for "GXE"
3	Car color	3 for "Gold"
4	Interior	4 for "Leather"
5	Radio	1 for "AM/FM"
6	Tire size	1 for "15 inch"

(b). Code for car model (a)

Fig. 5.23 Ploycode for a car model example

Table 5.8 Code scheme of Nissan passenger cars

Size	Package	Color	Interior	Radio	Tire size
(Digit 1)	(Digit 2)	(Digit 3)	(Digit 4)	(Digit 5)	(Digit 6)
1 for "Maxima"	1 for "GXE"	1 for "White"	1 for "Black"	1 for "AM/FM"	1 for 15"
2 for "Altima"	2 for "XE"	2 for "Black"	2 for "Gray"	2 for "CD"	2 for "17"
3 for "Sentra"	3 for "SE"	3 for "Gold"	3 for "Brown"	3 for "CD changer"	
	4 for "GLE"	4 for "Blue"	4 for "Leather"	4 for "Premium"	
		5 for "Red"			
		4 for "Dark Gray"			

six digits ($i = 6$), and each digit has a set of possible values from 0 to 9. The capacity of a monocode is found as $\sum_{i=1}^{6} 10^i = 1,111,110$; while the capacity of a polycode is found as the number of product variants in a polycode is $10 \times (i) = 60$.

5.4.2.5 Hybrid Codes

A hybrid code is for a hybrid code scheme where both monocode and polycode fully utilize their advantages. Figure 5.24 shows an example of a hybrid code scheme that integrates hierarchical and chain structures. The structure of a hybrid code can be customized to represent the attributes of products and manufacturing processes efficiently. The majority of product families are represented by hybrid codes. Popular hybrid code schemes are Opitz, Brisch System, CODE, CUTPLAN, DCLASS, Multi-Class, and the Part Analog System (Khan 2013). In the next section, the Opitz is introduced as an example of hybrid code schemes.

Optiz Code Scheme

Opitz was developed at the Technical University of Aachen in Germany (Haworth 1968). It was widely used as a hybrid code scheme to represent formed and machined product families. Opitz included the digitals for the information of products and manufacturing processes. As shown in Table 5.9, the Opitz code scheme includes three groups of digits, and the meanings of these digits are explained in Table 5.10 in detail.

The code scheme of Opitz is shown in Table 5.11 where the form code is monocode (digits 1–5) and the supplementary code is polycode (digits 6–9), and Table 5.12 shows the rules to assign values at these digits based on the given product attributes.

Example 5.3 Determine the five-digit Opitz code for the part shown in Fig. 5.25.

Fig. 5.24 Example of hybrid code for product families

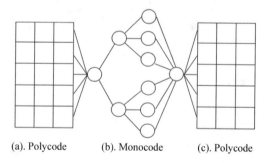

(a). Polycode   (b). Monocode   (c). Polycode

Table 5.9 Opitz code scheme with three groups of digits

Digits in Opitz Classification												
1	2	3	4	5	6	7	8	9	A	B	C	D
Form Code					Supplementary Code				Secondary Code			

Table 5.10 The meanings of digits in an Opitz code scheme

Digit	Description
1	rotational or non-rotational shapes; rotational shapes are further classified by length-to-diameter ratios and non-rotational shapes are further classified by length, width, and thickness.
2	features on external shapes.
3	internally machined features such as holes, threads, and other turning features.
4	machined surfaces such as flats and slots
5	special features such as auxiliary holes, gear teeth and others
6	overall dimensions of products
7	working materials such as steels, aluminums, and irons
8	original shapes of starting raw materials
9	tolerances of manufacturing processes.

Table 5.11 Opitze code scheme

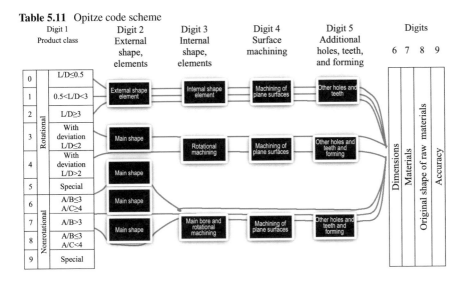

Solution. The attributes and dimensions of the part in Fig. 5.25 are used to assign the values of the digits in the Opitz code scheme as below.

(1)   The overall length (L) and diameter (D) of the part are 1.50 and 2.50, respectively; therefore, $L/D = 1.0$ and the first digit becomes '1'.

(2)   The part is rotational and has the steps at both ends but with one thread; therefore, the second digit is '5'.

(3)   The part has an internal shape with a through-hole; therefore, the third digit is '1'.

(4)   The part has no machined surface, therefore, the fourth digit is '0'.

(5)   The part has no other holes or gear teeth, the fifth digit is '0'.

Table 5.12 The rule to generate an opitz code

	1st digit		2nd digit			3rd digit			4th digit		5th digit
	Product class		External shape and elements			Internal shape and elements			Surface machining		Auxiliary holes and gear teeth
0	$L/D{\le}0.5$	0	Smooth, no shape elements		0	No hole no breakthrough		0	No surface machining	0	No auxiliary hole
1	$0.5{<}L/D{<}3$	1		No shape elements	1		No shape elements	1	Surface plane /curved	1	Axial, not on pitch circle diameter
2	$L/D{\ge}3$	2	Stepped one end or smooth	Thread	2	Stepped on one end	Thread	2	External plane surface, circular graduation	2	Axial on pitch circle diameter
3	With deviation $L/D{\le}2$	3		Groove	3		Groove	3	External groove and/or slot	3	Radial, not on pitch circle diameter
4	With deviation $L/D{>}2$	4	Stepped both ends	No shape elements	4	Stepped on both ends	No shape elements	4	External spline (polygon)	4	Radial, on pitch circle diameter
5	Special	5		Thread	5		Thread	5	External plane surface/slot spline	5	Axial and / radial and / other direction
6	$A/B{\le}3$ $A/C{\ge}4$	6		Groove	6		Groove	6	Internal plane surface or slot	6	Spur gear teeth
7	$A/C{\ge}4$	7	Functional cone		7	Functional cone		7	Internal spline (polygon)	7	Bevel gear teeth
8	$A/B{\le}3$ $A/C{<}4$	8	Operating speed		8	Operating speed		8	Internal or slot/ external polygon	8	Other gear teeth
9	$A/B{\le}3$	9	All others		9	All others		9	All others	9	All others

	6th digit		7th digit		8th digit		9th digit
	Diameter D or length of edge A (mm)		Material		Initial shape		Accuracy in coding digital
0	≤ 20	0	Grey cast iron	0	Round bar	0	No accuracy specified
1	>20 & ≤ 50	1	Nodular graphitic cast iron and malleable cast iron	1	Bright drawn round bar	1	2
2	>50 & ≤ 100	2	Steel < 42 kg/mm^2	2	Triangular, square, hexagonal, or other bars	2	3
3	>100 & ≤ 160	3	Steel $\ge$ 42 kg/mm^2	3	Tubing	3	4
4	>160 & ≤ 250	4	Steel 2+3 heat-treated	4	Angled U.-T. and similar sections	4	5
5	>250 & ≤ 400	5	Alloy steel	5	Sheet	5	2+3
6	>400 & ≤ 600	6	Alloy steel Heat-treated	6	Plates and slabs	6	2+4
7	>600 & ≤ 1000	7	Non-ferrous metal	7	Cast or forged component	7	2+5
8	>1000 & ≤ 2000	8	Light alloy	8	Welded group	8	3+4
9	> 2000	9	Other materials	9	Pre-machined component	9	(2+3)+4+5

Therefore, the five-digit Opitz code of the part in Fig. 5.25 becomes '15100'.

Other than the first nine digits as the form code and the supplementary code, a complete Optiz code scheme uses additional four digits to represent other manufacturing attributes. Each digit has the options of 10 different values. Therefore, the Opitz code scheme has the capacity of representing a large number of product families for a wide scope of applications.

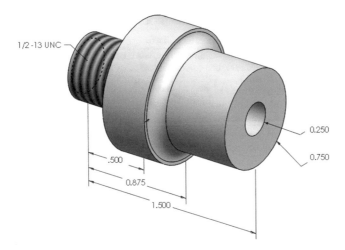

Fig. 5.25 Product drawing for Example 5.3

5.4.3 Production Flow Analysis

GT focuses on the classification of products; it would be better to consider products and productions simultaneously to analyze the similarities of products and machines. *Production flow analysis* (PFA) is used to model the correspondence of products and machines as relational matrices, and the relational matrices can be then analyzed to define workcells for product families. Each workcell consists of the machines for the manufacturing processes of one product family.

Instead of using product models in GT, PFA uses the production route sheets to group products and machines. The products in a family should have similarities in their production route sheets.

In defining workcells for products and machines, PFA follows the steps as below:

(1) determine the sequence of manufacturing processes and the routes of machines for products,
(2) group products and machines based on the similarities of the routes of manufacturing processes, the grouped products and machines are arranged as "packs"; each pack is modeled as an incidence matrix for product-machine relations,
(3) cluster the packs into the groups with similar routes, and
(4) define a workcell for each group of the machines.

Two popular algorithms to group products and machines are the *single-linkage clustering (SLC) algorithm* (SLCA) and *Rank-Order Clustering (ROC) Algorithm*. In the following, ROC by King (1980) is introduced to group products and machines based on a given product-machine matrix.

Assume that a product-machine relational matrix is given as $[M]_{n \times m}$; where n and m are the numbers of products and machines, respectively. ROC is used to sort rows and columns in $[M]_{n \times m}$ in Table 5.13.

Table 5.13 Main steps in a rank-order clustering (ROC) algorithm

Step	Task
(1).	Assign a binary weight to each column or row, and use the following equations to calculate a decimal weight for each row and column and calculate a decimal weight for each row and column using the formulas, ▪ Decimal weight for row $i = \sum_{p=1}^{m} b_{ip} 2^{m-p}$ ▪ Decimal weight for column $j = \sum_{p=1}^{n} b_{pj} 2^{n-p}$
(2).	Sort the rows in the order of decreasing decimal weight values,
(3).	Repeat the steps (1) and (2) for each column, and
(4).	Continue steps (1) to (3) until no switch is needed for any row or column.

Example 5.4 Define the workcells for the given product-machine relational matrix below.

Products

	M_{ij}	1	3	4	7	2	5	6	8
Machines	A	1	1			1			
	E				1				1
	C		1	1			1	1	
	F				1				1
	D			1	1		1	1	
	B	1	1			1			

Solution. Step 1: for column i ($i = 1, 2,..., n$), the binary weight 2^{n-i} is assigned as $(2^7, 2^6, \ldots 2^1, 2^0$, respectively. The decimal weight of *Row A* becomes $2^7(1) + 2^6(1) + 2^5(0) + 2^4(0) + 2^3(1) + 2^2(0) + 2^1(0) + 2^0(0) = 200$, and those for other rows are calculated as below:

Part

	M_{ij}	1	3	4	7	2	5	6	8	$\sum_{j=1}^{j=8} 2^{n-j} M_{i,j}$
Machine	A	1	1			1				200
	E				1				1	17
	C		1	1			1	1		102
	F				1				1	17
	D			1	1		1	1		54
	B	1	1			1				200
	$2^{(n-j)}$	2^7	2^6	2^5	2^4	2^3	2^2	2^1	2^0	

Step 2: sort the rows in the order of decreasing decimal weight values as *A, B, C, D, E,* and *F*.

Part

	M_{ij}	1	3	4	7	2	5	6	8	$\sum_{j=1}^{j=8} 2^{n-j} M_{i,j}$
Machine	A	1	1			1				200
	B	1	1			1				200
	C		1	1			1	1		102
	D			1	1		1	1		54
	E				1				1	17
	F				1				1	17
	$2^{(n-j)}$	2^7	2^6	2^5	2^4	2^3	2^2	2^1	2^0	

Step 3: Repeat step 1 for row j ($j = 1, 2,..., m$), the binary weight 2^{m-i} is assigned as $(2^5, 2^4, \ldots 2^1, 2^0$, respectively. The decimal weight of *Column 1* becomes $2^5(1)$

$+ 2^4(1) + 2^3(0) + 2^2(0) + 2^1(0) + 2^0(0) = 48$, and those for other columns are calculated as below:

Products

M_{ij}	1	3	4	7	2	5	6	8	$2^{(m-j)}$
A	1	1			1				2^5
B	1	1			1				2^4
C		1	1			1	1		2^3
D			1	1		1	1		2^2
E				1				1	2^1
F				1				1	2^0
$\sum_{i=1}^{i=6} 2^{m-i} M_{i,j}$	48	56	12	7	48	12	12	3	

(Machines)

The columns are then reordered in decreasing values from left to right as *3, 1, 2, 4, 5, 6, 7,* and *8*.

Step 4: Repeat steps 1, 2, and 3 to get the final result (no more switch when the steps are repeated) as.

Products

M_{ij}	3	1	2	4	5	6	7	8	$2^{(m-j)}$
A	1	1	1						2^5
B	1	1	1						2^4
C	1			1	1	1			2^3
D				1	1	1	1		2^2
E							1	1	2^1
F							1	1	2^0
$\sum_{i=1}^{i=6} 2^{m-i} M_{i,j}$	56	48	48	12	12	12	7	3	

(Machines)

Finally, three workcells should be defined, the first one consists of machines *A, B,* and *C,* for products *1, 2,* and *3;* the second one consists of machine *C* and *D* for products *4, 5,* and *6,* and the third one consists of *D, E,* and *F* for products *7* and *8.*

Note that ROC may not be able to generate workcells for some product-machine relational matrices; since it is not uncommon that the iterative process in ROC leads to an oscillation. This should be solved by introducing more machines of same types. Another scenario is that the finished clusters have an outlier or void; an outlier should be addressed by a machine replication, while no action is needed for a void; the product just skips any operation on the corresponding machine.

5.4.4 Cellular Manufacturing

GT or PFA has been applied to group products as families and sorts the machines as workcells. Moreover, the manufacturing resources should be well organized to plan, schedule, and control manufacturing processes efficiently. *Cellular manufacturing* is used to serve this purpose. The machines in a workcell should be organized at first, and it is referred to as a *cell formation problem* (CFP) in cellular manufacturing. CFP is generally a nondeterministic polynomial (NP) problem, and the solution to a CFP is the workcells, which is composed of heterogeneous machines for designated product families.

The required manufacturing capabilities of a workcell are defined based on the features of a composite product. *A composite product* can be *real* or *imaginal*; it is modeled as a collection of the primitives and features of all products in a family. Cellular manufacturing is designed to ensure that any feature on the composite product can be processed by the machines in the workcell. When a new product is assigned to the workcell, this product must show similarities with the composite part. The layout of CM is organized based on the sequence of manufacturing processes of the composite part.

Figure 5.26 shows an example of a composite product; it is an artificial product model with the machined features of the products in the family. Figure 5.26a–c show that the products in the family have the machined features of faces, chamfers, left shoulders, right shoulders, and slots. The composite product model in Fig. 5.26d has all of the aforementioned features.

A composite product model is used to select machines and tooling for a workcell. If a machine or tool is available to make any feature on a composite product, the workcell has the machines or tools for all the products in the family. Figure 5.27 shows an example of the product families (Fig. 5.27a), which can be derived from a composite part model. If the workcell is equipped with all manufacturing tools (Fig. 5.27c) to make the composite product, this workcell can be the solution to the manufacturing processes of the product family.

The design of cellular manufacturing can be viewed as a multi-objective optimization problem where the trade-offs are made among a set of conflicting design criteria. Strategosinc (2019) described the procedure of designing a cellular manufacturing system shown in Table 5.14; it begins with the determination of product families, then follows by the selection of machines and the design of cellular manufacturing system, and finally, the layout design of the system.

The procedure for the design of a cellular manufacturing system in Table 5.14 is simplified. In practice, many other factors such as the types, varieties, and volumes of products must be taken into considerations. By all means, the effectiveness of cellular manufacturing systems has been proven in manufacturing products with medium varieties and quantities. Cellular manufacturing can benefit companies to

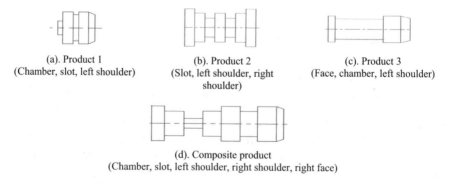

(a). Product 1
(Chamber, slot, left shoulder)

(b). Product 2
(Slot, left shoulder, right shoulder)

(c). Product 3
(Face, chamber, left shoulder)

(d). Composite product
(Chamber, slot, left shoulder, right shoulder, right face)

Fig. 5.26 Merging product features for a composite product

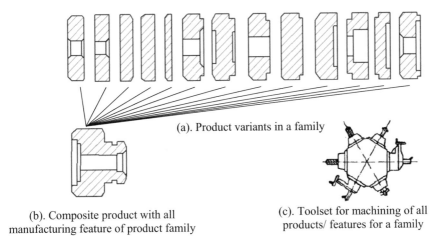

(a). Product variants in a family

(b). Composite product with all
manufacturing feature of product family

(c). Toolset for machining of all
products/ features for a family

Fig. 5.27 Composite product for the determination of machines and tools

improve system performance by reducing transportation, throughout time, lead time, and work-in-progress (WIP) and increasing the profit margin and quality of products (Hyer and Wemmerlov 2001).

5.5 Discrete Event Dynamic Systems

Manufacturing systems are typical *discrete event dynamic systems* (DEDSs) where state transitions are initiated by events that occur at discrete instants of time. *An event* refers to the start or the termination of an activity. Taking a manufacturing system as an example, an event can be the completion of a manufacturing process, a breakdown of machine, and the beginning of transportation, and many others. In addition, the intervals between two events might be deterministic or stochastic. Therefore, the complexity of planning, scheduling, and controlling a manufacturing system can be increased exponentially when the numbers of machines, tools, and parts and the interactions of these manufacturing resources increase. Many theories and computer-aided techniques have been proposed to model, analyze, and control DEDSs. In this section, prevalent Petri nets are introduced to model a manufacturing system at workcell level.

A DEDS consists of various system elements such as machines, tools, and products, the behaviors of these system elements must be controlled in way to optimize the performance of system, and system performances can be measured by qualitative and quantifiable indicators. Typical performance indicators of a manufacturing system include throughput, delivery time, inventory, utilization ratio, and the probabilities of malfunctions, breakdowns, and deadlocks. In a Petri net model, the behaviors of system elements are represented by the states and the transitions of states, and

Table 5.14 Procedure of designing a cellular manufacturing system (Strategosinc 2019)

Step	Main Activities
Step 1. The products are analyzed to form product families based on similarities. Each product family corresponds to a group of heterogamous machines. GT or PFA can be used to analyze and group products. The activities at this step yield the answers to some basic questions such as, 1) how to sort products as groups? (2) how to achieve better utilization rates of machines? (3) is manufacturing capability lacked or saturated?	
Step 2. Select manufacturing resources for product families The features of products are analyzed to determine manufacturing solutions including machines tools, and operations. The activities at this step yield the answers to some basic questions such as 1) how to manufacture a product? 2) what is the best sequence of manufacturing processes for a product? 3) what machine tools are needed for identified processes? 4) any human operation is needed to perform certain process?	

(continued)

the quantifiable indicators can be evaluated based on the dynamic changes of the properties of the states and the transitions.

Petri nets are featured with their capabilities in modeling concurrency, synchronization, mutual exclusion, conflict behaviors, and the representations of the states are more completely than analytical models but more structured and abstracted than simulations. A Petri net provides a graphical representation of a DEDS, which is indeed a mathematical model resembling the physical system. Moreover, Petri nets are very easy to be implemented as computer programs for the simulation and control

Table 5.14 (continued)

Step 3. Design a cellular manufacturing system that is a collection of workcells and accessary equipment such as storages and transportation equipment. The activities at this step yield the answers to some basic questions such as 1) what are the methods to transport materials from cell to cell? 2) how to balance the workloads on machines? 3) how to plan, schedule, and control production? 4) how to manage work-in-process workpieces? 5) how to assure the quality of products? 6) how to motivate human operators in cellular manufacturing system?	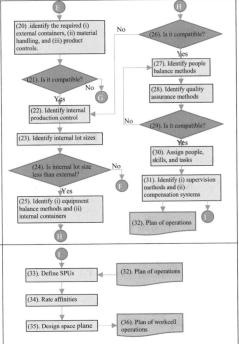
Step 4. Design the layout of the cellular manufacturing system, which is logical follow-up when the tasks in above three steps are accomplished thoroughly. A cellular manufacturing system includes 1) *inter-cell layouts* to arrange the machines and minimize the inter-cell movements of materials and 2) *intra-cell layout* to arrange workcells and minimize the transportation overall cost. The activities at this step yield the answers to two questions 1) how to optimize system performance by arranging machines and tools? (2) how to deal with external constraints and changes? (3) how to integrate manufacturing resources as a network?	

of DEDSs. A Petri net represents a DEDS as

$$PN = \{P, T, I, O, K, M\} \qquad (5.5)$$

where

P: the places for states of resources,

T: the transitions for state changes,

I: the inputs for the places to fire transitions,

O: the outputs for the places after firing transitions, and

K: the capacities of the places, and

M: the number of the tokens in each place.

In this section, a flexible manufacturing system in Fig. 5.28 is shown as an example of discrete event dynamic system (DEDS) to illustrate how the Petri net can be used to model and analyze DEDSs.

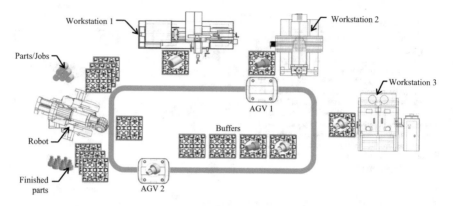

Fig. 5.28 A flexible manufacturing system (FMS) as a DEDS example

An FMS consists of material flow and tool flow, and the components in an FMS can be generally classified into two groups: *resources* to provide services and *jobs* to receive services. Table 5.15 shows the classification of service resources based on their primary functions, and Table 5.16 shows the classification of the workstations based on their accessory functions.

The states and events in the FMS can be modeled by the places and transitions of a Petri net. A *place* (*P*) represents the states of aggregated resources. For example, a machine at its idle state is a 1D place, a job is receiving the service on a machine is a two-dimensional (2D) place, a job is in transportation from one machine to another is a three-dimensional (3D) place.

Table 5.15 Types of service resources in an FMS (Bi et al. 2001; Zhang et al. 2000)

Types	Description	Examples
Workstation (*WS*)	A device to perform manufacturing process in the material flow or a device to change the state of tools in the tool flow	Machining, cleaning, inspecting and loading and unloading station in the material flow; The devices for tool changing and inspection in the tool flow
Buffer (*BF*)	A device to accommodate parts or tools	Pallets for parts, and central or local tool magazines for tools
Transportation (*CD*)	A device to change the physical location of parts or tools	Autonomous guided vehicles (AGV), tool changer, or other transportation equipment
Holding and Placing (*CT*)	A device to secure parts or tools in service	Fixtures, pallets, and tool holders.
Auxiliary Tooling (*AT*)	A device used to assist completion of a service	Manipulators, inspecting or cleaning tools.

Table 5.16 Classification of workstations

Type	1	2	3	4	5	6	7
Local buffer	No	No	No	Yes	Yes	Yes	Yes
Loading/unloading tool	No	Yes	Yes	No	No	Yes	Yes
Machining operation	Yes	No	Yes	No	Yes	No	Yes

A *transition* (**T**) is an event occurring to one or a few system resources; for example, a job arrives at the workstation, a machine completes one operation, and a machine is malfunctioned. Transitions in an FMS can be classified into the following four types,

one dimensional (**P**) $\oplus$ *one dimensional* **P**) $\rightleftarrows$ *two dimensional* (**P**)
one dimensional (**P**) $\oplus$ *two dimensional* (**P**) $\rightleftarrows$ *three dimensional* (**P**)
two dimensional (**P**) $\rightleftarrows$ *two dimensional* (**P**)
three dimensional (**P**) $\rightleftarrows$ *three dimensional* (**P**)

In addition, a transition can be *active* or *passive*. An active transition is fired from left to right by the control command, and a passive transition is fired from right to left when the system responds to the state change of system element. Upon the satisfaction of the firing conditions, an active transition is fired by a command issued by the FMS control system. Each active transition corresponds to a decision point in the control system, the decisions on the selections of parts and manufacturing resources are usually made based on predefined rules. Table 5.17 gives examples of typical decision points and rules for decision-making supports in controlling an FMS.

Figures 5.29 and 5.30 show the Petri net models for the material and tool flows of the FMS illustrated in Fig. 5.28. The places and transitions in the respective models are explained in the figures.

A Petri net model can be used to control an FMS or diagnose the malfunctions involved in the system operation; inappropriate commands such as those to cause the following scenarios can be identified in real-time operations (Bi et al. 2001).

(1) *Illegal command to activate a transition*: a transition can be fired only when all pro-places have the required number of tokens; if a command is issued when the firing condition is not satisfied, it is an illegal command.

(2) *Violation to synchronization*: the resources are shared by all transitions in an FMS, and one resource may be associated with a number of transitions at the same time; however, the state of the resource must be consistent with all the transitions.

(3) *Violation of designated rules at decision points*: at a decision point, the decision is made to select a resource, part, or tool from a list of the candidates. The selection is based on the decision-making rules shown in Table 5.17, a command to fire a transition must be aligned with the specified priority of manufacturing operations.

Table 5.17 Decision points and decision-making rules

Decision points \ Rules	Fixed	Random	Priority	Longest processing time	Shortest processing time	Maximum number of remained processes	Minimum number of remained processes	Lead-time	Average used time	Shortest path	First in first out
Part enters system	Yes	Yes	Yes					Yes			
Part chooses workstation	Yes	Yes	Yes						Yes		
Part chooses buffer	Yes	Yes								Yes	Yes
AGV chooses part		Yes								Yes	
Robot chooses tool		Yes								Yes	
Machine chooses tool									Yes	Yes	
Tool chooses AGV	Yes	Yes							Yes		Yes
Machine chooses part		Yes	Yes	Yes	Yes	Yes	Yes	Yes			Yes

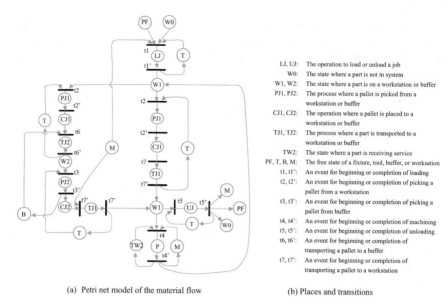

LJ, UJ:	The operation to load or unload a job
W0:	The state where a part is not in system
W1, W2:	The state where a part is on a workstation or buffer
PJ1, PJ2:	The process where a pallet is picked from a workstation or buffer
CJ1, CJ2:	The operation where a pallet is placed to a workstation or buffer
TJ1, TJ2:	The process where a part is transported to a workstation or buffer
TW2:	The state where a part is receiving service
PF, T, B, M:	The free state of a fixture, tool, buffer, or workstation
t1, t1':	An event for beginning or completion of loading
t2, t2':	An event for beginning or completion of picking a pallet from a workstation
t3, t3':	An event for beginning or completion of picking a pallet from buffer
t4, t4':	An event for beginning or completion of machining
t5, t5':	An event for beginning or completion of unloading
t6, t6':	An event for beginning or completion of transporting a pallet to a buffer
t7, t7':	An event for beginning or completion of transporting a pallet to a workstation

(a) Petri net model of the material flow   (b) Places and transitions

Fig. 5.29   Petri net of material flow of FMS (Bi et al. 2001)

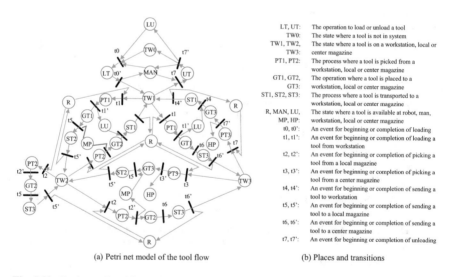

LT, UT:	The operation to load or unload a tool
TW0:	The state where a tool is not in system
TW1, TW2, TW3:	The state where a tool is on a workstation, local or center magazine
PT1, PT2:	The process where a tool is picked from a workstation, local or center magazine
GT1, GT2, GT3:	The operation where a tool is placed to a workstation, local or center magazine
ST1, ST2, ST3:	The process where a tool is transported to a workstation, local or center magazine
R, MAN, LU, MP, HP:	The state where a tool is available at robot, man, workstation, local or center magazine
t0, t0':	An event for beginning or completion of loading
t1, t1':	An event for beginning or completion of loading a tool from workstation
t2, t2':	An event for beginning or completion of picking a tool from a local magazine
t3, t3':	An event for beginning or completion of picking a tool from a center magazine
t4, t4':	An event for beginning or completion of sending a tool to workstation
t5, t5':	An event for beginning or completion of sending a tool to a local magazine
t6, t6':	An event for beginning or completion of sending a tool to a center magazine
t7, t7':	An event for beginning or completion of unloading

(a) Petri net model of the tool flow   (b) Places and transitions

Fig. 5.30   Petri net of tool flow of FMS (Bi et al. 2001)

(4)   *Illegal action against malfunctioned resources*: when a resource is broken down, any command to use the malfunctioned resource is illegal.

(5)   *Detection of deadlock*: a *deadlock* is a scenario where two or more parts are waiting for the resources, which are possessed by each other. A deadlock occurs

to a pair of parts, which may be the major issue to be addressed in resource-sharing systems such as DEDSs. A deadlock can be modeled and detected based on the following concepts.

Definition of *a waiting relation*. In a Petri net model $PN = \{P, T, I, O, K, M\}$, assume t_1, $t_2 \in T$ are two transitions, $s_0 \in P$ is the preplace of transition t_1 and the post-place of the transition t_2, and the token at the place s_0 is zero, i.e., $t_{1 \rightarrow} s_0$, $s_{0 \rightarrow} t_2$ and $m(s_0) = 0$. A waiting relation occurs to (t_1, t_2) under the mark M when the number of tokens of other preplaces of t_2 satisfies the firing condition of t_2 except for s_0, and it is denoted as $t_1 \overset{m}{\leftarrow} t_2$ and illustrated in Fig. 5.31a.

Definition of *a waiting loop*. In a Petri net model $PN = \{P, T, I, O, K, M\}$, assume there is a set of transitions $\forall T_n T$, where $T_n = \{t_1, t_2, \ldots t_n\}$, and the transitions in T_n satisfy the conditions of the waiting relations of,

$$\exists k \in (1, n-1), t_k \overset{m}{\leftarrow} t_{k+1} \ and \ t_n \overset{m}{\leftarrow} t_1 \tag{5.6}$$

Then, T_n is called a waiting loop under the mark M, and it is denoted as OT_n. Figure 5.31b shows a waiting loop with the minimum number of transitions. Note that $s_0 \in P$ is the preplace of transition t_1 and the post-place of the transition t_2, and $s_1 \in P$ is the preplace of transition t_2 and the post-place of the transition t_1 the tokens at s_0 or s_1 are insufficient to fire t_1 or t_2.

Definition of *deadlock field*. In a Petri net model $PN = \{P, T, I, O, K, M\}$, assume that there are m waiting loops $OT_{n_1}, OT_{n_2}, \ldots OT_{n_m}$, the set of m waiting loops ($T_d = OT_{n_1} \cup OT_{n_2} \ldots OT_{n_m}$) are called a deadlock field under mark M. The concept of deadlock field can be adopted to detect the following cases:

(1) If $T_d = null$, there is *no deadlock* under the mark M,
(2) If T_d *null* but $T_d < T$, there is *a local deadlock* under the mark M, and
(3) If $T_d = T$, there is *a whole deadlock* under the mark M.

Example 5.5 Use the Petri net to model the testbed for gearbox assembly in Fig. 5.31; the testbed was developed at the National Institute of Standards and Technology (NIST) to evaluate the grasping capabilities of end-effectors (Falco et al. 2015; Kootbally et al. 2016). A series of operations are to (1) open a gearbox kit, (2) move around parts, and (3) assemble them together as a functional gearbox. The testbed consisted of robots A and B, vision systems A and B, force sensors, trays and the assembling platform. Note that opening the gearbox kit and place parts on the platform are done

Fig. 5.31 A deadlock occurring to (t_1, t_2)

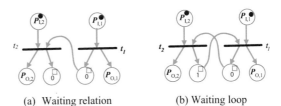

(a) Waiting relation   (b) Waiting loop

Table 5.18 Events for assembling processed of gearboxes (Bi et al. 2020)

Steps	Events
1.	Open gearbox kit and place parts on tray
2.	Move robot B, identify the base part, and transport it to buffer, retract robot B.
3.	Move robot A, identify the base part, and transport, and reposition it on assembly platform, retract robot A.
4.	Move robot B, identify the medium gear, and transport it to buffer, retract robot B
5.	Move robot A, identify the medium gear, transport, and assembly it on the base; robot B holds the base, retract robots.
6.	Place shaft A on the base part
7.	Move robot B, identify the large gear, and transport it to buffer, retract robot B.
8.	Move robot A, identify the large gear, transport, and assembly it on the shaft A; robot B holds the base, retract robots.
9.	Place shaft B on the base part
10.	Move robot B, identify the small gear, and transport it to buffer, retract robot B
11.	Move robot A, identify the small gear, transport, and assembly it on shaft B; robot B holds the base, retract robots.
12.	Move robot B, identify the gearbox cover, and transport it to buffer, retract robot B.
13.	Move robot A, identify gearbox cover, transport, and assembly it on sub-assembly; robot B holds the base, retract robots
14.	Robot A picks gearbox assembly, transport, place it to tray, retract robot A.

☐   1st event type   ▨   2nd event type   ▩   3rd event type

manually. robot *A* was equipped with a force sensor to assemble parts; while robot *B* was equipped with a vision system only to move and hold a part (Bi et al. 2020).

Solution. Table 5.18 shows the possible events which can be classified into three basic types, i.e., (1) an event initialized by human, (2) an event performed by a robot and a vision, (3) an event performed by a robot, a vision, and a force sensor.

Figures 5.32, 5.33, and 5.34 show the Petri net examples of three event types, respectively.

The sub-Petri net models for different events can be selected and assembled as a testing plan. In executing the testing plan, the states of resources and transitions are tracked to evaluate the performance of the assembly system.

5.6   Simulation of Discrete Event Dynamic Systems

Computer simulation is the most effective technique for the design and optimization of DEDSs. This section discusses how to use commercial software tools to simulate various DEDSs. Numerous software tools, such as *Anylogic, Enterprise Dynamics, Delmia, FlexSim,* and *Plant Simulation,* have been developed to simulate DEDSs (Wikipedia 2020c). Here, the *simulation modeling framework based on intelligent objects* (Simio) software by Simio LLC is used as an example for the simulation of DEDSs.

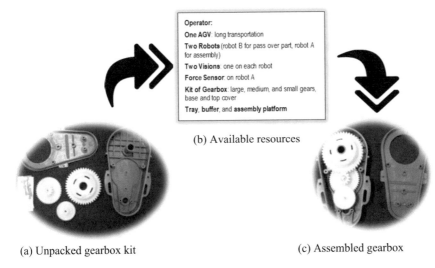

(b) Available resources

(a) Unpacked gearbox kit

(c) Assembled gearbox

Fig. 5.32 A testbed used to assemble a gearbox

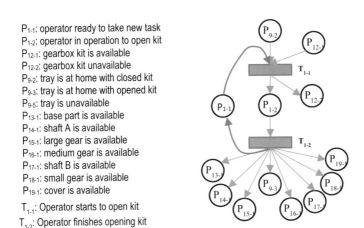

P$_{1-1}$: operator ready to take new task
P$_{1-2}$: operator in operation to open kit
P$_{12-1}$: gearbox kit is available
P$_{12-2}$: gearbox kit unavailable
P$_{9-2}$: tray is at home with closed kit
P$_{9-3}$: tray is at home with opened kit
P$_{9-5}$: tray is unavailable
P$_{13-1}$: base part is available
P$_{14-1}$: shaft A is available
P$_{15-1}$: large gear is available
P$_{16-1}$: medium gear is available
P$_{17-1}$: shaft B is available
P$_{18-1}$: small gear is available
P$_{19-1}$: cover is available

T$_{1-1}$: Operator starts to open kit
T$_{1-2}$: Operator finishes opening kit

Fig. 5.33 Event 1 (the 1st type): opening a gearbox kit

Simio is a graphic modeling tool that simplifies object representations and provides the flexibility of defining manufacturing processes without the need for programming (Simio 2020). Simio mainly uses the object-oriented modeling paradigm to define objects and processes; however, it supports other modeling paradigms such as event-based, process-oriented, or agent-based modeling techniques.

P_{2-1}: robot A ready to move new positon
P_{2-2}: robot A in motion
P_{2-3}: robot A arrives at given position
P_{2-4}: robot A holds at given positon
P_{2-5}: robot A retract to home positon
P_{7-1}: vision A ready to detect object
P_{7-2}: vision A searches object
P_{7-3}: vision A finds object/position
P_{3-1}: Robot A tool ready to pick an object
P_{3-2}: Robot A tool holds an object
P_{3-3}: An object is placed
P_{13-1}: base part is available
P_{13-2}: base part is unavailable

T_{3-1}: Robot A starts to move to buffer
T_{3-2}: Robot A arrives buffer
T_{3-3}: Vision A starts to search base part
T_{3-4}: Vision A finds base part
T_{3-5}: Robot tool A starts to pick base part
T_{3-6}: Robot A starts to move to assembly platform
T_{3-7}: Robot A arrives assembly platform
T_{3-8}: Vision A starts to search location for base part
T_{3-14}: Vision A finds location for base part
T_{3-9}: Robot A starts to move identified position
T_{3-10} Robot A arrives identified position
T_{3-11}: Robot tool A starts to place base part
T_{3-12}: Robot A stars to retract to home position
T_{3-13}: Robot A arrives home position

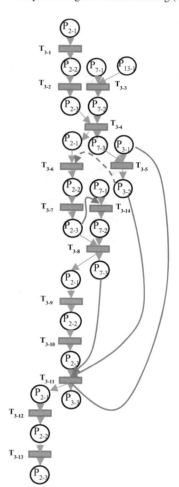

Fig. 5.34 Event 3 (the 2nd type): transporting a part

5.6.1   Modelling Paradigms

The supports for graphic modeling and animation are essential to computer simulation tools. *A graphics-based approach* simplifies the operations to model objects and processes, and *an animation-based approach* helps users in reviewing, visualizing, understanding, and verifying system behaviors efficiently.

Manufacturing systems are mostly *discrete event dynamic systems* (DEDSs). The changes of DEDS are made by asynchronous and discrete events occurring to objects over time. Therefore, early simulation tools such as Simulation Programming Language (Simscript) (Markowitz et al. 2020) and GASP (Hooper and Reilly 2020) are event-oriented; a DEDS is modeled as a series of the events that change system statuses dynamically. Accordingly, a process flow is modeled as a series of the

P₂₋₁: robot A ready to move new positon
P₂₋₂: robot A in motion
P₂₋₃: robot A at given position
P₂₋₄: robot A holds at given positon
P₂₋₅: robot A retract to home positon
P₇₋₁: vision A ready to detect object
P₇₋₂: vision A searches object/position
P₇₋₃: vision A finds object/position
P₃₋₁: Robot A tool ready to pick an object
P₃₋₂: Robot A tool holds an object
P₃₋₃: An object is placed
P₁₆₋₁: medium gear is available
P₁₆₋₂: medium gear is unavailable

P₅₋₁: robot B ready to move new positon
P₅₋₂: robot B in motion
P₅₋₃: robot B at given position
P₅₋₄: robot B holds at given positon
P₅₋₅: robot B retract to home positon
P₈₋₁: vision B ready to detect object
P₈₋₂: vision B searches object/position
P₈₋₃: vision B finds object/position
P₆₋₁: Robot B tool ready to pick an object
P₆₋₂: Robot B tool holds an object

T₅₋₁: Robot A starts to move to buffer
T₅₋₂: Robot A arrives buffer
T₅₋₃: Vision A starts to search medium gear
T₅₋₄: Vision A finds medium gear
T₅₋₅: Robot tool A starts to pick medium gear
T₅₋₆: Robot A starts to move to assembly platform
T₅₋₇: Robot A arrives assembly platform
T₅₋₈: Vision A starts to search location for medium gear
T₅₋₉: Visoin A finds the position
T₅₋₁₀: Robot A starts to move to identified position
T₅₋₁₁: Vision B starts to search sub-assembly
T₅₋₁₂: Robot B starts to move to sub-assembly
T₅₋₁₃: Vision B finds sub-assembly
T₅₋₁₄: Robot B tool starts to hold sub-assembly
T₅₋₁₅: Robot B arrives assembly platform
T₅₋₁₆: Robot arrives identified position
T₅₋₁₇: Robot A and B in assembling
T₅₋₁₈: Robot A starts to move to home position
T₅₋₁₉: Robot A arrives home position
T₅₋₂₀: Robot B starts to move to home position
T₅₋₂₁: Robot B arrives home position

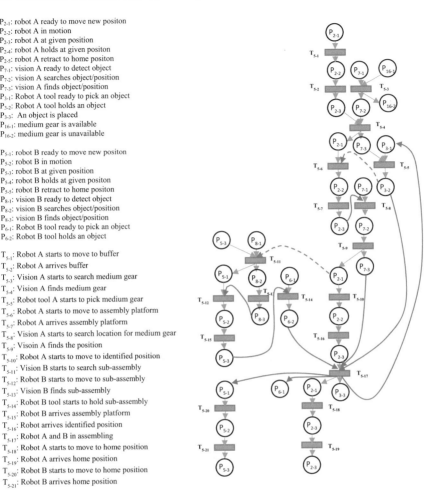

Fig. 5.35 Event 5 (the 3rd type): Performing assembling operation

movements occurring to system elements, and each movement changes the statuses of relevant system elements. From the perspective of processes, an event-based or process-oriented modeling technique is efficient and flexible in representing discrete events; however, the details of system elements are more or less ignored.

Simio has incorporated *object-oriented modeling* (OOM) in the event-based simulation, so that system elements can be defined in detail and the efficiency and flexibility of the modeling process can be preserved. In object-oriented modeling, both system elements and events can be defined as objects with attributes and states. Note that a manufacturing system consists of system elements such as machines, operators, robots, and transportation devices, and running manufacturing businesses corresponds to the events such as machining, assembling, transporting, and delivering; events are the results of the interactions of system elements.

In Simio, any type of system elements, such as a machine, robot, product, tool, and customer, can be defined as *an intelligent object* with given attributes and states. The state change of an intelligent object can be animated graphically for visualization purpose. Once a new intelligent object is defined, it would be stored in the design library and can be reused in any other models. Simio includes a design library with pre-defined objects for users.

5.6.2 Object Types and Classes

Other than the pre-defined objects in the design library, Simio provides the interface for users to build custom objects, and the common types of Simio objects are listed in Table 5.19 (Pegden 2020).

Objects in Table 5.19 can further be classified into six types shown in Fig. 5.36 (Thiesing and Pegen 2020). *A fixed* object has an immobile location in the system; the examples of fixed objects are stationary equipment such as machines, fueling, and stations; in Table 5.19, 'Source', 'Sink', 'Server', 'Combiner', 'Separator', 'Workstation', and 'Resource' are fixed objects. *An agent object* can move in the space freely. *An entity* is an instantiation from an agent class which can be moved from one object to another in the system. Examples of entities are the customers who wait in front of a service station and the workpieces in a production line. A *link* or *node* object is used to build a network in which entities move around to receive services. A *link* refers to a pathway for an entity to be moved from one place to another. A *node* corresponds to a starting or ending point in a link. In Table 5.19, 'Connector',

Table 5.19 Common types of Simio objects (Pegden 2020)

No.	Object	Description
1	Source:	A place where entities enter the system
2	Sink:	A place where entities level the system
3	Server:	A service provider with multiple channels and input/output queues.
4	Combiner:	A combiner of entities as batches.
5	Separator:	A separator of entities from a batches.
6	Workstation:	An operating place with three stages: setup, processing, and unloading.
7	Resource:	An object that can serve other objects.
8	Vehicle:	An object that can transport entities
9	Worker:	A human operator to move entities.
10	BasicNode:	A simple intersection of links.
11	TransferNode:	An intersection of links that is set as a destination or waiting stop on transporter.
12	Connector:	A connector of two nodes with no time delay.
13	Path:	A path between two nodes where an object travels with a given speed.
14	TimePath:	A path between two nodes with a specified time of traveling.
15	Conveyer:	An accumulating or non-accumulating conveyor device.

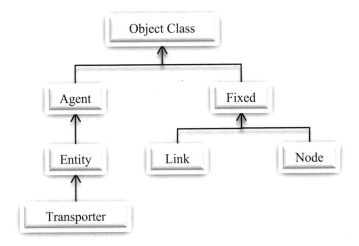

Fig. 5.36 Types of Simio objects

'Path', 'TimePath' and 'Conveyor' are link objects, and 'Basic Node' and 'Transfer Node' are node objects. A *Transporter* is a type of 'entity' to pick, carry, or drop object(s). The transportation equipment such as an AGV, crane, forklift, or truck can be defined as a transporter. In Table 5.19, *'Vehicle'* and *'Worker'* are modeled as transporter objects.

The behaviors of an object are governed by its properties. For example, a 'source' object has a property for the inter-arrival time, a 'server' object has a property to specify a processing time. Therefore, the intelligence of an object is modeled as the collection of the processes controlled by the defined properties.

5.6.3 Types of Intelligence

In defining an object, commonly used properties of an object are defined by clicking the checkboxes in the properties window. A basic object has some pre-defined events that are automatically triggered to change the status of object. Objects might need other properties such as *failures, status assignments, additional resource allocations, financials,* or *custom process logics,* these properties can be added from the fully exploded list of properties.

To insert an object in a Simio model, the object type is selected, and the object icon in the library is dragged and placed in *the Facility View*. The intelligence of a fixed object is modeled as one or more *processes*. A process is executed by an available token, and the token hold by the process is released when the process is terminated. A process itself can be treated as a sequence of the *process steps* that are triggered by events. Process steps are *stateless*; they have their input properties but no output or response. This implies that the process steps can be shared by an arbitrary number of objects. If the logic of a process is changed, the change occurs to any object with such process steps. The states for objects are recorded in elements. *Elements* may have both input and output properties.

A *process step* is a simple process such as holding a token in a time duration, holding or releasing a resource at a time moment, waiting an event to be activated, updating state of object, or making a decision of alternative paths. Some process steps such as 'delays' are common to any object including links, entities, transporters, agents, and groups. Other process steps are customized to certain types of objects such as 'pickup' or 'drop-off' steps to a *transporter* and 'engage' and 'disengage' steps to a *link*. Each object class has its own set of *events*. For example, a link object usually has the following events (1) an entity enters or leaves the link, (2) an entity fully merges to the link, and (3) an entity collides or separates from other entities in the link. By defining the intelligence on objects, processes, and events, the movements of the entities over the networked resources can be fully controlled.

5.6.4   Case Study

Simio provides many case studies of using Simio to simulate various DEDSs in aerospace engineering, business processes, supply chains, transportation, and manufacturing. Here, the simulation of an assembly workcell has cited an example to illustrate how a CAE tool such as Simio is used to model and simulate a DEDS (Simio 2020).

The entities in the workcells are *type-A parts*, *type-B parts*, *products* assembled from type-A and type-B parts. The objects include *two inventories* as the sources of type-A and type-B parts, respectively; *one preparation station, two assemble stations, one repairing station, one packing station*, and *two docks* to ship products. The process flows of the assembly workcell include (1) a pallet with type-A parts is picked by a forklift from one of two inventories for that type of parts; (2) a type-A part is sent to the preparation station; (3) a type-A part and a type-B part is assembled as one product; (4) a product with a defect is sent and repaired when the product is identified to have defects; (5) a product is packed; (6) a packed product is transported randomly to one of two docks for shipment. The simulation aims to predict the number of packaged products and repaired products over time.

Type-A parts. Type-A parts are picked from two inventories. At the beginning of the system operation, the pallet with part-A parts is taken from the first inventory; while the number of pallets in the first inventory is less than 2, a new pallet becomes randomly available at either of two inventories. This schedule is implemented by monitoring the number of pallets (*NumberInPrep*) by a *monitor element* whose threshold value is '2' and the crossing direction is 'negative'. When the state variable NumberInPrep is less than 2, the monitor element triggers an event to create a type-A part.

Pallets. Each pallet holds 10 type-A parts. The entities on a pallet are emptied at a 'separator' located at the beginning of the continuous assembly line. The type-A parts are sent by a 'conveyor', and a pallet is taken to a storage rack that is modeled as a 'sink'.

Preparation Station. The number of type-A parts at the preparation station is tracked by a user-defined state variable; the number is updated when a part enters or leaves the preparation station. When a part leaves, the designated assembly station is randomly selected from two assembly stations.

Assembly Station I and II. Two assembly stations are used to assemble type-A and type-B parts into products. One product has one type-A part (parent) and one type-B part (child). The assembly stations are modeled as 'combiners'. It is assumed that type-B parts are available by the sources that are linked to assembly stations.

Packaging Station and Repairing Station. 100% of the products from assembly station II and 70% of the products from assembly station I are directly sent to the packaging station and 30% of the products from assembly station I are sent to the repairing station, and they will be sent to the packaging station after they are repaired by workers.

Shipping. Packed products are randomly transported to two trucks that are modeled as sinks. The random selection is applied by (1) using 'random selection goal at the output of the package station, (2) setting the entity destination type as 'select from list', and (3) setting the 'node list name' as a list of input nodes of two trucks.

5.7 Computer Integrated Manufacturing

A manufacturing system involves numerous tasks and operations which require different information systems from different vendors for decision-making supports. As shown in Fig. 5.37, *computer integrated manufacturing* is an integrated solution of using computers to plan, schedule, and control manufacturing businesses in product lifecycles (Armagard 2020). The main sub-systems in CIM are *computer-aided design* (CAD), *computer-aided engineering* (CAE), *computer-aided manufacturing* (CAM), *computer-aided process planning* (CAPP), *computer-aided quality assurance* (CAQ), *production planning and control* (PPC), *enterprise resources planning* (ERP), and other high-level business systems such as *supply chain management* (SCM), and *customer relationship management* (CRM).

As shown in Fig. 5.38 (Wikipedia 2020d), CIM serves as the hub of an integrated solution for an enterprise to automate manufacturing businesses at all levels and domains in its production system. CIM deals with the integration, coordination, collaboration, and cooperation of functional units in an enterprise system. Therefore, three critical requirements of CIM are identified as follows (Wikipedia 2020d):

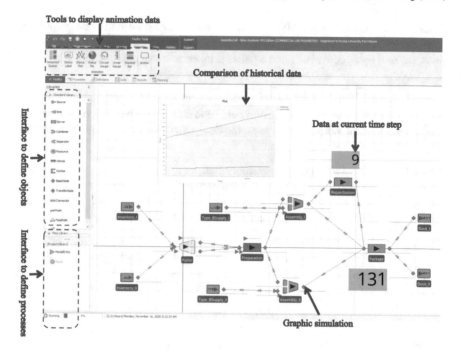

Fig. 5.37 Snapshot of DEDS simulation in Simio environment

Integration of sub-systems from different vendors. Different hardware and software systems are developed using different platforms, programming languages, and
standards. For example, automated machines such as robots, AGVs, conveyors,
and CNC use different protocols for communication. CIM must deal with the
heterogeneity of sub-systems in communication and interactions.

Human factors in process control. CIM is an ideal solution for full automation. In
reality, human intervenes are needed to deal with the changes and uncertainties in
both machine operations and decision-making supports. CIM must be able to provide
appropriate interfaces to (1) assist human operators in manufacturing facilities in
the material flow and (2) to allow engineers to tackle the circumstances which are
unpredicted by the automated information systems.

Data integrity. The higher the degree of automation a system is, the more crucial
the data integrity is to system control. When data is exchanged and shared between
two systems, it usually requires some additional human efforts to ensure that there
are proper safeguards for raw data to make effective interactions.

Similar to a modular system, CIM can be viewed as a philosophy in which the system
complexity and uncertainties system are tackled by selecting existing technical solutions and interacting with these solutions in different ways. Therefore, no universal
solution of CIM is available to manufacturing enterprises, especially for small- and

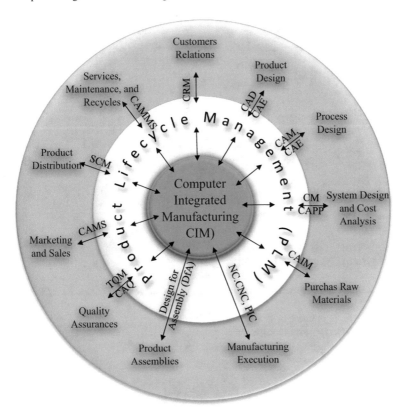

Fig. 5.38 Computer integrated manufacturing for system integration

medium-sized enterprises (SMEs) that have the limited resources in the investment of technological acquisition. Here, a case study is presented to show the manual intervention for data integrity in system integration.

Case Study 5.1. A manufacturer is a tier II supplier of thermoformed products in the automotive industry. The manufacturer takes the charge of designing and implementing manufacturing processes for the products that are designed by tier-I companies. The company needs to obtain product models from clients and using CAE to analyze the models to verify if all the product features can be manufactured. If all the product features can be manufactured, CAPP and CAM are used to set up fixtures, design molds, and run machine tools. Otherwise, the recommended changes are returned to clients for continuous improvement of products. As shown in Fig. 5.39, the system integration needs manual efforts for the communication of product models.

The company faces difficulty communicating with clients in data exchanges. As shown in Fig. 5.40, clients use different CAD tools such as Catia, Unigraphics, Creo, Solidedge to create their product modules; however, only SolidWorks is available at the company to review and edit CAD models. Note that product models must

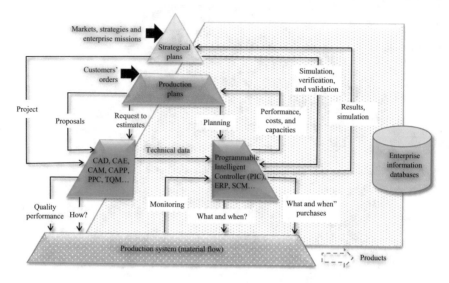

Fig. 5.39 Computer-integrated manufacturing for system integration (Wikipedia 2020d)

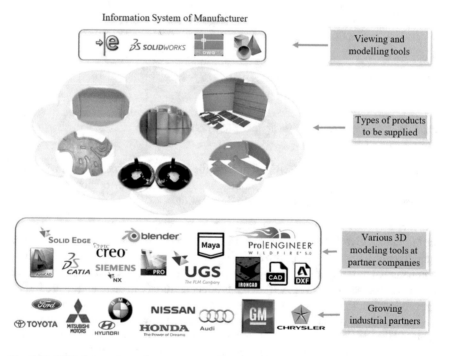

Fig. 5.40 CAD data exchange between a manufacturer and client companies in case study 5.1

Table 5.20 List of standard file formats of CAD models

3D file format	Type
STL	Neutral
OBJ	ASCII variant is neutral, binary variant is proprietary
FBX	Proprietary
COLLADA	Neutral
3DS	Proprietary
IGES	Neutral
STEP	Neutral
VRML/X3D	Neutral

be editable for the company to (1) obtain the dimensions relevant to thermoforming processes, (2) add new features required by thermoforming, (3) utilize product models to create molds and design fixtures, and (4) run CAE to ensure products can be manufactured with the required strengths and quality.

A CAD model can be generated in different file formats. Other than generating its native CAD format, a CAD tool is often compatible to a number of file formats such as .STL, .OBJ, .STEP, and .IGES; Table 5.20 shows some standard file formats for CAD models (Sketchfab 2020).

Even with the same geometric information, the CAD models in different file formats have different contents. If one CAD model is transferred from one software tool to another software tool, high-level information such as features, design intents, and constraints will be missed. SolidWorks is used by the company, and SolidWorks allows to import or export of over 30 different file formats (SolidWorks 2019) as listed in Table 5.21.

Figure 5.41 shows a comparison of the workflows with and without system integration. In Fig. 5.41 unicate the design changes with clients, create molds and make processing plans, and it led to a long lead time, high development time, and more importantly, missed business opportunities due to a long quote time and the uncertainty factors in quoting. In Fig. 5.40b, engineers are assisted greatly by the Solid-Works packages in recovering design intents and knowledge from imported CAD

Table 5.21 Compatible file formats in SolidWorks

3D XML	Autodesk Inventor	IGES*
3DS	Autodesk Mechanical	JPG
3MF	Desktop	OBJ
ACIS	CADKEY	Parasolid
Adobe Illustrator	CATIA Graphics	ProEngineer
AMP	**CGR (CATIA Graphics)***	Adobe Photoshop
PRC	SWG	PSD
PDF	DXF	Rhino
IFC	HCG	SAT
STEP*	**Unigraphics/NX***	SolidEdge
STL	VDA-FS	ECAD (IDF, ProStep)
TIFF	VRML	AEC (IFC format)

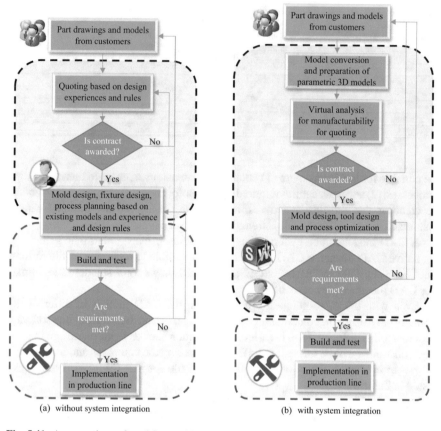

(a) without system integration   (b) with system integration

Fig. 5.41 A comparison of workflows with and without system integration

models, communicate the suggested design changes with client companies unambiguously and use the knowledge-based engineering method in CAM and CAPP. It is expected to reduce the lead time and development cost significantly.

Figure 5.42 shows the dataflow when product models are imported and exported from the information system at the company, and Table 5.22 shows the main steps to ensure that the information about features, design intents, and knowledge embedded in CAD models can be sustained to the maximum extent possible. Note that all of the computer aide functions in SolidWorks can be fully utilized to accomplish the tasks from step 2 to step 4.

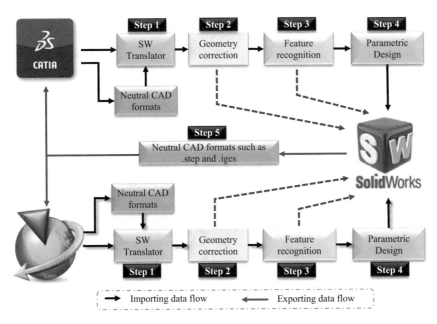

Fig. 5.42 Dataflow for importing or export product models

Table 5.22 Importing and exporting processes in solidworks

	Step	Task
Importing	1	Use the SolidWorks translator to open a file generated by the CAD package at the client company (such as Catia, Unigraphics/NX, .step, and .iges)
	2	Fix, delete, suppress errors on edges, faces, references, and datums to clean solids
	3	Run *feature recognition* to recover meaningful and correct features
	4	Use *the parametric design technique* to modify solids, add new features, and conduct engineering analysis
Exporting	5	Use the SolidWorks translator to export a file in the format compatible to the CAD package at the client company

5.8 Computer-Aided System Evaluation

The role of evaluation metrics in system designs can never be overestimated. An enterprise system requires the metrics to evaluate the performances of functional units at different domains and levels.

With an increasing concern on global warming, the scarce of natural resources, the pollution of eco-environment, enterprise values have been evaluated comprehensively from the perspective of economy, society, and environment, and the scope of manufacturing businesses has been expanded greatly to sustain manufacturing systems. As shown in Fig. 5.43, traditional manufacturing processes such as *design*,

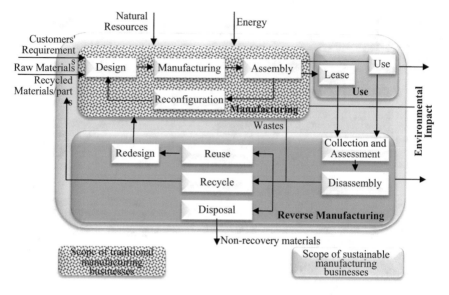

Fig. 5.43 Description of sustainable manufacturing system (Bi 2011)

manufacturing, and *assembly* used to be economic activities which are evaluated based on costs and profits; however, the scope of manufacturing processes was expanded greatly to *reuse*, *recover*, *redesign*, *recycle*, and *remanufacture*, and the impacts of manufacturing processes on economy, society, and environment have to be taken into consideration in determining enterprise values. Various standards (e.g. ISO 14,000 and ISO 14,064) have been developed to evaluate system sustainability and costs. Here, the techniques for computer-aided system evaluation are introduced.

5.8.1 Sustainability of Manufacturing Systems

Figure 5.44 represents the impacts of manufacturing businesses on three aspects of sustainability, i.e., economy, society, and environment. Manufacturing businesses are associated with the products at certain phases of product lifecycles, i.e. '*pre-manufacture*', '*manufacture*", '*use*", and '*post-use*'. The impact of a specific manufacturing process on three aspects is evaluated individually, and the overall impact of a manufacturing system on three aspects of sustainability is the sum of those of all manufacturing businesses in the system. However, it can be seen from Fig. 5.44 that the relations of manufacturing businesses with sustainability are so complicated to model them in analytical models; it would be more realistic to evaluate system sustainability using the statistical methods based on existing data. Fortunately, since

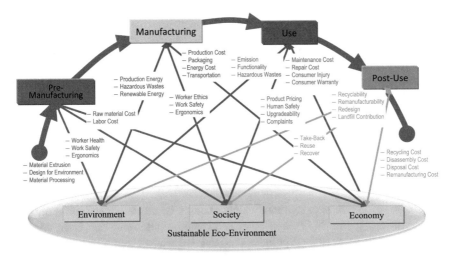

Fig. 5.44 Evaluation of sustainability in a manufacturing system (Bi 2011)

any manufacturing business is associated with products, the sustainability of a manufacturing system can be evaluated statistically based on the products that the system makes.

5.8.2 Main Indicators

Figure 5.43 showed that manufacturing businesses were expanded to the activities involved in entire product lifecycles; therefore, system sustainability should be evaluated in product lifecycles, and system sustainability is quantified by four main indicators as below.

5.8.2.1 Carbon Footprint

Carbon dioxide CO_2 and other gases with equivalent emissions are generated when natural resources such as fossil fuels burn in the atmosphere. CO_2 and its equivalence affect the average temperature of the earth, and continuous increase of average temperature causes *global warming* effects such as loss of glaciers, extreme weather conditions, and other environmental problems. *The carbon footprint* is defined to measure the impact of products on environment in the unit of a ton of carbon dioxide (CO_2).

5.8.2.2   Total Energy

Total energy is the total amount of direct energy consumption in making products, the unit of energy is in *mega-joules* (MJ). The total energy is calculated from (1) the upstream energy to acquire and process natural energy resources, (2) consumed energy by releasing or burning natural energy resources, (3) electrical power to make, transport, and use products. In addition, the efficiencies in energy conversion (e.g. power, heat, steam) must be taken into consideration.

5.8.2.3   Air Acidification

Air acidification relates to acid rain which is caused by the amount of sulfur dioxide (SO_2), nitrous oxide (N_2O), and other acidic emissions. Acid rain is an adverse environmental factor since it pollutes the land and water where plants and aquatic animals live; in addition, air acidification dissolves manmade building materials such as concrete slowly. Air acidification is measured in the unit of *kilogram sulfur dioxide equivalent* (SO_2–e).

5.8.2.4   Water Eutrophication

Water eutrophication measures the impact on the water ecosystem due to the overabundance of nutrients; this depletes the water of oxygen and results in the deaths of plants and animals. Nitrogen (N) and phosphorous (PO_4) in fertilizers are the main sources of water eutrophication. The unit of water eutrophication is *kilogram phosphate equivalent* (PO_4–e).

5.8.3   SolidWorks Sustainability

SolidWorks Sustainability allows engineers to estimate the environmental impacts of products via the lifecycle assessment (LCA) method. LCA is the method to assess the environmental impact of a product from the preparation of raw materials to production, distribution, use, disposal, and finally to recycles. LCA supports the comparison of optional materials in terms of environmental impacts, and LCA is used to assess system sustainability in terms of (1) the consumed energy and raw materials, (2) the generated emissions and wastes by manufacturing processes, (3) the potential environmental impacts, and (4) the options of reducing the environmental impact.

Figure 5.45 shows the framework of SolidWorks Sustainability. LCA assesses sustainability from the perspective of carbon footprints, total energy, air acidification, and water eutrophication, and the sustainability is quantified using the collected

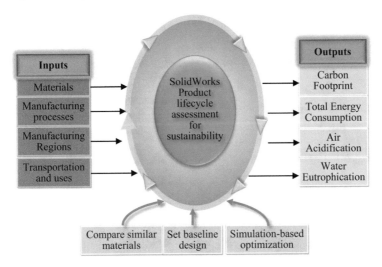

Fig. 5.45 SolidWorks Sustainability framework and tools

data of the environmental impacts by various materials in the library. Engineers can use it to select and compare raw materials since LCA supports simulation-based optimization.

5.8.3.1   Material Library

The impact of extraction and process of raw materials depends on the specific materials. Figure 5.45 shows the options of materials in the Material library. The materials are organized in two levels, the first level in Fig. 5.45a shows the classes of materials such as steel, iron, rubber, and woods. The second level in Fig. 5.45b shows the types of materials such as AISI 1020, AISI 304, and alloy steel in the steel class. Note that the types of materials correspond to the required treatments of materials.

5.8.3.2   Manufacturing Processes and Regions

The impact of manufacturing processes of products depends on the process types and the geographic locations where the processes are performed. Figure 5.46 shows the inputs for LCA to determine the impact of manufacturing processes on sustainability. The region input determines the consumed energy in kWh for performing such manufacturing processes.

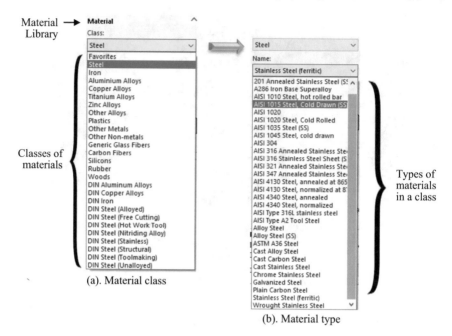

Material Library

Classes of materials

(a). Material class

Types of materials in a class

(b). Material type

Fig. 5.46   Classes and types of materials

5.8.3.3   Transportation and Use

Sustainability takes into account energy consumption by some non-value-added activities, especially, *transportation* and some activities that are not accomplished in production such as *product uses*. In addition, the energy consumptions of both these factors are affected by the regions where those activities take place. Figure 5.47 shows the interface to input the information for regions, transportation, and use of products.

Figure 5.48 shows an example of sustainability assessment of a product that includes (1) the quantified impact on carbon footprint, consumed energy, air acidification, and water eutrophication Fig. 5.48a and (2) the economic impact of products as the aspect of materials, manufacturing, use, and disposal.

5.8.3.4   Material Comparison Tool

The SolidWorks Sustainability supports engineers to compare raw materials from the perspective of the sustainability of products, a comparison is made between a baseline material and alternative materials, and Fig. 5.49 shows the interface to specify a baseline material for product. In comparison, the environmental impact

Fig. 5.47 Impact of manufacturing processes, regions, and transportation

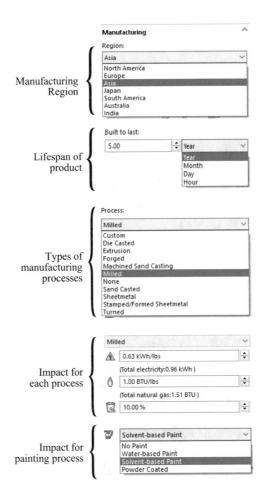

Manufacturing Region

Lifespan of product

Types of manufacturing processes

Impact for each process

Impact for painting process

can be converted to other quantities that are more familiar to people; for example, the carbon footprint can be interpreted into miles to drive a car, and this conversion can be made by using the online tool.

SolidWorks Sustainability is applicable to assembled products. The total environmental impact of product is the sum of those of individual parts in the product. Figure 5.50 shows the assessment of a structure that is assembled from structural elements. However, LCA of an assembly model is not supported by SolidWorks Sustainability, not the light version SustainabilityXpress.

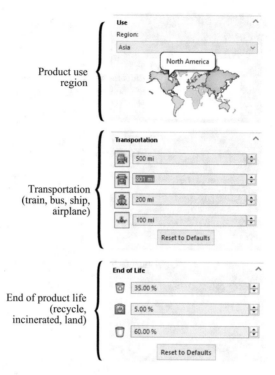

Product use region

Transportation (train, bus, ship, airplane)

End of product life (recycle, incinerated, land)

Fig. 5.48 Impact of transportation and use of product

(a). Environmental impact

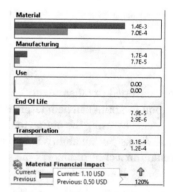

(b). Economic impact of product

Fig. 5.49 Example result of LCA

Fig. 5.50 Interface to specify a baseline material

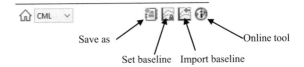

Save as   Set baseline   Import baseline   Online tool

5.8.3.5 Cost Analysis

Enterprises run manufacturing businesses for economic benefits; therefore, economic factors must be taken into the consideration in designing and operating manufacturing systems. In sustainable manufacturing, the costs of a manufacturing system must include direct and indirect costs, contingent liability, initial capitals, and tangible costs relevant to environmental impacts. The cost analysis of LCA includes overhead environmental costs. *SolidWorks Costing* supports the cost analysis with the consideration of environmental factors. The costing tool can be accessed in the '*Evaluation*' CommandGroup as shown in Fig. 5.51.

Figures 5.52 and 5.53 show the inputs for a cost analysis of product. In Fig. 5.52, the processing types and materials are specified, and in Fig. 5.53, the detail of each manufacturing process is provided. The details include the dimensions of raw materials and the quantity of products. Figure 5.54 shows an example result of the cost analysis of a product.

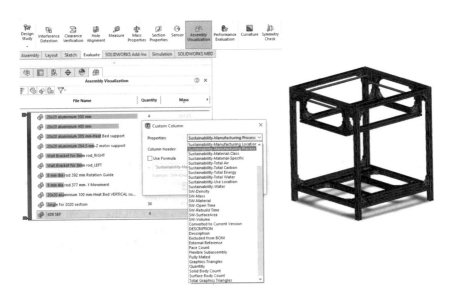

Fig. 5.51 Example of LCA for an assembly model

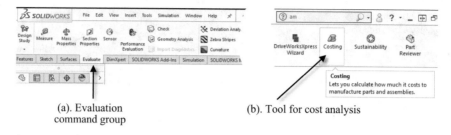

(a). Evaluation   (b). Tool for cost analysis
command group

Fig. 5.52 SolidWorks Costing for cost analysis

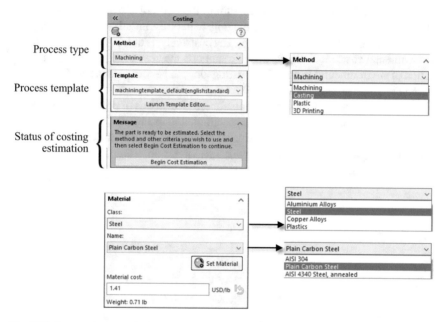

Fig. 5.53 Inputs of cost analysis—process type and material

5.9   Summary

A manufacturing system involves in numerous activities which are performed by
different hardware and software units from different vendors. Each functional unit
may be optimized for its purpose, but the manufacturing system has system-level
objectives. Therefore, an enterprise system must integrate all the system resources
seamlessly for coordination, collaboration, and cooperation of functional units. From
this chapter, engineers should learn some enabling technologies for organization,
integration, analysis, and evaluation of manufacturing systems. Especially, engineers

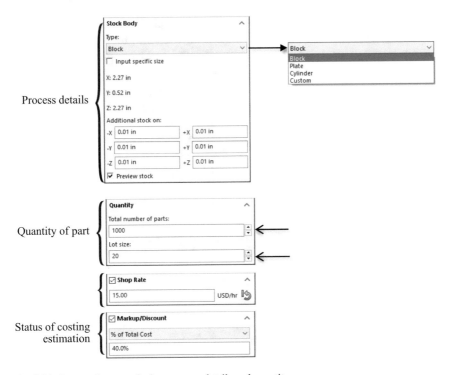

Fig. 5.54 Inputs of cost analysis—process details and quantity

are expected to gain understanding of computer integrated manufacturing (CIM), *cellular manufacturing*, *discrete event dynamic simulation*, LCA and *cost analysis* from the perspective of sustainability.

Design Problems

Problem 5.1 Given the product shown in Fig. 5.55, determine nine digits of 'form code' and 'supplementary code' using the Opitz coding system.

Fig. 5.55 Example result of cost analysis of product

Problem 5.2 Use Matlab or other languages to program the raking of clustering (ROC) algorithm, and use the program to find the machine groups for the following two machine-part incidence matrices. Alternatively, if you have limited programming skills, use ROC by hand to find the solution of GT.

Machine	'Part number'					
	1	2	3	4	5	6
M1		1		1		1
M2		1		1		1
M3	1		1		1	
M4	1		1		1	

(a) case 1

Machine ID	'Part Number'					
	1	2	3	4	5	6
A			1		1	
B		1	1			
C	1			1		
D		1	1		1	
E	1			1		1

(b) case 2

Problem 5.3 Consider a 6-part-and-9-machine problem shown in the following table, use ROC to form the part family and machine group.

Product

M_{ij}	1	2	3	4	5	6
A	1					1
B		1			1	
C		1	1		1	
D	1			1		
E	1			1	1	
F			1		1	
G	1			1	1	1
H		1			1	
I	1			1		

Machine

Problem 5.4 Consider a 20-part-and-10-machine problem shown in the following table, use ROC to form the part family and machine group.

Product

M_{ij}	1	2	3	4	5	6	7	8	9	10	11	12	13	14	15	16	17	18	19	20
A				1		1					1							1		
B		1			1						1				1		1			
C		1			1		1				1									
D			1							1				1						
E										1				1						1
F							1					1				1			1	
G			1									1				1			1	
H	1			1			1									1			1	
I						1			1				1		1					
J						1			1					1			1			

Machine

Problem 5.5 Create a composite product for a product family with the following instances (Fig. 5.56).

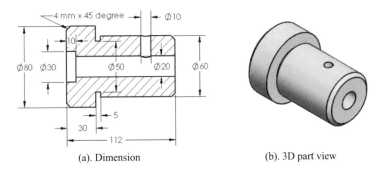

(a). Dimension   (b). 3D part view

Fig. 5.56 Product drawings for Problem 5.1

Problem 5.6 Create a similar model to analyze the operation of a 3D printing lab. The lab is equipped with four 3D printers, i.e., Dreamer-Flashforge, SeeMeCNC, Dreamel, and Dreamel digiLab. The lab provides 3D services for students in the courses of *'Solid Modelling'*, *'CAD/CAM applications'*, *'senior design projects'*, and *'course projects'* in other mechanical courses. Each student is required to submit one printing job for each course. The complexity of parts varies from one course to another. The simulation aims to evaluate if available resources are sufficient to meet students' needs; in other words, the utilization rates of the 3D printers will be evaluataed. Note that the frequency of 3D printing orders from a certain class is proportional to the average enrolments in each class in a specified period of time. Therefore, the needs of printing jobs are assumed in Table 5.23 (Fig. 5.57).

Design Project

Project 5.1 Use one of the products you modeled in Chap. 2 (e.g. Problems 2.5, 2.6, 2.7, and 2.8), use the SolidWorks Sustainability and SolidWorks Costing to evaluate its environmental impact and cost, and modify the materials, processes, materials, and business regions to reduce the environmental impact and cost.

Table 5.23 Predicted needs of printing jobs for students in different courses at lab

Courses	Solid Modelling (ME160)	CAD/CAM Applications (ME546)	Senior Design (ME487)	Other Courses
Average enrolments in semester	24	20	15	30% of 150
Probability (%)	22.86%	19.05%	14.29%	42.86%
Printing Time (hours)	(1, 3)	(5, 10)	(5, 30)	(5, 10)

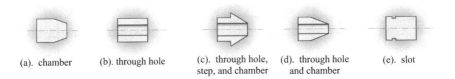

(a). chamber   (b). through hole   (c). through hole, step, and chamber   (d). through hole and chamber   (e). slot

Fig. 5.57 Product variants for Problem 5.5

References

Andrew L (2019) What is the difference between discrete and process manufacturing? https://www.sensrtrx.com/difference-between-discrete-and-process-manufacturing/

Al-Habahbah OS (2015) Chapter 5 industrial control systems. https://eacademic.ju.edu.jo/o.hababhbeh/My%20Documents/Industrial%20Automation%20By%20Dr%20Osama%20AlHababbeh/9-%20Ch_5_Industrial%20Control%20Systems.pptx

Armagard (2020) Computer-integrated manufacturing explained clearly. https://www.armagard.com/ip54/computer-integrated-manufacturing-explained-clearly.html

Askin RG, Standridge CR (1993) Modeling & analysis of manufacturing systems. Wiley. ISBN 0-417-51418-7

Bi ZM (2011) Revisiting system paradigms from the viewpoint of manufacturing sustainability. J Sustain 3(9):1323–1340

Bi ZM, Zhang WJ, Li Q (2001) A software environment for flexible manufacturing system control software testing. Proc Inst Mech Eng Part B: J Eng Manuf 215(B):339–352

Bi ZM, Miao ZH, Zhang B, Zhang WJ (2020) A framework for performance assessment of heterogeneous robotic systems. IEEE Syst J. https://doi.org/10.1109/JSYST.2020.2990892

Cruz-Mejia O, Vilalta-Perdomo E (2018) Merge-in-transit retailing: a micro-business perspective. Univers Empresa 20(34):83–101. https://doi.org/10.12804/revistas.urosario.edu.co/empresa/a.5500

El-Sherbeeny am(2016) Introduction and overview of manufacturing. https://fac.ksu.edu.sa/sites/default/files/1_introduction_sep07_13_ams_2.pdf

Falco J, Wyk KV, Liu S, Carpin S (2015) Facilitating replicable performance measures via benchmarking and standardized methodologies. IEEE Robot Autom Mag 22(4): 125–136.

Fang W, Guo Y, Liao W, Huang S, Yang N, Lui J (2020) A parallel gated recurrent units (P-GRUs) network for the shifting lateness bottlenech prediction in making-to-order production system. Comput Ind Eng 140(2020):106246

GAO (2001) A pratical guid to dederal enterprise architecture. Chief Information Officer Council, Version 1. https://www.gao.gov/assets/590/588407.pdf

Haworth EA (1968) Group technology—using the Optiz system. Prod Eng 47(1):25–35

Hyer N, Wemmerlov U(2001) Reorganizing the factory: competing through cellular manufacturing, the 1st version. Productivity Press, Portland, USA, ISBN-10:1563272288.

Hyer N, Wemmerlov U (2002) Reorganizing the factory, competing through cellular manufacturing. Productivity Press, Portland, OR, 2001. ISBN-10:1563272288

HIBA (2019) Operations management. https://www.slideshare.net/Joanmaines/process-and-layout-strategies

Hooper J, Reilly K (2020) The GPSS—GASP combined (GGC) system. https://www.semanticscholar.org/paper/The-GPSS%E2%80%94GASP-combined-(GGC)-system-Hooper-Reilly/ce6a58307f1fb8052b28762a9db8216ff7b18242

Kaminsky P, Kaya O (2009) Combined make-to order/make-to-stock supply chains, IIE Trans 41(2009): 103–119.

Khan N (2013) Computer integrated manufacturing. https://www.slideshare.net/NoumanKhan2/9-oct-2013-lec-13-1415161718

King JR (1980) Machine-component grouping in production flow analysis: an approach using a rank order clustering algorithm. Int J Prod Res 18(2):213–232

Kirby K (2019) Focused factories and group technology. https://web.utk.edu/~kkirby/IE527/Ch9.pdf

Kiran DR (2019) Chapter 18: plant layout, production planning and control, pp 261–278. https://doi.org/10.1016/B978-0-12-818364-9.00018-4

Knowledgiate Team (2017) Advantages and disadvantages of continuous production systems. https://www.knowledgiate.com/wp-content/cache/wp-rocket/www.knowledgiate.com/advantages-disadvantages-continuous-production-system/index.html_gzip

Kootbally Z (2016) Industrial robot capability models for agile manufacturing. Ind Robot, 43(5):481-494.

Markowitz HM, Hausner B, Karr HW (2020) Simsript—a simulation programming language. https://www.rand.org/pubs/research_memoranda/RM3310.html

Pegden D (2020) An introduction to Simio for beginners. https://www.simio.com/resources/white-papers/Introduction-to-Simio/Introduction-to-Simio-for-Beginners.pdf

Pritchett AR, Lee S, Huang D, Goldsman D (2000) Hybrid-system simulation for national airspace system safety analysis. In: Joines JA, Barton RR, Kang K, Fishwick PA (eds), Proceedings of the 2000 winter simulation conference. https://www.researchgate.net/publication/221526881_Hybrid-System_Simulation_for_National_Airspace_System_Safety_Analysis

SAC (2018) Alignment report for reference architectural model for industrie 4.0/ intelligent manufacturing system architecture. https://sci40.com/files/assets_sci40.com/img/sci40/Alignment%20Report%20RAMI.pdf

Simio (2020) The story of Simio. https://www.simio.com/about-simio/

Strategosinc (2019) Design of workcelland micro layouts—cellular manufacturing, https://www.strategosinc.com/celldesign.htm

Sketchfab (2020) 3D file format. https://help.sketchfab.com/hc/en-us/articles/202508396-3D-File-Formats

Solidworks (2019) SMG export options. https://help.solidworks.com/2019/english/SolidWorks/sldworks/r_SMG_export_options.htm

Thiesing RM, Pegen CD (2020) Introduction to Simio. https://informs-sim.org/wsc13papers/includes/files/407.pdf

Viriyasitavat W, Xu L, Bi ZM (2019a) rmSWSpec: real-time monitoring of service workflow specification language for specification patterns. IEEE Trans Ind Inf 15(7):4021–4032

Viriyasitavat W, Xu L, Bi ZM (2019b) The extension of semantic formalization of service workflow specification language. IEEE Trans Ind Inf 15(2):741–754

Vollmann TE, Berry WL, Whybark DC, Jacobs RF (2004) manufacturing planning and control systems for supply chain management, 5th edn. McGrawHill Professional, New York

Weber A (2004) The pros and cons of assembly cells. https://www.assemblymag.com/articles/83136-the-pros-and-cons-of-cells

Williams TJ (1994) The Purdue enterprise reference architecture and methodology (PERA). https://citeseerx.ist.psu.edu/viewdoc/download?doi=10.1.1.194.6112&rep=rep1&type=pdf

Wikipedia (2020a) NIST enterprise architecture model. https://en.wikipedia.org/wiki/NIST_Enterprise_Architecture_Model

Wikipedia (2020b) Fixture (tool). https://en.wikipedia.org/wiki/Fixture_(tool).

Wikipedia (2020c) List of discrete event simulation software. https://en.wikipedia.org/wiki/List_of_discrete_event_simulation_software

Wikipedia (2020d) Computer integrated manufacturing. https://en.wikipedia.org/wiki/Computer-integrated_manufacturing

Zeng BC (2009) Group technology (GT) in manufacturing. https://www.me.nchu.edu.tw/lab/CIM/www/courses/Computer%20Integrated%20Manufacturing.htm

Zhang WJ, Li Q, Bi ZM, Zha XF (2000) A generic petri net model for flexible manufacturing systems and its use for FMS control software testing. Int J Prod Res 38(5):1109–1132

Chapter 6
Digital Manufacturing (DM)

Abstract This chapter discusses the applications of digital technologies in manufacturing systems: (1) the main functional requirements are discussed to construct enterprise system architecture; (2) a new system architecture is proposed for the applications of digital manufacturing; (3) two examples of digital technologies, i.e., reverse engineering (RE) and direct manufacturing (DM) are discussed in details; and (4) some case studies are briefed to show the diversified applied researches in using digital technologies in manufacturing.

Keywords Digital manufacturing (DM) · Digital twin (DT) · Reverse engineering (RE) · Additive manufacturing (AM) · Direct manufacturing (DM) · Cyber-physical systems (CPS) · Cloud computing (CM)

6.1 Introduction

A digital twin (DT) refers to a computer representation of (1) potential and physical assets, devices, peoples, places, processes, and systems and (2) the dynamics of how the networked things operate and interact over time. The computer representation for a DT focuses on the connections of physical and virtual models and how both models are interacted to optimize the performances of physical systems. Therefore, DT replies greatly on the cutting-edge information technologies (IT) such as *Internet of Things* (IoT), *artificial intelligence* (AI), *machine learning, cyber-physical systems* (CBSs), and *big data analytics* (BDA). The state of a virtual model in a DT system is updated continuously based on the real-time data feedback from its physical twin (Wikipedia 2020; Parrott and Warshaw 2020).

This chapter focuses on the application of digital technologies in manufacturing. The idea of DT was proposed by Gelernter (1991) that the physical world was modeled, simulated, and evaluated in the virtual reality. The DT concept was coined by Grieves 11 years later, and it was introduced as an underlying principle for product lifecycle management (PLM) (Geieves and Vickers 2006; Catapult 2018). However, due to the high requirement of computing and network capabilities, early DT technologies were not prevalent tools until the General Electric (GE) managed to use DT in designing and manufacturing aircraft structures, vehicles, and turbines (Glaessgen

© The Author(s), under exclusive license to Springer Nature Switzerland AG 2021   389
Z. Bi, *Practical Guide to Digital Manufacturing*,
https://doi.org/10.1007/978-3-030-70304-2_6

and Stargel 2012; Tuegel et al. 2011; GE 2020). In this chapter, *firstly*, the functional requirements (FRs) of modern manufacturing systems are discussed. *Secondly*, DT is introduced as the solution to design and operation of manufacturing systems, and DT architecture is presented to illustrate main system components and their relations. *Thirdly*, the system architecture of DT is introduced and some critical enabling technologies are discussed. *Fourthly*, reverse engineering (RE) and additive manufacturing (AM) are introduced as the examples of computer implementation of digital technologies. *Finally*, a number of case studies have been presented to illustrate the ongoing applied studies in digital technologies.

6.2   Functional Requirements (FRs) of Digital Twin

In a complex manufacturing system, different information systems are used to support the decision-making supports at different domains and stages such as designs, productions, assemblies, and marketing and sales. These information systems have to be integrated as a cross-functional enterprise system to share data seamlessly, coordinate business processes, and plan and schedule enterprise resources. DT is an implementation of an enterprise system; therefore, DT relies on various digital manufacturing technologies.

Digital manufacturing technologies are the collection of computer-aided tools for modeling, simulation, visualization, and optimization of products, processes, and manufacturing systems. Early digital technologies, such as computer-aided design (CAD), computer-aided manufacturing (CAM), product data management (PDM), and total quality Management (TQM), are stand-alone systems; however, modern enterprises prioritize the needs of integration, coordination, collaboration, and inter-operation in a digital twin (Siemens 2020). The applications of digital technologies have been widely explored in manufacturing industry. In fact, the industry evolution is driven by the emerging digital technologies. For example, the fourth industrial revolution (i.e., Industry 4.0) aims at the data transparency, interoperability, predictivity, and adaptability via newly developed technologies such as the Internet of Things (IoT), blockchain technology (BCT), and big data analytics (BDA) (Berger 2019).

Materials in a physical world are similar to the data in a virtual world. All of the functional units in a physical manufacturing system are used to transfer raw materials into finish products; similarly, all of the functional modules in virtual manufacturing system (i.e., DT) are used to generate, collect, process, distribute, transfer, share, mine, or utilize data. As shown in Fig. 6.1 (Bi and Wang 2020), with an increasing complexity of a physical manufacturing system, the volume of data in DT will be increased exponentially along the lifecycle of products. Other than its volume, the data in DT can also be characterized by high heterogeneity and diversity since the data can be sensed signals, events, methods, plans, commands, knowledge, rules, or models for different manufacturing resources at different spatial and temporal schemes.

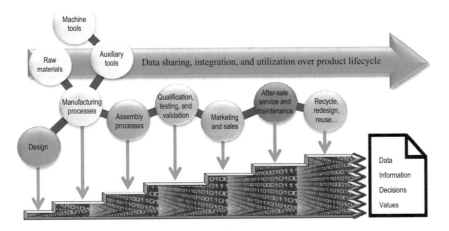

Fig. 6.1 The data growth in product lifecycle

In addition, the functions of DT are far beyond the data collection and sharing. A digital environment may suffer from a *data-rich but information-poor* (DRIP) syndrome. When the volume, velocity, and variety of data from design, manufacture, assembly, and production are increased exponentially, the system becomes incapable of processing and utilizing data for decision-making supports. Therefore, it is essential for DT to incorporate advanced digital technologies to process, utilize, and mine data to reach right decision for right physical count-partners promptly. The digital technologies in DT are expected to fulfill the main functional requirements (FRs) as follows.

6.2.1 Data Availability, Accessibility, and Transparency

An information system counts on correct, reliable, abundant, and updated data. DT must have the information infrastructure to provide right data at right time to right place whenever the data is needed. All manufacturing resources at different levels and domains should be modeled in the digital world to ensure the accessibility, availability, and transparency of real-time data (Mäkiö-Marusik et al. 2019). For example, when a CAM package is used to program a machining process, the part CAD model must be available since the data of the machined features on the part determines the types of machines and tools, cutting paths, and processing parameters. Moreover, virtual models of physical systems make it possible to (1) integrate external and internal manufacturing resources, (2) reconfigure systems when an enterprise's mission is adjusted, and (3) incorporate new manufacturing technologies in an existing information infrastructure (Buttner and Muller 2018).

6.2.2   Integration

Many technological solutions may be available to one functional unit, and an enterprise system usually consists of technological solutions from different vendors. As a customized and holistic system, an enterprise system must integrate all of functional modules seamlessly to support the interaction, interoperation, coordination, and collaboration of physical systems. Therefore, system integration includes the integrations of data, platforms, business processes, and hardware and software applications at various levels and domains (Fenner 2003).

(1)   **Data integration**. In an enterprise system, right decisions rely on reliable and abundant data. Data integration ensures the data availability, so that it can be utilized in various business processes in product lifecycles (Tao et al. 2019). The data in an enterprise system must be shared, maintained, and integrated through the network. When it is needed, data should be distributed and accessed across different applications. Some mechanisms that are used to support data integration include *component object model* (COM), *distributed component object model* (DCOM), *common object request broker architecture* (CORBA), *enterprise date integration* (EDI), *Java remote method invocation* (JavaRMI), and *extensible markup language* (XML).

(2)   **Integration of platforms**. A manufacturing system consists of different business units for purchasing, designing, manufacturing, assembling, logistics, human resources, marketing, and sales and an enterprise system integrates the processes, software, and tools from different vendors and platforms. These system resources must be integrated to support the interactions and interoperation in the heterogeneous environment safely and efficiently.

(3)   **Integration of business processes**. In an enterprise, the required business processes are defined, and the methods for data exchanges of business processes are specified. Business process integration includes *process management*, *process modeling*, and *process workflow*. The solution to the integration includes the tools to integrate the procedures, organizations, inputs and outputs, and required tools. The integration of business processes will streamline business operations.

(4)   **Integration of applications**. Applications in an enterprise are implemented as black boxes; however, applications have their interfaces to be integrated with other applications for high-level system objectives. Both internal and external applications must be considered. For example, an enterprise's relations with the manufacturing business environment must be taken into consideration; *a customer relationship management* (CRM) system can be integrated with the backend applications of the enterprise to support the interoperations cross the boundaries of enterprises.

6.2.3 Coordination, Collaboration, and Cooperation

Functional units in the material flow of a manufacturing system are locally optimized. When low-level units are integrated for system-level objectives, the enterprise system must be able to provide decision-making supports on the coordination, collaboration, and cooperation of system elements to optimize system performances. Conventionally, a steady hierarchical structure is used to organize system elements in layers and domains, so that the scope of each decision-making task becomes manageable. However, a steady structure is impractical when the business environment is highly turbulent, and the boundaries of manufacturing businesses or the environment are vague and dynamic.

6.2.4 Decentralization

Manufacturing businesses to produce parts, assemble parts into components and products, and distribute products to customers are operated in a geographically distributed environment. To achieve the flexibility and adaptability of the manufacturing system, decision-making on organizing, planning, scheduling, and controlling of business processes are decentralized. Therefore, DT must support the decentralized controls of manufacturing processes so that these businesses can be operated at geographically distributed locations through the material and information flows.

6.2.5 Reconfigurability, Modularity, and Composability

In a dynamic environment, the optimization of a manufacturing system must take sustainability into consideration; a manufacturing system must be reconfigurable to accommodate the changes and uncertainties over time. Reconfigurability is the capability of changing system configurations to meet new needs. Reconfigurability and flexibility deal with the changes in different ways in sense that both of hardware and software components are changed in a reconfigurable system, while only the software components are changed in a flexible system. Therefore, reconfigurability can be further characterized by modularity, compatibility, universality, mobility, and scalability (Wiendahl et al. 2007; Buttner and Muller 2018).

Modularity is implemented by selecting different system elements, and assembling them in different manners to build a system configuration for a set of specific task requirements; a set of system elements can serve for different processes and products. A modular system is a collection of independent, reusable system elements. Composability is a measure for the capability of system elements to be connected with others. Taking an example of one computer-aided design (CAD) package in DT, its composability is measured by the number of compatible computer-aided tools that can import and export data from the package directly.

6.2.6 Resiliency

Any system element has the probability of failure; in addition, an integrated manufacturing system is vulnerable to the uncertainties, changes, and disruptions occurring to its supply chains. Resiliency is the measure of the system capability to sustain system operations and recover the system from an abnormal state to a normal state. The resiliency relates to *adaptability, agility, redundancy*, and *learning capability* of a manufacturing (Kusiak 2019). When the manufacturing environment is dynamic, DT should be resilient to cope with the unexpected disruptions in the dynamic business environment.

6.2.7 Sustainability

Sustainability is the measure of the long-term survivability of a manufacturing system. The sustainability is evaluated at the three aspects, i.e., economy, society, and environment. The impact of taking the sustainability as one of system FRs has been discussed in depth (Bi 2011). Simply speaking, this would greatly expand the scope of manufacturing businesses since the activities in the whole lifecycles of products have to be covered. DT should have the enabling technologies to provide decision-making supports on expanded manufacturing processes; in addition, DT should be able to evaluate system sustainability from the perspective of economy, society, and environment (Gregori et al. 2017).

6.3 System Architecture

Making complex products involves numerous direct manufacturing processes and indirect business operations; DT provides a platform to network and integrate all of the computer-aided tools to support these manufacturing businesses. However, any system resource has its limitation in dealing with the scale, the complexity, and the uncertainty of system; *system architecture* aims to maximize the system capability in dealing with the complexity and uncertainties of system. In the digital world, the complexity of a system can be measured by entropy. *Entropy* is a thermodynamic property; it refers to an amount of system energy that is not yet transferred into works. *The Shannon entropy* is used to quantify the randomness, the disorder, and the complexity of system (Sönmez and Koç 2015) as

$$H(X) = \sum_{x \in \chi} p(x) log_2\left(\frac{1}{p(x)}\right) \tag{6.1}$$

where $H(X)$ is the Shannon entropy for the measure of uncertainty or information; χ is the set of all possible events in a system; $p(x)$ is the probability of event x where $x \in \chi$; and log_2 is a logical operation over the probability $p(x)$.

The higher $H(X)$ is, the more complex the manufacturing system is, and the larger amount of information or uncertainty the manufacturing system has. Equation (6.1) shows that the system entropy is mainly determined by two factors. *The first factor* is the number of possible events. A manufacturing system has many types of events, for example, the interaction of two or more system elements is an event, the case when a machine is malfunctioned is an event, and the state change of system element is an event. The system entropy increases when the number of system elements is increased. *The second factor* is the probability of having a certain event in a system. The higher possibility of occurrence of an event, the less uncertainty or a lower entropy this event contributes to the system.

Conventionally, system architecture is introduced to reduce the complexity and uncertainty of manufacturing systems. As we discussed in Chap. 5, many system architectures have been proposed. For example, some popular system architectures include Open System Architecture for Computer-Integrated Manufacturing (CIMOS), GRAI-Integrated Methodology (GRAI-GIM), Purdue Enterprise Reference Architecture (PEPA), and Enterprise Architecture by the National Institute of Standards and Technology (EA-NIST) (Williams 1994). However, adopting a conventional system architecture in enterprises has exposed the following major drawbacks:

(1)   Hierarchical or grid architecture may cause the delays for system elements to respond to the changes. Taking an example of a hierarchical architecture, the directions of the communication for plans, schedules, controls, and executions are specified from high-level functional units to low-level functional units in a sequence, and the directions of the communications for data collections and feedbacks are specified from low-level to high-level functional units. Moreover, the functional units at different levels are updated in different time durations; this scarifies the flexibility of low-level functional units to accommodate the changes promptly.

(2)   System architecture represents system elements and their interactions; it actually brings some artificial barriers to the direct communications of system elements at different levels or domains even though all system elements are networked.

(3)   To develop system architecture, the boundary for system elements and manufacturing businesses is clarified. However, the system boundary becomes ambiguous when the enterprise needs to collaborate with business partners; in addition, an enterprise has to adjust its business spectrum continuously to adopt to the dynamic changes. Existing system architectures are not effective when an enterprise deals with the changes and dynamics in the environment.

(4)   System architecture is designed for given products and manufacturing processes within a period. As shown in Fig. 6.2, it is evolved along with the changes of the business environment continuously. System architecture will

Development of new enterprise architecture (EA)

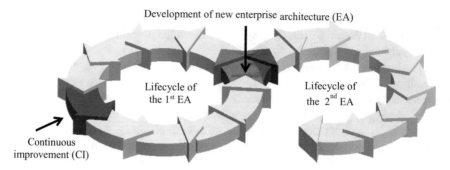

Continuous
improvement (CI)

Fig. 6.2 Evolution of system architecture in its lifecycle

be terminated when continuous improvement (CI) is not able to fill the gap of system architecture and the mission of enterprises. This leads to an abrupt shift to new system architecture.

Conventional system architectures reduce the complexity and uncertainties by regulating the interactions of system elements by standards and rules. This limits the capabilities of a manufacturing system at the aspects of the delays caused by indirect communications, artificial barriers to respond changes, and discontinuities of updates of enterprise systems. Existing system architectures were able to manage system complexity by sacrificing system flexibility to certain degree. They are effective only when the business environment is relatively steady.

The higher the system entropy is, the higher is uncertainty level of information, and more possible interactions occur to system. From this perspective, a networked but free system architecture can support the interactions of system elements to the greatest extent. It enhances the flexibility and adaptability for a manufacturing system to deal with uncertainties and changes promptly. Figure 6.3 shows a free architecture of a digital twin (Berman and Bell 2007; Pati and Bandyopadhyay 2017). All of functional units in the digital world are networked, and they can communicate and interact directly with each other; this offers the flexibility, adaptability, and agility to the physical twin. Based on their roles in system, digital tools can be classified into five types, i.e., *digital engineering, manufacturing operations, digital customer engagement, business intelligence* (BI), and *digital technology platform.*

(1) *Digital engineering* deals with engineering problems involved in the material flow such as designs of products and manufacturing processes and selections of manufacturing equipment. The solution to an engineering problem must be verified in the digital world before it is released for implementation. The computer-aided technologies discussed in previous chapters such as CAD, CAE, CAM, and DEDS simulation are used for digital engineering.

(2) *Manufacturing operations* refer to the executions of manufacturing processes such as the value-added processes such as metal forming, material removing processes, injecting molding, and additive manufacturing and non-value-added

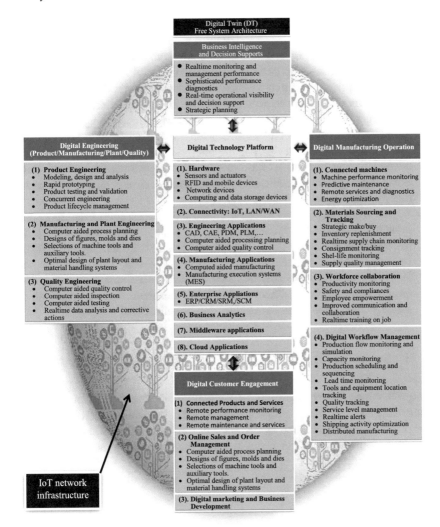

Fig. 6.3 Free system architecture of DT

processes such as inspecting and transportation. The executions of manufacturing operations are the outcomes of the direct interactions of digital and physical twins; the performances of a manufacturing system such as cost, delivery time, and productivity depend on the enabling technologies for manufacturing operations. The examples of the digital tools to support the decision-making activities of manufacturing operations are FMS, MRP-I, MRP-II, MES, and SCM.

(3)   *Customer engagement* deals with the relations to customers since the businesses in an enterprise are driven by customers' needs. Products must be

designed and manufactured to meet customers' needs to the great extent. *Customer relation management* (CRM) is used to integrate an enterprise with suppliers and end users in the distributed environment through supply chains. In addition, information and communication technologies (ICTs) allow end users to be fully engaged in product lifecycle from the product design stages.

(4)   *Business intelligence* deals with the strategical changes of enterprises for the sustainability in the competitive environment. When everything is networked, an enterprise is capable of fully utilizing the external resources over the Internet to catch emerging businesses, and the enterprise can enhance its flexibility and adaptability by forming workflows with business partners dynamics. Traditional digital tools for business intelligence are return of investment (ROI) and quality function deployment (DFD).

(5)   *Digital technology platforms* serve as the information infrastructure to integrate heterogeneous digital technologies for various applications in the physical world. A number of advanced technology platforms have been developed recently. For example, the Industrial Internet of Things (IIoT) supports both *machine-to-machine* (MTM) and *machine-to-man* communications; *Social, Mobility, Analytics and Cloud* (SMAC) provides the technology platform for the implementation of digital strategies in organizations.

6.4   Example of Digital Engineering—Reverse Engineering

Figure 6.3 shows that so many digital technologies are applied in digital twins; the introduction of all these technologies is certainly beyond the limit of coverage. Some digital technologies such as CAD, CAE, CAM, and DEDS have been introduced in previous chapters; in Sects. 6.4 and 6.5, reverse engineering (RE) and direct manufacturing (DM) are discussed as the examples of digital engineering, respectively.

6.4.1   Forward Engineering (FE) and Reverse Engineering (RE)

Engineering is a process of accumulating knowledge and information data about materials, parts, products, processes, and systems. Depending on the method by which the knowledge and information is accumulated, an engineering process can be either of *forward engineering* (FE) or *reverse engineering* (RE). As shown in Fig. 6.4, forward engineering begins with a conceptual design at the abstract level and concludes at a detailed physical solution. In contrast, reverse engineering begins with a physical model and completes with a virtual model that represents the knowledge and information of the physical model. Reverse engineering is an effective digital technology to (1) create a virtual model of a physical object, product, process,

Fig. 6.4 Forward
engineering (FE) and reverse
engineering (RE)

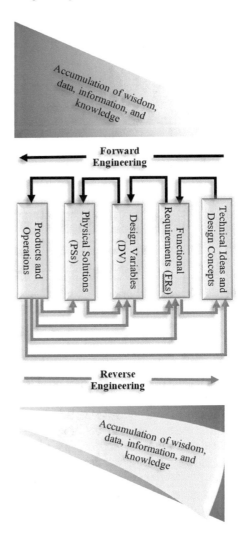

or system; (2) explore geometric dimensions, information, and knowledge based on physical objects; and (3) explore the structure (system elements and relations) of an assembled product and system. A reverse engineering process can shorten the technical gaps of an enterprise and business competitors in new technology developments.

RE aims to recover, capture, and reuse knowledge, wisdom, and information embedded in existing objects, products, or systems. However, RE is effective only when the physical model to be reverse engineered shows the economic benefits based on the return of investment (ROI). Since ROI depends on the expected product volumes on markets, the cost of an RE project has to be analyzed to justify if mimicking existing innovations is a right option for new product development. RE is

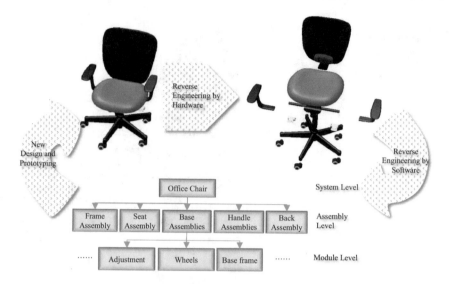

Fig. 6.5 Example of reverse engineering at different levels

applicable to any of design aspects; however, only the RE applications in geometric designs of products are interested; computer-aided reverse engineering (CARE) tools are introduced to (1) create virtual models from physical objects and (2) explore the design intents of original physical objects such as design variables, constraints, and features.

RE can be applied to assembled products, and Fig. 6.5 illustrates an RE process used to create a digital model of an office chair. To create a virtual model, (1) the chair is disassembled, and parts are scanned to acquire point clouds from exterior surfaces; (2) the acquired data is processed to recover the surfaces of the virtual model; (3) the solid model can then be defined using the boundary surfaces that are defined at the previous step; (4) the solid model is then followed by a re-conception process to discover design intents and generate a high-level parametric representation of models; and (5) the part models are then assembled as a complete chair.

6.4.2 Procedure of Reverse Engineering

An RE project involves four critical steps as shown in Fig. 6.6. *Firstly*, a physical object is scanned to acquire point cloud dataset from exposed surface or internal solids. Commonly used scanning devices are laser scanners, cameras, coordinate measuring machine (CMM), and computed tomography (CT). *Secondly*, the point cloud dataset is processed to generate a polymesh as the boundary surfaces of physical objects. *Thirdly*, the polymesh is cleaned to define a watertight volume of solid. *Fourthly*, the solid model is analyzed to identify features, design intents,

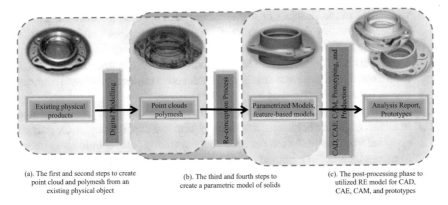

(a). The first and second steps to create point cloud and polymesh from an existing physical object

(b). The third and fourth steps to create a parametric model of solids

(c). The post-processing phase to utilized RE model for CAD, CAE, CAM, and prototypes

Fig. 6.6   Main tasks in a reverse engineering project

and constraints for a parametric model. *Finally*, the reverse engineered model can be utilized in computer-aided engineering analyses or applications.

6.4.3   Reverse Engineered Modeling

One of the key tasks in an RE project is to construct a surface from point cloud. To construct a watertight surface for a solid model, points are acquired by three-dimensional sensors such as stereo cameras and laser sensors, and the point cloud is then analyzed by computers to generate the boundary surfaces of solids. Figure 6.7 shows the surface reconstruction process, which consists of some main tasks described below (Bi and Kang 2014).

(1)   *Data acquisition.* Three-dimensional scanners are used to capture points over visible surfaces of a physical object. When the object cannot be fitted in

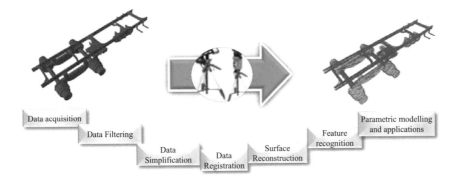

Fig. 6.7   Critical tasks in surface reconstruction (Bi and Kang 2014)

workspace of the scanners or surface points are collected from different orientations, the points should be acquired from different views and segments to ensure the sufficient information for a watertight volume.

(2) *Data Filtering.* Raw data from sensors includes noise, redundant, distorted, or even invalid data; it must be filtered and cleaned to remove noise and redundant points for the collected data.

(3) *Data Registration and Integration.* When the data acquisition process involves in a relative motion between the sensing system and physical object, the datasets obtained from multiple views or continuous scanning paths should be registered to merge datasets in a unified coordinate system. Registration is to determine the transformation of a dataset between two views, and integration is to create single part model from multiple sources of the datasets.

(4) *Surface reconstruction.* A point cloud from visible surface, or a volumetric dataset, is analyzed to construct boundary surfaces of the object.

(5) *Data simplification and smoothing.* For a dataset with a large amount, or there are some special requirements about the smoothness of a constructed surface, surface data is simplified to reduce the size of the dataset and smooth the boundary surface for better quality. Data simplification and smoothing are extremely critical when a reverse engineered model is used in other engineering designs such as finite element analysis and collision detection.

(6) *Feature recognition.* A part model consists of a number of features, which are assembled by logical operators such as addition, subtraction, and union. It is desirable to recognize the features, such as the machined features of holes, bosses, and fillets, from a solid model.

(7) *Parametric modeling and applications.* An RE process aims to recover insights, wisdom, and design intents of an engineering design from existing physical objects. The parametrization of a solid model helps to catch the knowledge and design intents. In addition, the obtained solid model can be utilized for other engineering purposes such as rapid prototyping, mold design, and computer-aided manufacturing.

6.4.4 Computer-Aided Reverse Engineering (CARE)

Many computer-aided tools have been developed for *computer-aided reverse engineering* (CARE), for example, the commercial and open-source CARE tools that are shown in Fig. 6.8. In comparison, commercial CARE tools offer better functionalities, while open-source tools are flexible to be integrated with other computer programs for continuous developments. Many software tools, such as Inventor, Sketchup, and Blender, have commercial and open-source licenses.

A CARE tool is mainly used to accomplish two tasks (1) convert point cloud into a polymesh model and (2) create a watertight solid model, and explore design features and design intents for a parametric model. Here, the Autodesk Recap and

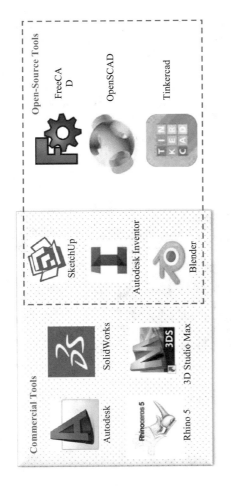

Fig. 6.8 Examples of computer-aided reverse engineering (CARE) tools

the ScanTo3D are used to illustrate how the CARE tools are applied to accomplish these tasks.

6.4.4.1   Creating Polymesh Model

The raw data acquired by the triangulation, time-of-flight, or interferometry method is in the format of point cloud, and a hardware system with such a capability is usually equipped to produce point clouds directly. However, the raw data from photogrammetry is 2D images, some sophisticated tools, such as Autodesk Recap Pro, are required to convert 2D images into polymesh models. As shown in Fig. 6.9a, Recap accepts three types of raw data, point cloud, image files from mobile devices, and photos. The data processing service can be accessed online as shown in Fig. 6.9b; the outcome of image processing is the polymesh model, which can be exported as in the format of .rcs, .obj, .rcm, .fbx, or .ipm. Figure 6.10 shows three steps of using Recap

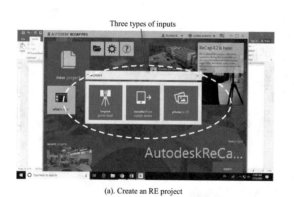

(a). Create an RE project   (b). Interface for cloud-computing

Fig. 6.9   Cloud reverse engineering service by Autodesk Recap Pro

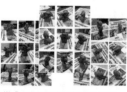

(1). Prepare a number of photos, which are taken from different locations, directions, and heights of the object.

(2). Create a new project in ReCap Pro, select the mesh quality, the export file types, upload the photos, and confirm the submission.

(3). Allow the cloud service to complete surface reconstruction, download a polymesh model in the .rcm format. The generated model can be viewed online and exported to other computer-aided tools.

Fig. 6.10   The steps to create a polymesh model in Recap Pro

(a). Tools for editing a polymesh model

(b). Exporting a polymesh model

Fig. 6.11 Editing and exporting a polymesh model in Recap Pro

Pro to convert a set of image files into a polymesh model. Recap also provides some basic functions to analyze, edit, and refine a polymesh model as shown in Fig. 6.11a before the model is exported in a format of .obj, .stl, or .fbx (see Fig. 6.11b). Create an object from photos

The quality of the polymesh model depends on the completeness and resolution of photos. In photogrammetry, the location and orientation of a camera are determined by matching pixels in a group of photos. Ideally, 80% of the pixels of one photo should be covered by other photos. The following guides should be applied to prepare high-quality photos for a reverse engineering project (Autodesk 2020):

(1) **Lighting conditions**. Lights should be placed to avoid shadows in photos. In an indoor environment, flash should not be used, and diffuse lights should be used without a shadow on object. In an outdoor environment, photos should not be taken under direct lights to avoid a strong contrast in photos.

(2) **Cameras**. Use a lens that produces a good sharpness and avoid the focal change when the photos are taken. Stabilize the camera using support and clicker, expand the wide depth of field to reduce amount of light, and use a camera with a high optical resolution as possible.

(3) **Photoing strategies**. Place the object completely in the view window but let the object occupy the window as large as possible. Avoid a background with complex textures rather than monochromatic ones. Avoid moving any objects in the scene in catching photos. Take photos from all directions of object to ensure a feature of interest is covered by multiple photos. At the height, take the photos around the object at each of 5–15°.

6.4.4.2   Generate Parametric Models

A polymesh model only includes the geometric information, while the ultimate goal of a reverse engineering process is to explore features, intents, and knowledge of physical products. Therefore, polymesh models should be further analyzed and processed to create parametric models with identified parameters, features, and design intents. Figure 6.12 shows the procedure where *the SolidWorks ScanTo3D* was used to create

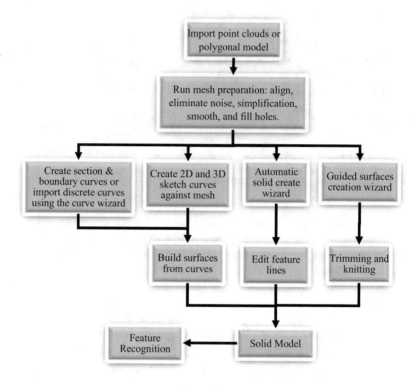

Fig. 6.12 Parametrize a solid model using Solidworks ScanTo3D

a parametric model from a point cloud or polymesh model. *Firstly*, a data file is imported as an application of ScanTo3D. *Secondly*, the 'mesh preparation' tool is used to correct, simplify, align, and smooth data; if no watertight boundary surface is available, further trim extra faces and fill holes. *Thirdly*, the parametric modeling tools are used to define sections, edges, surfaces, and generate a legal boundary surface for a solid model. *Fourthly*, the 'feature recognition' tool is used to identify possible features such as holes, cylinders, extrusions, and fillets.

6.5 Example of Digital Engineering—Direct Manufacturing

Direct manufacturing provides an alternative to make unique products or products with a low volume. Direct manufacturing has been identified as one of the emerging digital technologies that will have a great impact on manufacturing industry. *Direct manufacturing* is a type of additive manufacturing in which physical products are made directly from their digital models without custom manufacturing tools.

Figure 6.13 shows three basic types of fabrication processes in creating geometric shapes of parts. *A formative process* in Fig. 6.13c uses the cavity in a mold assembly to define part geometry. When the material is added into the cavity, it is solidified or sintered into a desired shape. *A subtractive manufacturing process* in Fig. 6.13a cuts away unwanted materials to reshape a part by the cutting tool that moves along the part; the part geometry is determined by the cutting tool and its motion with respect to the part. A subtractive process is performed at a machine tool such as a lath, drill, or mill. *An additive manufacturing* (AM) *process* in Fig. 6.13a makes a part by adding the materials to a solid layer by layer gradually. A machine and programming tool of additive manufacturing processes can be used to make any geometries of parts without using customized tools.

Many technologies were developed for additive manufacturing, and Fig. 6.14 shows that there are seven basic types of additive manufacturing: vat photo-polymerization, powder bed fusion, binder jetting, material jetting, sheet lamination, material extrusion, and directed energy deposition. The characteristics of these technologies have been briefly explained in the figure.

A power bed fusion process in Fig. 6.15 is used as an example to illustrate the additive manufacturing approach. The power particle is supplied by a supply mechanism on the left layer by layer. Powder particles at each layer are fused and melted at a high degree close to the melting point of power materials, and the thermal energy is typically from a laser or electron beam. The portion of the heated powder particles are sintered and fused as the bonded solid over the part. Once the additive process is finished, unheated and loose powder particles can be blown or blasted away. The

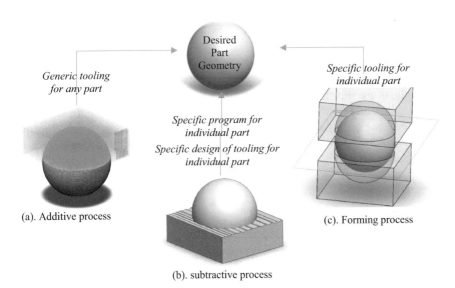

(a). Additive process

(b). subtractive process

(c). Forming process

Fig. 6.13 Three basic methods to fabricate parts in manufacturing

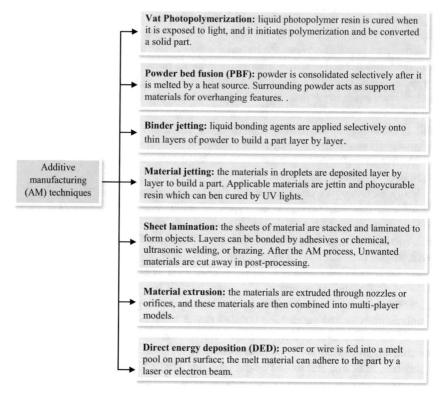

Vat Photopolymerization: liquid photopolymer resin is cured when it is exposed to light, and it initiates polymerization and be converted a solid part.

Powder bed fusion (PBF): powder is consolidated selectively after it is melted by a heat source. Surrounding powder acts as support materials for overhanging features. .

Binder jetting: liquid bonding agents are applied selectively onto thin layers of powder to build a part layer by layer.

Additive manufacturing (AM) techniques

Material jetting: the materials in droplets are deposited layer by layer to build a part. Applicable materials are jettin and phoycurable resin which can ben cured by UV lights.

Sheet lamination: the sheets of material are stacked and laminated to form objects. Layers can be bonded by adhesives or chemical, ultrasonic welding, or brazing. After the AM process, Unwanted materials are cut away in post-processing.

Material extrusion: the materials are extruded through nozzles or orifices, and these materials are then combined into multi-player models.

Direct energy deposition (DED): poser or wire is fed into a melt pool on part surface; the melt material can adhere to the part by a laser or electron beam.

Fig. 6.14 Types of additive manufacturing

materials that are suitable to a power bed fusion process include metal, metal alloys, plastics, ceramic powders, and sand.

Despite of the fact that additive manufacturing techniques are highly diversified, the procedure of making products by direct manufacturing is straightforward and Fig. 6.16 shows seven main steps of a direct manufacturing process. In addition, the hardware and software resources in accomplishing the tasks in these steps are also introduced in the figure.

6.5.1   Prepare Digital Models

Direct manufacturing begins with a virtual model of product. A virtual model may be for a newly designed product from scratch or an existing product for reverse engineering. If the target is an existing physical object, reverse engineering techniques in Sect. 6.4 are used to collect data from the physical object, process data, and construct it into a sold model. If the target is a new product, it is assumed that the design is detailed

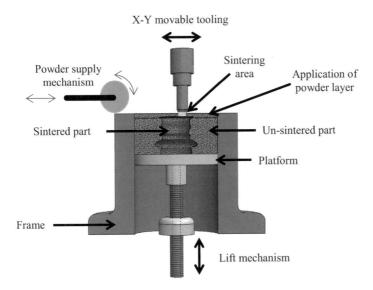

Fig. 6.15 Powder bed fusion (PBF) process

enough to determine geometric shapes of products. In addition, design for manufacturability has to be considered since not all of the features, materials, and functions can be made with the required accuracy appropriately. Many computer-aided and manufacturing tools such as commercial SolidWorks, SolidEdge, Unigraphics, and free computer-aided packages such as FreeCAD, 3D Builder, and LibreCAD can be used to create virtual models of products.

6.5.2 Prepare STL Files

A rapid prototyping machine accepts a product model only in the tessellated format. Tessellation is a method to manage polygon datasets, and the .stl format is special tessellated type for rapid prototyping. The abbreviation of '.stl' stands for *stereolithography*, *standard triangle language*, and *standard tessellation language*. A .stl file can be '*binary*' or '*ASCII*'. A binary file is more concise and commonly used unless the virtual model has to be reviewed and changed manually. Note that a virtual model in .stl includes the information of boundary surfaces of an object only.

A .stl file should be exported only when the product design is finalized since different formats correspond to different contents even though the geometric information is the same. A .stl file cannot be changed easily. Figure 6.17 shows a comparison of a SolidWorks .sldprt file and .stl file for a part with the same geometry. Figure 6.17a is parametrized in which both of the 2D sketch and the revolving axis can be changed easily at the feature level. However, the .stl file in Fig. 6.17b is a triangulated surface model; it only includes a list of triangles and associated vertices.

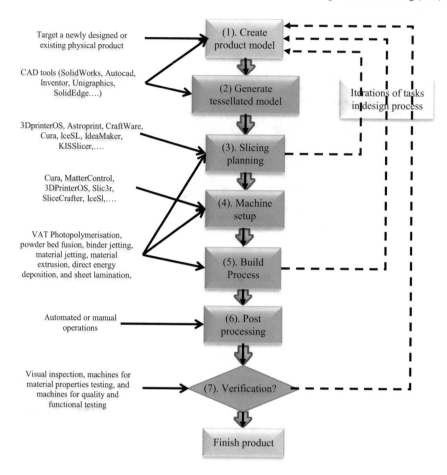

Fig. 6.16 Procedure of AM processes

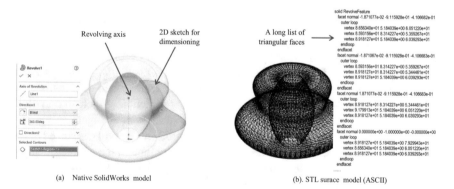

Fig. 6.17 The same geometry with different contents in .sldprt and.stl formats

Vertices are specified by the coordinates in a Cartesian system. Before the model is ready for a rapid prototyping process, engineers are encouraged to preserve as the high-level information (such as features and parameters) as possible to improve designs in an iterative process.

Due to the importance and popularity of .stl format to additive manufacturing, nearly all CAD packages are capable of exporting native solid models as the models in a .stl format. As an example, Table 6.1 shows the interface, the steps, and the settings in SolidWorks when a .sldprt model is exported as a .stl is sile.

When a product is involved in some curvy features, the resolution determines the smoothness of curvy features unless the resolution of a .stl file exceeds that of the rapid prototyping process. The resolution also affects the file size of a virtual model. Figure 6.18 gives an example of a solid model as well as two .stl files with coarse

Table 6.1 The interface, steps, and settings to export a.stl file from SolidWorks

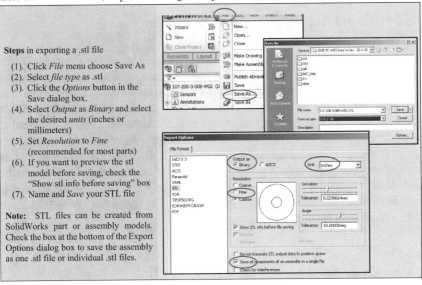

Steps in exporting a .stl file

(1). Click *File* menu choose Save As
(2). Select *file type* as .stl
(3). Click the *Options* button in the Save dialog box.
(4). Select *Output* as *Binary* and select the desired *units* (inches or millimeters)
(5). Set *Resolution* to *Fine* (recommended for most parts)
(6). If you want to preview the stl model before saving, check the "Show stl info before saving" box
(7). Name and *Save* your STL file

Note: STL files can be created from SolidWorks part or assembly models. Check the box at the bottom of the Export Options dialog box to save the assembly as one .stl file or individual .stl files.

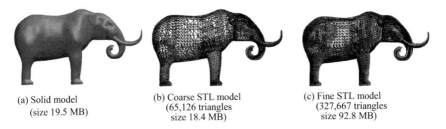

(a) Solid model
(size 19.5 MB)

(b) Coarse STL model
(65,126 triangles
size 18.4 MB)

(c) Fine STL model
(327,667 triangles
size 92.8 MB)

Fig. 6.18 Resolution affects the smoothness of curvy features and file sizes

Fig. 6.19 Visualize a rapid prototyping process in Simplify3D

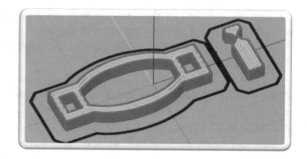

and fine resolutions. The file size can be increased greatly with an increase of the number of triangles in the model.

6.5.3 Slicing Algorithms and Visualization

A rapid prototyping process builds a part layer by layer. A slicing algorithm is used to decompose a volume into a set of layers with a specified resolution. At each layer, the algorithm determines the moving paths to cover the cross-sectional area. The slicing algorithm generates the results of (1) a set of toolpaths for the given .stl model, (2) estimated processing time and material consumption based on the specified percentage of infills, and (3) the constructions of supporting materials when one or a few of features on part are not self-supported (All3DP 2019). It is desirable that the results from slicing algorithms are visualized and reviewed graphically to ensure that the toolpaths or supports are reasonable. As shown in Fig. 6.19, many software tools, such as Simplify3D, can be used to visualize printing paths and layered structures from a rapid prototyping process.

6.5.4 Setup Machine

The quality of a rapid prototyped product is determined not only based on the tool-paths from a slicing algorithm, but also on various design factors relevant to an additive manufacturing process such as machine type, material type, and process parameters. A control program has to be tailored to certain RP machine. Figure 6.20 shows a list of major design parameters involved in a *fused deposition modeling* (FDM) process (Ha 2016). Even though the RP machine can be run to generate a control program using the default settings with few manual intervenes, the optimized quality of part relies on users' understanding on various process parameters and the experience in setting up the machines.

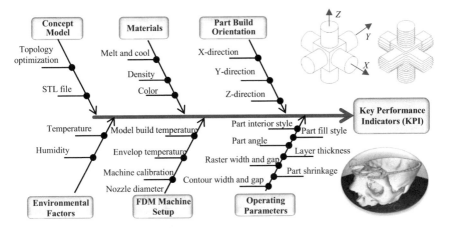

Fig. 6.20 Main design factors of a rapid prototyping process (Bi and Wang 2020)

6.5.5 Building Process

A control program is a set of instructions to prepare for tooling, move and follow a toolpath for operating, and operate auxiliary tools based on the specified settings of the process parameters. After the slicing algorithm processes the .stl file and generates a control program, the RP machine is calibrated and setup appropriately, and the rapid prototyping process can be performed to build a part layer by layer. Depending on the part geometry, the operating speed of end-effector tools, and the effectiveness of the slicing algorithm, the machining time varies from one part to another.

6.5.6 Post-processing

When the building process is terminated, the part can be taken from the rapid prototyping machine. The post-processing is often needed to clean debris and remove a supportive structure from the part. Occasionally, secondary processes such as machining are used to create some machined features or improve geometric and dimensioning tolerances on existing features.

Rapid prototyping offers some significant advantages in reducing the cost of tooling and shortening the time of product development. However, RPs have their disadvantages on quality and strengths of products due to (1) the applicable raw materials having limited strengths and (2) staircase effects in a layer-by-layer operation. Table 6.2 shows the ranges of surface quality and minimum possible thickness from different additive manufacturing. The surface roughness could be as poor as (9–40 μm) from fused deposition modeling (Campbell et al. 2002; Kumbhar and Mulay 2018).

Table 6.2 Surface roughness from different AM processes (Campbell et al. 2002; Kumbhar and Mulay 2018)

No.	Process type	Minimum layer thickness (mm)	Surface roughness (Ra) in μm
1	Stereolithography (SLA)	0.100	2 ~ 40
2	Selective laser sintering (SLS)	0.125	5 ~ 35
3	Fused deposition modelling (FDM)	0.254	9 ~ 40
4	Material extrusion (3D printing)	0.175	12 ~ 27
5	Laminated object manufacturing (LOM)	0.114	6 ~ 27
6	Material jetting	0.100	3 ~ 30

The surface finish of a RP part may be improved in two ways: (1) optimizing orientation of toolpath and reducing thickness between layers and (2) applying secondary processes such as milling, laser surface finishing operations, or abrasive flow machining.

6.5.7 Verification and Validation

Most of the rapid prototyped parts are for verification and validation purpose. As the last step, physical parts are tested and evaluated to see if continuous improvement is necessary. An iterative process applies until the performance of the prototyped part meets the design requirements satisfactorily. Simple validation such as part defects can be manually inspected; some quantifications such as surface roughness, hardness, and tensile strengths need sophisticated testing machines. Recent efforts on verification and validation were emphasized on micro-level structures, residual stresses, fatigue strengths, and mechanical properties (Kim et al. 2015).

6.6 Studies in Customizing Digital Manufacturing Technologies

There is no universal solution of DT, which is applicable to any manufacturing enterprises. Some customized efforts of research and development are usually required to tailor general digital manufacturing solutions to certain applications. In this section, some case studies of the research and development are introduced to show the diversity of the applied studies in digital manufacturing.

6.6.1 Ubiquitous Sensing

Digital manufacturing is data driven, and the smartness and intelligence of the manufacturing system rely on the abundance and sufficiency of data collected from system

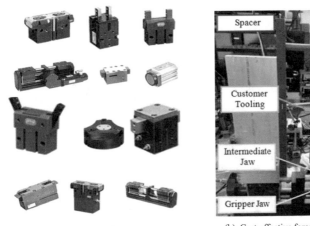

(a). various grippers   (b). Cost-effective force sensing

Fig. 6.21   Force sensors for smart grippers (Bi et al. 2018)

elements. From this aspect, more sensors are used in a manufacturing system, the better chance the right data is collected, and the faster and better decisions can be made for the operations of manufacturing systems. A large amount of research and development is needed to have cost-effective instrumenting solutions for smart things in manufacturing systems.

Figure 6.21 shows an example of developing smart force sensors for robotic grippers. As smart things, industrial robots are widely applied in manufacturing. However, the controls of industrial robots are mostly based on kinematic models, and this poses the challenge to (1) optimize the performance of a robot since any motion is caused by force and (2) use them in complex environment such as a human-machine coexistence environment. To implement dynamic control for robots, the external forces occurring to system elements should be measured in real time and the instrumenting solutions are demanded for robotic modules such as grippers in Fig. 6.21a. A cost-effective sensing solution was developed to measure gripping forces, and the collected data can be utilized to support the human-robot interaction in open working environment.

6.6.2   Holistic Multi-sensing Solution for Real-Time Controls

Many manufacturing operations require the coordination, collaboration, and cooperation of multiple system elements. Taking an example of an automated guided vehicle (AGVs) system in an assembly plant, the information of delivering tasks, the states of AGVs, jobs, and relevant manufacturing resources are collected and fed to the control system as inputs to plan, schedule, and control AGVs. The system may

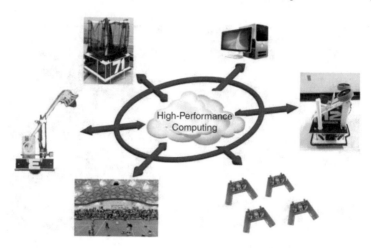

Fig. 6.22 Holistic sensing solutions for real-time decision-making supports (Bi et al. 2017)

be complicated when the controls are decentralized since system elements coordinate, collaborate, and cooperate with each other dynamically. In addition, the data about relevant environmental factors should be collected and made available to these system elements, and the real-time data will be fused, processed, and utilized to support decision-making activities. Such types of research are highly demanded in distributed and decentralized manufacturing systems; enterprise systems should be able to collect environmental data from multiple sources, fuse, process, and utilize data to achieve a high-level intelligence of decision-making units.

Figure 6.22 shows an example of decentralized control systems in which system components coordinated and collaborated to accomplish respective tasks for optimized system performances. It was a football robot team consisting of different types of robots equipped with various sensors including range sensors, vision sensors, and force sensors. With the real-time data collected from robots, robot members coordinated and collaborated at system level. In developing such systems, engineers need to select right sensors to detect events or measure physical quantities, and the information systems should be able to utilize data for decision-making.

6.6.3 Methods of Coping with Big Data

To shorten the development time, virtual design must deal with the ever-increasing volume, velocity, and variety of big data. In general, a big data means that the required computation to an engineering problem is far beyond the capability of available computing resources. With an increase of system complexity, more and more engineering problems face the challenges of limited computing resources; data analysis

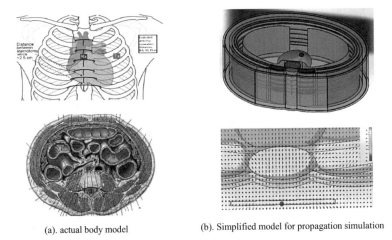

(a). actual body model   (b). Simplified model for propagation simulation

Fig. 6.23 Developing of cost-effective solution for ubiquitous sensing (Särestöniemi et al. 2019)

must be innovative to narrow down big data into the dataset with a manageable size, so that a feasible solution can be found in a reasonable timeframe.

Figure 6.23 gives an example of using an innovative approach to deal with the big data in simulating the signal propagation in human body (Särestöniemi et al. 2019). A human body is formed by skin, fat, bones, and sterns, and these organic matters have different dielectric properties that affect signal propagating. Using a realistic body model shown in Fig. 6.23a leads to the scenario of big data in numerical simulation. However, the models should be greatly simplified by approximating the volumes of different material properties as parametrized bodies. Accordingly, the computing time for an acceptable solution was reduced significantly.

6.6.4 Methods of Data Mining

In an IoT-based system network, numerous smart things have their local control units to optimize the performances. However, the coordination and cooperation of these functional units are required to optimize system-level objectives. System-level decisions are made based on the real-time data collected from smart things. However, certain data may be relevant or irrelevant, and interpretable or uninterpretable to a system-level objective such as variety, quality, lead time, and cost. Effective methods are demanded to sort and process data and make a useful dataset manageable in decision-making.

Figure 6.24 shows that the axiomatic design theory (ADT) is used to process the data collected from low-level functional models and map the data into the category related to system-level objectives to make bid data manageable (Cochran et al. 2017).

Fig. 6.24 Axiomatic design theory (ADT) for data mining (Cochran et al. 2017)

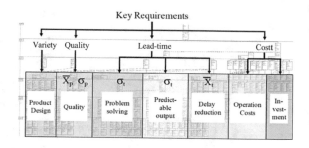

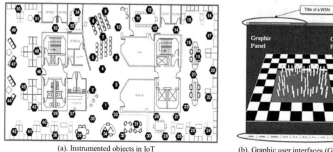

(a). Instrumented objects in IoT   (b). Graphic user interfaces (GUIs) for data visualization

Fig. 6.25 Data visualization of human-machine interaction (Bi et al. 2016)

6.6.5 Data Visualization Methods for Human-System Interaction

Humans still play important roles in digital manufacturing; in particular, humans are usually needed to oversee the operations of manufacturing systems in identifying abnormal situations and taking immediate actions. Effective interactive tools are needed for humans to understand the system states using the real-time data collected from manufacturing plants. Figure 6.25 shows an integrated surveillance system, which is capable of accessing, retrieving, and visualizing real-time data based on the selection criteria by users.

6.6.6 Data-Driven Decision-Making Units

Conventionally, the decision-making units at one-level of system architecture get the inputs from a upper level unit and send the outputs to lower level unit(s). The decision-making units do not require the flexibility to deal with the changes and uncertainties beyond fixed interactions. However, IoT allows the direct interactions of functional units, and the capabilities of these functional units should be expanded in terms of flexibility and adaptability to deal with changes and uncertainties. Figure 6.26

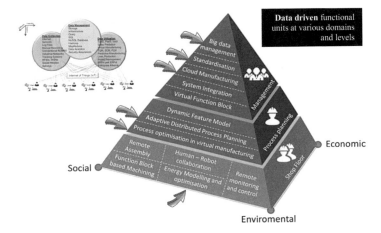

Fig. 6.26 Data-driven functional units (Bi and Wang 2013)

describes a scenario of an intelligent system, in which all of functional units are data driven, so that a functional unit in the system can directly interact and respond to another unit at any level and domain (Bi and Wang 2013).

6.6.7 Methods for Workflow Compositions

IoT brings the possibilities for small- and medium-sized companies to utilize external manufacturing resources to catch emerging business opportunities. A virtual enterprise can be formed to execute an engineering project. However, a virtual enterprise differs from physical enterprises in sense that manufacturing resources are across the boundaries of physical enterprises, and additional businesses are involved in selecting partners to configure workflows; executing engineering projects collectively; and ensuring the privacy, security, safety, and reliability of business operations. Figure 6.27 shows a research example on the algorithms for the selection and composition of services for virtual enterprises.

6.6.8 Standardization of Specifications

In the material flow of a manufacturing system, the complexity of products or manufacturing processes is mainly tackled by selecting different types of manufacturing resources and assembling them in different ways. Manufacturing resources are selected if their capabilities satisfy the functional requirements in applications. It becomes critical to standardize the measures and descriptions of manufacturing resources, so that a manufacturing system can be configured optimally. As shown

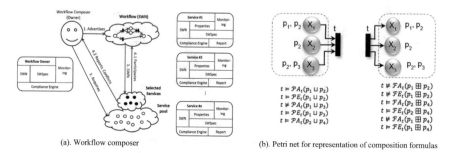

(a). Workflow composer   (b). Petri net for representation of composition formulas

Fig. 6.27   Service selection and composition in virtual enterprise (Viriyasitavat et al. 2019a, b, c)

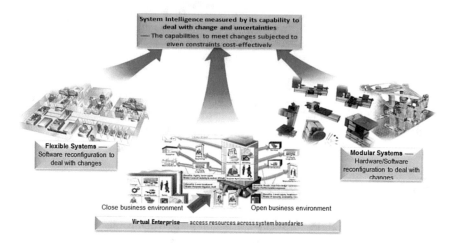

Fig. 6.28   Standardization of functional requirements of robotic modules (Bi et al. 2020)

in Fig. 6.28, Bi et al. (2020) discussed the challenges in standardizing functional requirements of robotic systems in testing and measurement.

6.7   Summary

The idea of digital twin (DT) I discussed to have the first-time-right mapping from a virtual twin and physical twin. DT emphasizes the mirror relation of a digital twin and physical twin. However, the chapter focuses on the digital technologies to narrow down the solutions of digital twins in vast design space. To this end, the functional requirements of DT are discussed at all main aspects of system decision-making, a free architecture is proposed, and a number of emerging digital technologies have been discussed. From this chapter, engineers are expected to know how to (1) use RE in digitizing physical objects; (2) use DM to create physical objects directly based on digital models; and 3) understand main areas of applied researches in digital manufacturing.

Design Projects

Project 6.1. Find a malfunctioned machine or machine elements you can reach such as the broken dehumidifiers shown in Fig. 6.29, disassembly the machine to investigate the parts and the assembly, and understand how the machine should function, then create a digital model for the machine, demonstrate how the mechanical system is supposed to work, and make your recommendation how the machine could be fixed.

Project 6.2. Find a physical object you can reach, such as some decorative objects shown in Fig. 6.30; create its solid model by using reverse engineering technique; determine its physical properties such as mass, volume, and inertial properties; and run feature recognition on the solid model to see if some parameters and features can be recovered from constructed model. The following steps can be used as the guide in using a cellphone or digital camera to collect data and using Autodesk Recap Pro and Solidworks ScanTo3D to reconstruct a solid model and feature recognition.

(1) Select a physical object to be modeled in your RE project. Place an object on a flat surface. Make sure the object can be taken photos from all sides and angles.
(2) Use a cellphone or digital camera to take over 50 photos from different positions and angles.
(3) Create a user account at https://www.autodesk.com, download and install ReCap Pro, create an RE project under in ReCap Pro, upload all photos, convert them into a mesh file, download .rcm file, and use ReCap to export .rcm file as an .obj file or .stl file for its use in Solidworks.

Fig. 6.29 Examples of used products in design project 6.1

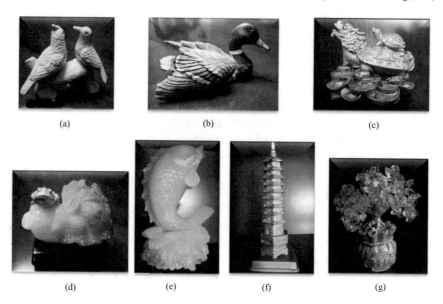

Fig. 6.30 Examples of physical objects in design Project 6.2

(4)   Use Solidworks—ScanTo3D to clean the imported mesh file, fix the mesh, create a surface model, fix the surface model, and generate a solid model from the watertight surface model.

(5)   For the solid model, assign the materials and evaluate mass and volume properties.

(6)   Document your project idea, procedure, the challenges and solution, and result of reverse engineering.

(7)   3D print a physical model using the solid model.

References

All3DP (2019) Want to get the best results from your 3D printer? https://all3dp.com/1/best-3d-sli cer-software-3d-printer/#what

Autodesk (2020) What makes photo good for photogrammetry? https://blogs.autodesk.com/recap/ what-makes-photos-good-for-photogrammetry/

Berger R (2019) Digital factories, the renaissance of the U.S. automotive industry. http://secure.rbj. net/collaborative-design-and-planning-for-digital-manufacturing-1st-edition.pdf

Berman SJ, Bell R (2007) Digital transformation—creating new business models where digital meets physical. https://www-07.ibm.com/sg/manufacturing/pdf/manufacturing/Digital-transformation.pdf

Bi ZM (2011) Revisiting system paradigms from the viewpoint of manufacturing sustainability. J Sustain 3(9):1323–1340

Bi ZM, Kang B (2014) Sensing and responding to the changes of geometric surfaces in flexible manufacturing and assembly. Enterprise Inf Syst 8(2):225–245

Bi ZM, Liu YF, Krider J, Buckland J, Whiteman A, Beachy D, Smitch J (2018) Real-time force monitoring of smart grippers for Internet of Things (IoT) applications. J Indus Inf Integration 11:19–28

Bi ZM, Miao ZH, Zhang B, Zhang WJ (2020) A framework for performance assessment of heterogeneous robotic systems. IEEE Syst J. https://doi.org/10.1109/JSYST.2020.2990892

Bi ZM, Wang G, Xu L (2016) A visualization platform for internet of things in manufacturing applications. Internet Res 26(2):377–401

Bi ZM, Wang G, Xu LD, Thompson M, Mir R, Nyikos J, Mane A, Witte C, Sidwell C (2017) IoT-based system for communication and coordination of football robot team. Internet Res 27(3):162–181

Bi ZM, Wang L (2013) Manufacturing paradigm shift towards better sustainability in cloud manufacturing. Springer, London. ISBN: 978-1-4471-4934-7, http://doi.org/10.1007/978-1-4471-4935-4_5

Bi ZM, Wang XQ (2020) Computer aided design and manufacturing (CAD/CAM). John Wiley & Sons, Inc, Hoboken, NJ USA. IBSN-13:9781119534211

Buttner R, Muller E (2018) Changeability of manufacturing companies in the context of digitalization. Procedia Manuf 17(2018):539–546

Campbell RI, Martorelli M, Lee HS (2002) Surface roughness visualization for rapid prototyping models. Comput Aided Des 34(2002):717–725

Catapult (2018) Feasibility of an immersive digital twin: the definition of a digital twin and discussions around the benefit of immersion. https://www.amrc.co.uk/files/document/219/1536919984_HVM_CATAPULT_DIGITAL_TWIN_DL.pdf

Cochra D, Ki YS, Fole J, Bi ZM (2017) Use of the manufacturing system design decomposition for comparative analysis and effective design of production systems. Int J Product Res 55(3):870–890

Fenner J (2003) Enterprise application integration technique. http://www-icm.cs.ucl.ac.uk/staff/W.Emmerich/lectures/3C05-02-03/aswe21-essay.pdf

GE (2020) GE digital twin—analytic engine for the digital power plant, https://www.ge.com/digital/sites/default/files/download_assets/Digital-Twin-for-the-digital-power-plant-.pdf

Geieves M, Vickers J (2006) Digital twin: mitigating unpredictable, undesirable emergent behavior in complex systems (excerpt). https://www.researchgate.net/publication/307509727_Origins_of_the_Digital_Twin_Concept/link/57c6f44008ae9d64047e92b4/download

Gelernter DH (1991) Mirror worlds: or the day software puts the universe in a shoebox—how it will happen and what it will mean. Oxford University Press, Oxford, New York. ISBN 978-0195079067. OCLC 23868481

Glaessgen EH, Stargel DS (2012) The digital twin paradigm for future NSA and US air force vehicles. https://ntrs.nasa.gov/archive/nasa/casi.ntrs.nasa.gov/20120008178.pdf

Gregori F, Papetti A, Pandolfi M, Peruzzini M, Germani M (2017) Digital manufacturing systems: a framework to improve social sustainability of a production site. Procedia CIRP 63(2017):436–442

Ha S (2016) 3D printing/process parameters. https://worldmaterialsforum.com/files/Presentations/WS1-1/WMF%202016%20-%20WS%201.1%20-%20Sung%20Ha%20Final.pdf

Kim DB, Witherell P, Lipman R, Feng SC (2015) Streamlining the additive manufacturing digital spectrum: a systems approach. Additive Manuf 5(2015):20–30

Kumbhar NN, Mulay AV (2018) Post processing methods used to improve surface finish of products while are manufacturing by additive manufacturing technologies: a review. J Instit Eng (India) Ser C 99(4):148–487

Kusiak A (2019) Fundamentals of smart manufacturing: a multi-thread perspective. Annual Rev Control. https://doi.org/10.1016/j.arcontrol.2019.02.001

Mäkiö-Marusik E, Colomboa AW, Mäkiö J, Pechmann A (2019) Concept and case study for teaching and learning industrial digitalization. Procedia Manuf 31(2019):97–102

Parrott and Warshaw (2020) Industry 4.0 and the digital twin. https://www2.deloitte.com/us/en/insights/focus/industry-4-0/digital-twin-technology-smart-factory.html

Pati A, Bandyopadhyay PK (2017) Digital manufacturing: evolution and a process oriented approach to align with business strategy. Int J Econ Manag Eng 11(7):1746–1751

Särestöniemi M, Pomalaza-Ráez C, Bi ZM, Kumpuniemi T, Kissi C, Sonkki M, Hämäläinen M, Iinatti J (2019) Comprehensive study on the impact of sternotomy wires on UWB WBAN channel characteristics on the human chest area. IEEE Access 7(1):74670–74682

Siemens (2020) Digital manufacturing—a holistic approach to the complete product lifecycle. https://www.plm.automation.siemens.com/global/en/our-story/glossary/digital-manufacturing/13157

Sönmez OE, Koç VT (2015) On quantifying manufacturing flexibility: an entropy based approach. In: Proceedings of the world congress on engineering 2015 Vol II WCE 2015, July 1–3 2015. London

Tao F, Zhang M, Nee AYC (2019) Digital twin and big data in digital twin driven smart manufacturing, pp 183–202

Tuegel EJ, Ingraffea AR, Eason TG, Spottswood (2011) Reengineering aircraft structural life prediction using a digital twin. Int J Aerospace Eng. Article ID 154798

Viriyasitavat W, Xu L, Bi ZM, Hoonsopon D (2019a) Blockchain technology for applications in internet of things—mapping from system design perspective. IEEE Internet of Things J 6(5):8155–8168

Viriyasitavat W, Xu L, Bi ZM (2019b) rmSWSpec: real-time monitoring of service workflow specification language for specification patterns. IEEE Trans Indus Inf 15(7):4021–4032

Viriyasitavat W, Xu L, Bi ZM (2019c) The extension of semantic formalization of service workflow specification language. IEEE Trans Indus Inf 15(2):741–754

Wiendahl HP, ElMaraghy HA, Nyhuis P, Zäh MF, Wiendahl HH, Duffie N, Brieke M (2007) Changeable manufacturing - classification, design and operation. CIRP Ann 56(2):783–809

Williams TJ (1994) The Purdue enterprise reference architecture and methodology (PERA). http://citeseerx.ist.psu.edu/viewdoc/download?doi=10.1.1.194.6112&rep=rep1&type=pdf

Wikipedia (2020) Digital twin. https://en.wikipedia.org/wiki/Digital_twin

Printed in the United States
by Baker & Taylor Publisher Services

By Corollary 4, any edge 3-admissible graph has vertices of degree 2 and 3 in each induced C_4's and K_4, resp. Hence, Construction 2 presents a way to break C_4's and K_4's into P_5's and K_3's, resp., in order to present a stronger necessary condition in Lemma 4.

Construction 2 *Let G be a graph that satisfies: G does not have induced C_k nor K_k, for $k \geq 5$, as induced subgraphs; for each induced C_4 there is a vertex of degree two in G; and for each induced K_4 there is a vertex of degree three in G. We construct a graph H from G as follows:*

1. *each induced $C_4 = a, b, c, d, a$, for $d_G(a) = 2$, is transformed into a $P_5 = a, b, c, d, a'$ by adding a new vertex a' and the edge da', and removing the edge da;*
2. *each induced $K_4 = \{a, b, c, d\}$, for $d_G(a) = 3$, is transformed into three complete graphs K_3 by adding a new vertex a' and: removing edge ba; adding edges ba' and ca'.*

Lemma 3 *A graph G is edge 3-admissible if and only if the graph H from Construction 2 is edge 3-admissible.*

Proof If G is edge 3-admissible, then all edges of an edge tree 3-spanner of G are used to obtain a spanning tree of H and we do not increase the edge stretch index from G to H, because, by construction, we are not increasing a maximum path between any two adjacent vertices of G in H. If H is edge 3-admissible, then all edges of an edge tree 3-spanner of H are used for a spanning tree of G and, since we are identifying vertices that belong only to C_4's or K_4's in G, such identification does not affect cycles that give the edge tree 3-spanner of H and does not increase such index of G by the used edges of H. □

A *k-tree* is a graph obtained from a K_{k+1} by repeatedly adding vertices in such a way that each added vertex v has exactly k neighbors defining a clique of size $k + 1$. A *partial k-tree* is a subgraph of a k-tree [9].

Lemma 4 *Let G be an edge 3-admissible graph. If H is the graph obtained from G in Construction 2, then H is a chordal partial 2-tree graph.*

Proof If G is edge 3-admissible with $X \in \{C_4, K_4\}$ as an induced subgraph, then, by Corollary 4, X must have at least one vertex a such that $N(a) \subseteq X$. Based on that, in Construction 2 we obtain a graph without C_4's nor K_4's. Since, by Lemma 3, the transformed graph H from an edge 3-admissible graph G is also edge 3-admissible, we have that the length of any clique is at most 3 and it does not have C_k, for $k \geq 4$. Since chordal graphs with maximum clique of length 3 are partial 2-tree [9], we have that H is a chordal partial 2-tree graph. □

By Lemma 4, edge tree 3-spanner graphs are formed by 2-trees where either an edge or a vertex connects two 2-trees. Hence, for the former case such edge is a bridge and for the later case it is a cut vertex of the graph. Lemmas 5 and 6 present conditions that force spanning trees correspond to edge 3-admissible graphs.

Lemma 5 *Given an edge 3-admissible graph G and two 2-trees A_1 and A_2 connected by a bridge uv, such that $|V(A_i)| > 3$ for $i \in \{1, 2\}$, then for any edge 3-spanner T, uv is a pendant vertex in $T[A_1 \cup \{u, v\}]$, i.e. $d_{T[A_1 \cup \{u,v\}]}(uv) = 1$.*

Proof Assume $u \in A_1, u, x, y$ is a triangle and $v \in A_2$. Suppose $d_{T[A_1 \cup \{u,v\}]}(uv) \geq 2$, hence xy must be adjacent to either ux or to uy in T. W.l.o.g., let xy be adjacent to uy, then there is an edge wx in A_1 which implies the distance between wx and xy to be equal to 4 by a path through uv, a contradiction. $\square$

Each bridge forces a unique way to obtain an edge tree 3-spanner of G. Hence, by Lemma 5, assume G is 2-*edge connected*, i.e. there is not a bridge in G. Otherwise, we consider each connected component separately after the bridges removal of G.

Now, consider the case that G has a cut vertex. Let a *windmill graph* $Wd(3, n)$ be the graph constructed for $n \geq 2$ by identifying n copies of K_3 at a universal vertex. Since an edge 3-admissible graph is partial 2-tree, we have that if there is a cut vertex u in G, then $G[N_G[u]]$ contains a windmill graph $Wd(3, d)$, for $2 \leq d \leq \frac{d_G(u)}{2}$. Let a *diamond graph* be a K_4 minus an edge. Each K_3 of a windmill centered in u has two vertices of degree 2, or it has a cut vertex of G distinct of u, or it belongs to a diamond graph of G.

Lemma 6 *Let G be 2-edge connected graph with a cut vertex u and edge 3-admissible. If the associated windmill graph $Wd(3, n)$ centered in u satisfies $n \geq 3$, then u belongs to at most 2 diamonds in G.*

Proof Assume that u is center of the windmill graph $Wd(3, 3)$ and it belongs to 3 diamonds D_1, D_2 and D_3 in G. We prove that G is not edge 3-admissible, and then it implies that if G is edge 3-admissible, then u does not belong to more than 3 diamonds for every $n \geq 3$, either, because the hereditary property proved in Lemma 2.

Note that $L(H)$, for $H = Wd(3, 3) \cup D_1 \cup D_2 \cup D_3$, is composed by a K_6 and the addition of three other subgraphs, named B_1, B_2 and B_3, constructed by a join between a vertex and a C_4. Moreover, each edge of a perfect matching of the K_6, $\{e_1, e_2, e_3\}$, is identified to an edge of B_1, B_2 and B_3 that belongs to the C_4s, resp. Suppose that $L(H)$ is 3-admissible, hence for any tree 3-spanner T of $L(H)$ we have that $T \cap L(H)$ is a fl-centered bi-star, for f and l being any two K_6's vertices. Since any vertex of the K_6 belongs to exactly one of the other three subgraphs added to it, i.e. each K_6's vertex belongs to either B_1, B_2 or B_3, then at least two adjacent vertices of $L(H)$ are adjacent to leaves of the fl-centered bi-star, implying $\sigma'(H) = 4$. $\square$

If there is a vertex u that belongs to $Wd(3, 2)$ then there are two solutions in $T \cap Wd(3, 2)$, less than isomorphism. Consider a $Wd(3, 2)$ such that $V(Wd(3, 2)) = \{u, v, w, v', w'\}$ such that u, v, w and u, v', w' induce K_3's. Note that an edge tree 3-spanner $T \cap Wd(3, 2)$ can be formed as follows: Case 1: $\{uv, uw\}$, $\{uv, vw\}$, $\{uv, uv'\}$, $\{uv', uw'\}$, $\{uv', v'w'\}$; Case 2: $\{uv, uw\}$, $\{uv, vw\}$, $\{uv, uv'\}$, $\{uv, uw'\}$, $\{uv', v'w'\}$. Any other edge tree spanner of $Wd(3, 2)$ is not edge tree 3-spanner.

Although a $Wd(3,2)$ graph centered in u may have two spanning trees, if each triangle also belongs to a diamond, let D_1 and D_2 be such diamonds with vertices $V(D_1) = \{u, v, w, x\}$ and $V(D_1) = \{u, v', w', x'\}$, then the previous Case 1 is the unique edge tree 3-spanner for $T \cap Wd(3,2)$, less than isomorphism.

Furthermore, let $H = Wd(3,2) \cup D_1$ be formed by a $Wd(3,2)$ centered in u with vertices $V(Wd(3,2)) = \{u, v, w, v', w'\}$ such that vw belongs to the diamond D_1 with vertices $V(D_1) = \{v, w, s, t\}$, then we have that H is not edge 3-admissible, which can be verified by conditions above and a simple case analyses.

Hence, we have presented necessary conditions of a 2-edge connected graph G satisfying Construction 2 to be edge 3-admissible when it has a cut vertex.

Now, consider G a biconnected graph. Theorem 2 characterizes such graphs. The *diameter* of a graph G is the greatest distance between any pair of vertices, and is denoted by $D(G)$.

Theorem 2 *Given G a biconnected graph with $D(G) \le 3$. We have that $\sigma'(G) \le 3$ if and only if either there is distance two dominating edge $e_1 = uv$ or for any edges $e_1 = uv$, $e_2 = uw$, and $e_3 \notin N(u) \cup N(v) \cup N(w)$, e_3 is adjacent to edges only of $N(v)$ (or equivalently, only of $N(w)$).*

Proof If G has a dominating edge, for $D(G) \le 3$, then $\sigma'(G) \le 3$ by a uv centered bi-star. Or, if any edge is not dominated by e_1 but it is adjacent to edges only of $N(v)$, then in the solution spanning tree such vertex is adjacent to a leaf of v and it does not turn $\sigma'(G) \ge 4$ because it is not adjacent to leaves of u. Assume that G is edge 3-admissible, there is not a distance two dominating edge and there is an edge e_3, such that $e_3 \notin N(u) \cup N(v) \cup N(w)$ that is adjacent to edges of $N(v)$ and $N(w)$. In this case e_3 is connected to leaves of the two centers of the bi-star in $L(G)$, which implies that T' is not edge 3-admissible, a contradiction. $\square$

Note that Theorem 2 gives another argument on the lower bound of Corollary 3, since a K_n does not satisfy conditions of Theorem 2.

Corollary 5 *Edge 3-admissibility is polynomial-time solvable.*

4 Edge Stretch Index for Split and Generalized Split Graphs

Since $\sigma'(G) \le 4$ for graphs with a distance two dominating edge (Theorem 1), the polynomial time algorithm for edge 3-admissible of Corollary 5 also works for these graphs and their subclasses, such as split graphs, join graphs and P_4-tidy graphs. I.e., we know whether these graphs have $\sigma'(G) = 2$, $\sigma'(G) = 3$ or $\sigma'(G) = 4$.

Corollary 6 *Edge t-admissibility is polynomial-time solvable for split graphs, join graphs and P_4-tidy graphs.*

As presented in Corollary 6, we are able to determine the edge stretch index for split graphs. Split graphs can be generalized as the (k, ℓ)-graphs, which are the

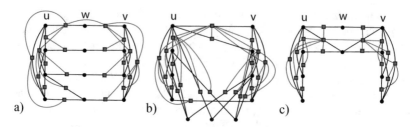

Fig. 5 Cases of $(1, 2)$-graphs and the corresponding edge tree spanners. (**a**) an edge 5-admissible graph. (**b**) and (**c**) are edge 4-admissible graphs

graphs that the vertex set can be partitioned into k stable sets and ℓ cliques. The (k, ℓ)-graphs are also denoted as the generalized split graphs [5].

In [4], the dichotomy P *versus* NP-complete on deciding the stretch index for (k, ℓ)-graphs was partially classified. One of the open problems regarding MSST is to establish the computational complexity for $(1, 2)$-graphs. Next, we prove that the edge stretch index for $(1, 2)$-graphs can be determined in polynomial time.

We denote a $(1, 2)$-graph as a graph $G = (V, E)$ where V is partitioned into $V = \mathcal{K}_1 \cup \mathcal{K}_2 \cup S$, such that each $\mathcal{K}_i$ induces a clique and S is a stable set.

Lemma 7 *If G is a $(1, 2)$-graph, then G is edge 5-admissible.*

Proof Since G is connected, there is a path between a vertex $u \in \mathcal{K}_1$ and $v \in \mathcal{K}_2$ by an edge uv or by a $P_3 = u, w, v$. Figure 5 depicts the cases of $(1, 2)$-graphs and their edge 5-tree spanners. In Fig. 5a there is an induced C_6 by two vertices of each clique and two vertices of S, implying a non-edge in any tree, hence $\sigma'(G) \leq 5$. □

Theorem 3 *A $(1, 2)$-graph $G = (\mathcal{K}_1 \cup \mathcal{K}_2 \cup S, E)$ has $\sigma'(G) \leq 4$ if and only if G has a distance two dominating edge or two adjacent distance two dominating edges that are adjacent to at least one edge of each pair of edges incident to a vertex of S such that one endpoint of an edge of this pair is in $\mathcal{K}_1$ and another one in $\mathcal{K}_2$.*

Proof From Lemma 1, if G has a distance two dominating edge, then G is edge 4-admissible. Moreover, if G has two distance two dominating edges e_1 and e_2 adjacent to at least one edge of each pair of edges incident to a vertex of S such that one endpoint of an edge of this pair is in $\mathcal{K}_1$ and an endpoint of the other edge is in $\mathcal{K}_2$, one obtain an edge tree 4-spanner T of G by selecting any spanning tree of $L(G)$ that maximizes the degrees of these two distance two dominating edges in T.

Conversely, for the sake of contradiction assume that G does not have such distance two dominating edges and T is an edge tree 4-spanner of G. Since G is connected, there is a vertex of S adjacent to both $\mathcal{K}_1$ and $\mathcal{K}_2$ and we can select these two edges of S to be two distance two dominating edges of G. Therefore, for all distance two dominating edges e_1 and e_2 of G we have two edges e_i and e_f incident to a vertex of S such that these edges are both not adjacent to e_1 and e_2. Therefore, in the best case scenario these two edges are adjacent to edges e'_1 and e'_2

adjacent to e_1 and e_2. However, we have a path in T e_i e'_1 e_1 e_2 e'_2 e_f with these two edges e_i and e_f sharing an endpoint, which implies that T is not an edge 4-tree spanner of G. □

Corollary 7 *Edge t-admissibility is polynomial-time solvable for* $(1, 2)$*-graphs.*

5   Edge 8-Admissibility Is **NP**-Complete for Bipartite Graphs

Next, we present a polynomial time transformation from 3-SAT [6] to edge 8-admissibility for $(2, 0)$-graphs, i.e. bipartite graphs.

Construction 3 *Given an instance* $I = (U, C)$ *of* 3-SAT *we construct a graph* G *as follows. We add a* P_2 *with labels* x *and* x' *to* G. *For each variable* $u \in U$ *we add a* C_8 *to* G *with three consecutive vertices labeled as* u, m_u, *and* $\overline{u}$ *and the other five consecutive vertices labeled as* u_1 *to* u_5. *For each* $u_i, i = 1, \dots, 5$, u *and* $\overline{u}$ *we add a pendant vertex. For each variable* $u \in U$ *we add the edge* xm_u *to* G. *For each clause* $c_1 = (u, v, w) \in C$, *we add two vertices vertex* c_1 *and* c'_1 *to* G *and the edges* $c_1 c'_1$, $c_1 u$, $c_1 v$, *and* $c_1 w$. *For each variable* $u \in U$ *we add a* P_4 *to* G *with endpoints labeled* p_{u1} *and* p_{u4} *and the edges* $p_{u1}x$ *and* $p_{u4}m_u$.

Figure 6 depicts an example of a graph obtained from a 3-SAT instance.

The key idea of the proof of Theorem 4 is that, for each variable $u \in U$, we have exactly one edge in the edge tree 8-spanner T which is near to x and u or $\overline{u}$. We relate this proximity to a true assignment of that literal. Next, we require that at least one edge incident to each clause to be connected to a true literal. Otherwise, if they are all false literals, we end up with two of the edges incident to that clause being vertices of $L(G)$ with distance at least 9 in T.

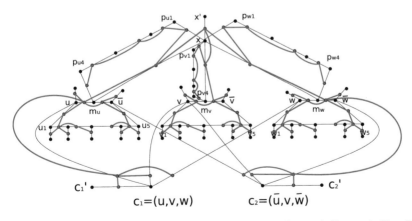

Fig. 6 Graph obtained from Construction 3 on the instance $I = (\{u, v, w\}, \{(u, v, w), (\overline{u}, v, \overline{w})\})$ and an edge tree 8-spanner of it in red

Theorem 4 *Edge 8-admissibility is NP-complete for bipartite graphs.*

Proof By construction, G is bipartite. Moreover, not only the problem is in NP, but also the size of the graph G, obtained from Construction 3 on an instance $I = (U, C)$ of 3-SAT, is polynomially bounded by the size of I. We prove that G is edge 8-admissible if and only if there is a truth assignment to I. Consider a truth assignment of $I = (U, C)$. We obtain an edge tree 8-spanner T of G as follows (see Fig. 6).

Add to T the edges: $\{x'x, xm_u \mid u \in U\}$; $\{xm_u, m_u u \mid u \in U \text{ and } u \text{ is true}\}$ or $\{xm_u, m_u\bar{u} \mid u \in U \text{ and } \bar{u} \text{ is true}\}$; $\{um_u, \bar{u}m_u \mid u \in U\}$; For each clause select a true literal and add to T: $\{c'c, uc \mid c \text{ is a clause with the selected true literal } u\}$;

$\{uc, um_u \mid c \text{ is a clause with the selected true literal } u\}$;

$\{\bar{u}c, \bar{u}m_u \mid c \text{ is a clause with the selected true literal } \bar{u}\}$;

$\{uc, vc \mid c \text{ is a clause with the selected true literal } u \text{ and } v \text{ is other literal of } c\}$;

For each variable $u \in U$ add to T the edges: $\{m_u p_{u_4}, p_{u_4} p_{u_3}\}$; $\{p_{u_4} p_{u_3}, p_{u_3} p_{u_2}\}$; $\{p_{u_3} p_{u_2}, p_{u_2} p_{u_1}\}$; $\{p_{u_2} p_{u_1}, p_{u_1} x\}$; $\{p_{u_1} x, xm_u\}$; $\{um_u, uu_1\}$; $\{\bar{u}m_u, \bar{u}u_5\}$; $\{uu_1, u_1 u_2\}$; $\{u_3 u_4, u_4 u_5\}$; $\{u_4 u_5, \bar{u}u_5\}$; and each pendant G is added to a solution tree as Fig. 6

Consider an edge tree 8-spanner T of G (resp. tree 8-spanner of $L(G)$), we present a truth assignment of $I = (U, C)$. First we claim that for each variable $u \in U$, there is exactly one of these two edges in T: $\{xm_u, um_u\}$ and $\{xm_u, \bar{u}m_u\}$. Assume that both edges are in T. There are in $L(G)$ two adjacent vertices $u_i u_{i+1}$ and $u_{i+1} u_{i+2}$ of the cycle C_9 of variable u with distance 9 in T, a contradiction. Now, assume that both edges are not in T. We consider two cases. If there are no edges $p_{u_4} m_u, um_u$ or $p_{u_4} m_u, \bar{u}m_u$, then there are in $L(G)$ two adjacent vertices $p_{u_4} m_u$ and um_u (or $\bar{u}m_u$) with distance at least 9 in T, since it is necessary to make a path passing through xx', a contradiction. Otherwise, there is an edge $p_{u_4} m_u, um_u$ or $p_{u_4} m_u, \bar{u}m_u$. In both cases, let $c_1 = (u, v, w)$ be a clause that contains u, there are in $L(G)$ two adjacent vertices $c_1 v, vv_1$ that have distance at least 9 in T, a contradiction.

Hence, relate the edge $\{xm_u, um_u\}$ or $\{xm_u, m_u\bar{u}\}$ in T for each variable $u \in U$ to a true assignment to the literal u or $\bar{u}$. Assume that there is a clause with three false literals $c_3 = (x, y, z)$. No matter how we connect the vertices $c_3' c_3, c_3 x, c_3 y$ and $c_3 z$ in T, two of them have distance at least 9 in T, a contradiction. Therefore, each clause has at least one true literal, and this is a truth assignment of I. □

Construction 3 can be adapted in order to prove that edge $2k$-admissibility is NP-complete, for $k \geq 5$. It can be obtained by subdividing the edge $m_u x$ and the cycles corresponding to each variable u.

6 Concluding Remarks

We have obtained the edge stretch index of some graph classes, or equivalently, the stretch index of line graphs, such as gridline graphs (line graphs of bipartite graphs); complement of Kneser graphs $KG_{n,2}$ (line graphs of complete graphs); and line graphs of (k, ℓ)-graphs. Although deciding the 3-admissibility is open for more than 20 years, we characterize the edge 3-admissible graphs in polynomial time, and we also prove that edge 8-admissibility is NP-complete, even for bipartite graphs. Hence, some open questions arise, such as determine the computational complexity of edge t-admissibility for $4 \leq t \leq 7$, and $t = 2k + 1, k \geq 4$.

Acknowledgments This study was partially supported by Coordenação de Aperfeiçoamento de Pessoal de Nível Superior—Brasil (CAPES)—Finance Code 001.

References

1. Bhatt, S., Chung, F., Leighton, T., Rosenberg, A.: Optimal simulations of tree machines. In: 27th Annual Symposium on Foundations of Computer Science, pp. 274–282. IEEE, Piscataway (1986)
2. Cai, L., Corneil, D.G.: Tree spanners. SIAM J. Discrete Math. **8**(3), 359–387 (1995)
3. Couto, F., Cunha, L.F.I.: Tree t-spanners of a graph: minimizing maximum distances efficiently. In: 12th COCOA, Lecture Notes in Computer Science, vol. 11346, pp. 46–61 (2018)
4. Couto, F., Cunha, L.F.I.: Hardness and efficiency on minimizing maximum distances for graphs with few P4's and (k, ℓ)-graphs. Electron. Notes Theor. Comput. Sci. **346**, 355–367 (2019)
5. Couto, F., Faria, L., Gravier, S., Klein, S.: Chordal-(2, 1) graph sandwich problem with boundary conditions. Electron. Notes Discrete Math. **69**, 277–284 (2018)
6. Garey, M.R., Johnson, D.S.: Computers and Intractability: A Guide to the Theory of NP-Completeness. W. H. Freeman Co., New York (1979)
7. Giakoumakis, V., Roussel, F., Thuillier, H.: On P4-tidy graphs. Discr. Math. Theoretical Comput. Sci. **1**, 17–41 (1997)
8. Godsil, C., Royle, G.: Kneser graphs. In: Algebraic Graph Theory, pp. 135–161. Springer, New York (2001)
9. Heggernes, P.: Treewidth, partial k-trees, and chordal graphs. INF334-Advanced algorithmical techniques, Department of Informatics, University of Bergen (2005)
10. Jamison, B., Olariu, S.: P-components and the homogeneous decomposition of graphs. SIAM J. Discrete Math. **8**(3), 448–463 (1995)
11. Peleg, D., Ullman, J.D.: An optimal synchronizer for the hypercube. SIAM J. Comput. **18**(4), 740–747 (1989)